Biochemical and Environmental Bioprocessing
Challenges and Developments

Biochemical and Environmental Bioprocessing
Challenges and Developments

Edited by
M. Jerold and V. Sivasubramanian

CRC Press
Taylor & Francis Group
Boca Raton London New York

CRC Press is an imprint of the
Taylor & Francis Group, an **informa** business

CRC Press
Taylor & Francis Group
6000 Broken Sound Parkway NW, Suite 300
Boca Raton, FL 33487-2742

© 2020 by Taylor & Francis Group, LLC
CRC Press is an imprint of Taylor & Francis Group, an Informa business

No claim to original U.S. Government works

Printed on acid-free paper

International Standard Book Number-13: 978-0-367-18739-2 (Hardback)

This book contains information obtained from authentic and highly regarded sources. Reasonable efforts have been made to publish reliable data and information, but the author and publisher cannot assume responsibility for the validity of all materials or the consequences of their use. The authors and publishers have attempted to trace the copyright holders of all material reproduced in this publication and apologize to copyright holders if permission to publish in this form has not been obtained. If any copyright material has not been acknowledged, please write and let us know so we may rectify in any future reprint.

Except as permitted under U.S. Copyright Law, no part of this book may be reprinted, reproduced, transmitted, or utilized in any form by any electronic, mechanical, or other means, now known or hereafter invented, including photocopying, microfilming, and recording, or in any information storage or retrieval system, without written permission from the publishers.

For permission to photocopy or use material electronically from this work, please access www.copyright.com (http://www.copyright.com/) or contact the Copyright Clearance Center, Inc. (CCC), 222 Rosewood Drive, Danvers, MA 01923, 978-750-8400. CCC is a not-for-profit organization that provides licenses and registration for a variety of users. For organizations that have been granted a photocopy license by the CCC, a separate system of payment has been arranged.

Trademark Notice: Product or corporate names may be trademarks or registered trademarks, and are used only for identification and explanation without intent to infringe.

Library of Congress Cataloging-in-Publication Data

Names: Jerold, M., editor. | Sivasubramanian, V., editor.
Title: Biochemical and environmental bioprocessing : challenges and developments / M. Jerold and V. Sivasubramanian, editors.
Description: First edition. | Boca Raton, FL : CRC Press/Taylor & Francis Group, 2020. | Includes bibliographical references and index.
Identifiers: LCCN 2019008604 | ISBN 9780367187392 (hardback : acid-free paper)
Subjects: LCSH: Green chemistry. | Biochemical engineering. | Bioremediation. | Natural products--Biotechnology.
Classification: LCC TP155.2.E58 B565 2020 | DDC 660.028/6--dc23
LC record available at https://lccn.loc.gov/2019008604

Visit the Taylor & Francis Web site at
http://www.taylorandfrancis.com

and the CRC Press Web site at
http://www.crcpress.com

Contents

Foreword vii
Preface ix
Editors xi
Contributors xiii

Chapter 1 Biosorption – An Elective Strategy for Wastewater Treatment: An Eco-Friendly Approach 1

A. Saravanan, P. Senthil Kumar and P. R. Yaashikaa

Chapter 2 Recent Advancements and Perspectives on Biological Degradation of Azo Dye 17

Rajmohan K. S., Ramya C. and Murali Mohan Seepana

Chapter 3 Performance Analysis of Anaerobic Digestion of Textile Dyeing Industry Effluent in a Modified Sequential Batch Reactor 37

S. Venkatesh Babu and M. Rajasimman

Chapter 4 Waste Sea Shells for Biodiesel Production – Current Status and Future Perspective 53

Subramaniapillai Niju and M. Balajii

Chapter 5 An Intensified and Integrated Biorefinery Approach for Biofuel Production 87

Devadasu Sushmitha and Srinath Suranani

Chapter 6 Hydrothermal Carbonization for Valorization of Rice Husk 105

B. Sai Rohith, Naga Prapurna, Kuldeep B. Kamble, Rajmohan K. S. and S. Srinath

Chapter 7 Production of Biofuels from Algal Biomass 123

Murali Mohan Seepana, M. Jerold and Rajmohan K. S.

Chapter 8	Diffusion Limitations in Biocatalytic Reactions: Challenges and Solutions 139	
	Carlin Geor Malar, Muthulingam Seenuvasan and Kannaiyan Sathish Kumar	
Chapter 9	Recent Advancements and Applications of Nanotechnology in Expelling Heavy Metal Contaminants from Wastewater 151	
	Muthulingam Seenuvasan, Venkatachalam Vinothini, Madhava Anil Kumar and Ayyanar Sowmiya	
Chapter 10	Organic Flocculation as an Alternative for Wastewater Treatment 163	
	Devlina Das	
Chapter 11	Power Production in Microbial Fuel Cells (MFC): Recent Progress and Future Scope 179	
	E. Elakkiya and Subramaniapillai Niju	
Chapter 12	Synthesis, Characterization and Antimicrobial Properties of CuO-Loaded Hydrophobically Modified Chitosans 203	
	P. Uma Maheswari, K. Sriram and K. M. Meera Sheriffa Begum	
Chapter 13	Fucoxanthin: Biosynthesis, Structure, Extraction, Characteristics, and Its Application 221	
	K. Anjana and K. Arunkumar	

Index 251

Foreword

I am so delighted to write the Foreword for *Biochemical and Environmental Bioprocessing: Challenges and Developments*. Bioremediation is gaining more importance due to environmental pollution. There are several methods for the removal of toxic substances, but in many cases huge amounts of byproducts are released as wastes. Hence, there is a sustainable bioremediation technique to overcome such difficulties. This book brings in energetic ideas for the current generation of scientists and engineers. Perhaps "waste is said to be waste" for the common person, but to bioprocess engineers it is feedstock for the production of value-added products. For example, cane molasses is waste produced after extracting sugar in the sugar industry. Surprisingly, the fermentation technology can be adopted for the production of ethanol by using suitable microorganisms. Similarly, the same raw material is used for the production of antibiotics. Therefore, the two common examples show potent applications of bioprocessing technology for valorizing the waste materials (solid/liquid) for the sustainable production of value-added products in clean and green methods. This book would be useful to students, academicians, researchers and engineers for updating their knowledge and understanding the technological background on each and every process involved in environmental bioprocessing. I am so happy to see the content selection and organization of the chapters by the editors. I extend my best wishes to both of the editors for their marvelous and sincere effort in bringing such a wonderful book for the present and future generations of researchers. Indeed, the topics are very meticulously planned and written by eminent academicians and researchers who are working in the field of biotechnology and bioprocessing. I believe this book is not only going to be a high-ranking textbook but also a practical handbook. The readers will definitely enjoy the recent perspectives on biochemical engineering waste management. I anticipate this book will serve as invaluable knowledge material for research communities interested in protecting the environment.

Dr. Selvaraju Narayanasamy
IIT Guwahati

Preface

An emerging field in applied science, bioprocess engineering is the amalgamation of science and engineering. Basically, the life science aspects are translated into applied products for the commercialization process. Today, bioprocessing has become an essential part of various industries such as food, chemical, pharmaceutical and environmental industries. In microbial bioprocessing, plant and animal cells are used as a major component for the production of biological products. In environmental components, the toxic products are neutralized by means of microbial bioprocessing. In microbial bioprocessing, enzymes play an important role in the bioconversion of substrate into products. Microorganisms are key players which can be either aerobic or anaerobic. In some cases, they can be facultative organisms. The reliability of the bioprocess depends upon the usage of either wild type or recombinant type strains.

In the present scenario, the environment is disturbed by various xenobiotic compounds. Industries especially discharge various toxic and other byproducts into the atmosphere. Such pollutants are accumulating and cause environmental problems. Biotechnology has a wide scope in degradation and bioconversion. In order to make the process into a realistic approach, there is a need for large-scale bioreactors. Hence, the integration of bioprocessing and biochemical engineering surely can be a prospect for the successful implementation of life science fundamental research. This book, *Biochemical and Environmental Bioprocessing: Challenges and Developments*, is the work of eminent researchers and academicians in the field of biotechnology and chemical engineering. This book was initiated with a vision to project the importance of biochemical engineering for sustainable pollution management. Green technology has a significant role in the development of eco-friendly bio-products. Essentially, there is a great demand for the production in terms of energy efficiency and cost effectiveness. Therefore, there is continuous research in the bioprocess to make the process simpler to cut down the cost. This book contains 13 chapters related to wastewater treatment, bioenergy production, synthesis of nanoparticles, microbial fuel cells, biopolymer, algal biomass, kinetics, modeling and so on. This book is a consolidated research output by various interdisciplinary researchers. Hence, one can easily correlate the fundamentals and applied knowledge for the valorization of waste into value-added bio-products. Basically, environmental sustainability is to reduce the impact on the environment by the concepts: reduce, reuse and recycle. However, due to the population explosion and industrial expansion, it is very difficult to safeguard the environment. Thus, it is highly essential to implement bioremediation to avoid contamination. In addition, the waste has to be used as feedstock for the conversion into useful products. Hopefully, this book will address the problems pertaining to the environmental issues by innovative and advanced bioprocess techniques. This book is meticulously organized to give comprehensive ideas to solve the challenging environmental issues by benign green technologies. Moreover, this book can be used by students, researchers, academicians and engineers who are working in any area of research. The multidimensional perspective of this book will enable readers to explore a wide range of recent trends in bioremediation.

"Science of today is the technology for tomorrow."

Enjoy reading the book!

Best wishes,

Dr. M. Jerold
Prof. V. Sivasubramanian

Editors

Dr. Manuel Jerold is an Assistant Professor in Department of Biotechnology at the National Institute of Technology Warangal, Telangana State, India. Dr. Jerold has immense knowledge in the field of bioprocessing and biochemical engineering. He has industrial, teaching and research experience. He works on many research areas such as fermentation, phytoremediation, biological wastewater treatment, biosorption, microbial fuel cells, nanotechnology, biosurfactant, biocatalyst, biofuels, and adsorption of heavy metals, etc. His research is highly focused on problems related to environmental issues, especially liquid and solid waste management. He has been awarded with research seed money for projects on biodiesel production using enzymatic catalysts by the National Institute of Technology Warangal. He has identified macroalgae as feedstock for the production of bioethanol at low cost. He has extensively worked on dye removal using algae for the prospect of the textile industry. He has developed a biocomposite for the removal of dye with maximum biosorption capacity. As an interdisciplinary area researcher, he is working on hot topics related to sustainable energy production from microbial fuel cells, algal lipid, biohydrogen from wastewater and so on. He is also involved in the extraction of various phytochemicals from algae. In addition to his research, Dr. Jerold is involved in teaching undergraduate engineering students. He has taught various subjects related to biotechnology and biochemical engineering, bioreactor design and analysis, downstream processing, protein engineering, bioprocess heat transfer, biofuel technology and engineering, bioprocess instrumentation, environmental biotechnology, herbal biotechnology, introduction to life science, engineering biology and transport phenomena in bioprocess system. He has taught various laboratory courses for Bachelor of Technology in Biotechnology students such as bioprocess and bioreaction engineering lab, transport phenomena in bioprocess system lab, heat and mass transfer, instrumental methods of analysis, fluid mechanics lab and environmental and pollution control lab. He is a highly committed teacher and a researcher who works enthusiastically on cutting-edge research areas of biotechnology. He has been awarded the Dr. D.G.S. Dhinakaran fellowship for his postgraduate education and MHRD Fellowship by the Government of India for his doctoral studies. In addition, he has been honored with gold medals for his higher academic credentials by the university at his higher studies. He was the state rank holder during his pre-university level. Dr. Jerold has published 24 research and review papers in international and national peer-reviewed journals. In addition, Dr. Jerold has authored five book chapters. He is a reviewer for various peer-reviewed journals at the international and national level.

Professor Velmurugan Sivasubramanian is Professor and Former Head of the Department of Chemical Engineering at the National Institute of Technology Calicut (NITC) in Kozhikode, India. He has five years of industrial experience after his graduation. He has been involved in teaching fluidization engineering, environmental engineering, wastewater engineering, energy management, safety and hazards control, principles of chemical engineering, operations management, total quality management for master of technology degree in chemical engineering, petroleum refining and petrochemicals, pharmaceutical technology and environmental science and engineering; chemical reaction engineering, mechanical operations, mass transfer, chemical technology, environmental studies, safety in chemical process industries, energy management, process economics and industrial management, petrochemicals, downstream processing/bioseparations and operations research for a Bachelor of Technology in Chemical Engineering and Biotechnology; and bioreactor engineering, fluidization engineering, biological wastewater treatment, environmental biotechnology, safety management in process industries, fire engineering and explosion control and fire modeling and dynamics for the doctoral program since 2001. He received four awards: the Dr. Radhakrishnan Gold Medal Award, the Jawaharlal Nehru Gold Medal Award and the National Citizenship Gold Medal Award, from Global Economic Progress and Research Association (Thiruvannamalai, Tamil Nadu), and the Universal Achievers Gold Medal Award from the Universal Achievers Foundation (Salem, Tamil Nadu) for his outstanding individual achievement in education and contribution to education and national development. He has developed and commercially exploited wastewater treatment using novel bioreactors such as the inverse fluidized bed reactors (IFBRs), photobioreactors (PBRs) and self-forming dynamic membrane reactors for the remediation of dyes, pharmaceuticals and heavy metals. His team has successfully installed a novel magnetic biocomposite for the treatment of dye effluent. Recently, his team developed a 4 m^3 floating drum biodigester for the production of biogas, which is used as a fuel in the NITC hostel mess. Prof. Sivasubramanian has published several research papers in international and national peer-reviewed journals and conferences in the area of environmental and chemical engineering for the treatment of effluents from various industries. His team has earned nine best paper awards at national and international conferences. He is editor for six peer-reviewed journals, and he also serves as reviewer for more than 50 journals.

Contributors

K. Anjana
Department of Plant Science, School of Biological Sciences
Central University of Kerala
Kasaragod, India

K. Arunkumar
Department of Plant Science, School of Biological Sciences
Central University of Kerala
Kasaragod, India

S. Venkatesh Babu
Department of Petroleum Engineering
JCT College of Engineering & Technology
Coimbatore, India

M. Balajii
Department of Biotechnology
PSG College of Technology
Coimbatore, India

K. M. Meera Sheriffa Begum
Department of Chemical Engineering
National Institute of Technology
Trichy, India

Ramya C.
Department of Biochemistry
National University of Singapore
Singapore, Singapore

Devlina Das
Department of Biotechnology
PSG college of Technology
Coimbatore, India

E. Elakkiya
Department of Biotechnology
PSG College of Technology
Coimbatore, India

M. Jerold
Department of Biotechnology
National Institute of Technology Warangal
Warangal, India

Kuldeep B. Kamble
Department of Chemical Engineering
National Institute of Technology Warangal
Warangal, India

Rajmohan K. S.
Department of Chemical Engineering
National Institute of Technology Warangal
Warangal, India

Madhava Anil Kumar
Analytical and Environmental Science Division & Centralized Instrument Facility
CSIR-Central Salt & Marine Chemicals Research Institute
Bhavnagar, India

Kannaiyan Sathish Kumar
Department of Chemical Engineering
SSN College of Engineering
Chennai, India

P. Senthil Kumar
Department of Chemical Engineering
SSN College of Engineering
Chennai, India

P. Uma Maheswari
Department of Chemical Engineering
National Institute of Technology
Trichy, India

Carlin Geor Malar
Department of Biotechnology
Rajalakshmi Engineering College
Chennai, India

Subramaniapillai Niju
Department of Biotechnology
PSG College of Technology
Coimbatore, India

Naga Prapurna
Department of Chemical Engineering
Chaitanya Bharathi Institute of Technology
Hyderabad, India

M. Rajasimman
Department of Chemical Engineering
Annamalai University
Annamalai Nagar, India

B. Sai Rohith
Department of Chemical Engineering
Chaitanya Bharathi Institute of Technology
Hyderabad, India

A. Saravanan
Department of Biotechnology
Rajalakshmi Engineering College
Chennai, India

Muthulingam Seenuvasan
Department of Chemical Engineering
Hindusthan College of Engineering and Technology
Coimbatore, India

Murali Mohan Seepana
Department of Chemical Engineering
National Institute of Technology Warangal
Warangal, India

Ayyanar Sowmiya
Department of Petrochemical Engineering
SVS College of Engineering
Coimbatore, India

K. Sriram
Department of Chemical Engineering
National Institute of Technology
Trichy, India

Srinath Suranani
Department of Chemical Engineering
National Institute of Technology Warangal
Warangal, India

Devadasu Sushmitha
Department of Chemical Engineering
National Institute of Technology Warangal
Warangal, India

Venkatachalam Vinothini
Department of Petrochemical Engineering
SVS College of Engineering
Coimbatore, India

P. R. Yaashikaa
Department of Chemical Engineering
SSN College of Engineering
Chennai, India

1 Biosorption – An Elective Strategy for Wastewater Treatment
An Eco-Friendly Approach

A. Saravanan, P. Senthil Kumar and P. R. Yaashikaa

CONTENTS

1.1 Introduction 2
1.2 Water Pollutants 2
    1.2.1 Heavy Metals and Their Toxicity 2
1.3 Conventional Methodologies 3
    1.3.1 Coagulation and Flocculation 4
    1.3.2 Ion Exchange 4
    1.3.3 Precipitation 5
    1.3.4 Membrane Filtration 5
    1.3.5 Electrochemical Processes 5
1.4 Biosorption Process 6
1.5 Biosorption Mechanism 7
    1.5.1 Bacteria 8
    1.5.2 Fungi 8
    1.5.3 Algae 8
1.6 Factors Affecting the Biosorption Process 9
    1.6.1 pH 9
    1.6.2 Temperature 9
    1.6.3 Characteristics and Concentration of Biomass 10
    1.6.4 Initial Metal Ion Concentration 10
1.7 Biosorbents 10
    1.7.1 Techniques for Biosorbent Characterization 11
1.8 Biosorption Isotherms... 12
    1.8.1 Langmuir Isotherm 12
    1.8.2 Freundlich Isotherm 13
    1.8.3 Constraint of Freundlich Adsorption Isotherm 13
1.9 Kinetics Study... 13
1.10 Conclusion 14
References... 15

1.1 INTRODUCTION

A number of industries – for example, material, paper and mash, printing, press steel, coke, oil, pesticide, paint, dissolvable, pharmaceutics, wood-saving synthetic compounds – expend substantial volumes of water and naturally based synthetics. These synthetic substances demonstrate an inspiring distinction in concoction synthesis, sub-atomic weight, lethality, and so forth. Effluents of these ventures may likewise contain undesired amounts of toxins (Aksu, 2005).

Ecological contamination by overwhelming metals presents a genuine risk to human wellbeing due to hemotoxicity. Unlike natural poisons, substantial metals are not biodegradable and have the capacity to amass in life forms as harmful or cancer-causing. Research has demonstrated that metals are a standout amongst the most well-known reasons for illnesses in living beings, and expanded consideration is paid to their assurance and expulsion from tainted wastewater. The increase in the concentration of metal ions beyond the permissible limits causes various cytological and physiological effects (Babel and Kurniawan, 2003; Krstic et al., 2018).

1.2 WATER POLLUTANTS

Water pollutants can be classified as agricultural, industrial, sewage, radioactive, and thermal pollutants. Fertilizers, pesticides, manures, etc. contribute to pollutants from the agricultural sector. These pollutants, when dissolved in groundwater or surface water, result in eutrophication and water pollution, thus making it unfit for use. These chemicals are toxic to animals and aquatic organisms. Effluents discharged from different industries such as leather and tannery, dairy, electrochemical plating, etc., release toxic heavy metals such as copper, lead, chromium, mercury, arsenic, chromium, and zinc, along with poisonous organic and inorganic waste materials. Sewage pollutants include contaminants from both domestic and municipal wastes such as food waste, household waste, etc. The harmful bacteria and fungi present in these wastes spread various diseases such as hepatitis, polio, and sometimes cancer. Thermal and radioactive wastes come under physical pollutants. Radioactive waste releases due to leaching of minerals into the human body through food. This causes cancer and poses a serious threat. Thermal pollutants signify the impairment of value and decay of amphibian and earthly condition by different industrial plants such as nuclear, atomic, coal-terminated, and thermal plants, as well as oil field generators, industrial facilities, and factories. Thus, all these pollutants are spread through various sources, posing serious threat to human life and other creatures on earth.

1.2.1 Heavy Metals and Their Toxicity

Substantial metal contamination is one of the major natural issues today. A large portion of overwhelming metal particles are lethal to living beings. These metal particles are non-degradable and remain in the earth. In this way, the disposal of substantial metal particles from wastewater is essential to secure general wellbeing. Modern effluents are a noteworthy reason for an overwhelming metal focus;

these effluents originate from numerous ventures, for example, consumption of water funnels, misuse of dumping, electroplating, electrolysis, electro-osmosis, mining, surface completing, vitality and fuel creating, manure, pesticide, iron and steel, cowhide, metal surface treating, photography, aviation and nuclear vitality establishments, and so forth (Wang and Chen, 2009; Abbas et al., 2014). Along these lines, the evacuation and recuperation of overwhelming metals from gushing streams are basic to the insurance of the earth.

1.3 CONVENTIONAL METHODOLOGIES

Growth in the populace and urbanization by the nineteenth century made it important for a central answer for the issue of waste administration to be found. New strategies assembled wastewaters in extraordinary trenches, redirecting the loss from towns and releasing it into streams. The release of untreated wastewater in streams and different waterways that were utilized by residents for drinking water and washing led to the spread of ailments, particularly cholera, which took a large number of lives. The production of the principal magnifying instrument and perceptions of microbes brought about another phase in wastewater administration. Reports of the time demonstrated that microbiological procedures could be in charge of natural compound oxidation. Populace development and financial advancement have expanded water requests, bringing about the arrangement of vast volumes of wastewater with synthetically complex creations. The nature of common water is subject to the level of wastewater treatment; the more progressive the treatment, the more exorbitant the procedure. The modern methods of wastewater treatment are comprised of three principle stages: primary, secondary, and tertiary treatment. The quantity of stages depends on the degree of poison expulsion and the components through which toxins are expelled. Primary or essential treatment is intended to expel natural and inorganic solids from wastewater. Amid the secondary treatment, fine suspended solids, scattered solids, and broken up organics are expelled by volatilization, biodegradation, and consolidation into slop. The motivation behind the tertiary (propelled) treatment is to enhance the nature of the water that is released in common waters, utilizing an assortment of natural, physical, and substance treatment approaches (Wiesmann, 2007). Contingent upon the utilized techniques, the level of water decontamination can reach a maximum of 95% and may be 98–99%. The synthesis and properties of wastewater, which are released in common water bodies, are imperative in light of their impact on water environment conditions. The accompanying necessities ought to be addressed:

1. Wastewater must not be dangerous with respect to water tenants.
2. Parameters, for example, pH, temperature, and the convergence of substantial metals and biogenic substances, ought to be sufficiently low so that the waterway arrangement of self-cleaning is not hurt or obliterated.

Physico-chemical strategies, for example, coagulation, flocculation, precipitation, adsorption, particle trade, electro-dialysis, and layer partition, can be connected in wastewater treatment plans (Carolin et al., 2017; Thekkudan et al., 2017).

1.3.1 Coagulation and Flocculation

Coagulation and flocculation are a basic piece of drinking water treatment, and in addition, wastewater treatment. Coagulation and flocculation are fundamental procedures in different orders. In consumable water treatment, illumination of water utilizing coagulating specialists has been polished from antiquated circumstances. Coagulation is additionally vital in a few wastewater treatment activities. A typical case is concoction phosphorus expulsion, and another, in over-burdened wastewater treatment plants, is the act of synthetically improving essential treatment to decrease suspended solids and natural burdens from essential clarifiers (Kimura et al., 2013).

Coagulant synthetics with charges opposite to those of the suspended solids when added to the water kill the negative charges on undisolvable solids. Once the charge is killed, the little suspended particles are equipped for staying together. These marginally bigger particles are called miniaturized scale flocs and are not unmistakable to the naked eye. The water encompassing the recently shaped miniaturized scale flocs ought to be clear. If not, coagulation and a portion of the charged particles have not been killed. Maybe more coagulant synthetics ought to be included (Xu et al., 2012).

Flocculation, a delicate blending stage, expands the molecule estimate from submicroscopic microfloc to noticeable suspended particles. Microfloc particles impact, making them cling to deliver bigger, noticeable flocs called stick floc. Floc measure keeps on working with extra crashes and connection with included inorganic polymers (coagulant) or natural polymers. Macroflocs are framed, and high sub-atomic weight polymers, called coagulant helps, might be added to enable scaffolding, to tie and reinforce the floc, including weight and increment settling rate. When floc has achieved its ideal size and quality, water is prepared for sedimentation.

The main impediments that may block an entirely physical–synthetic solution for wastewater treatment are the issues related with the profoundly putrescible slop created and the high working expenses of compound expansion. Nonetheless, a significant part of the ebb and flow interests in physical–concoction treatment originate from wastewater's appropriateness for treatment under crisis measures; for occasional applications, to dodge abundant wastewater releases amid storm occasions; and for essential treatment before organic treatment, where the above inconveniences happen to lesser effect.

1.3.2 Ion Exchange

Ion exchange frameworks, for the most part, contain particle exchange pitches which are worked on a cyclic premise. Water goes through the basis until the point when it is soaked; the water leaving the tar contains more than the coveted centralization of the particles that must be evacuated. The tar is then recovered by discharging the sap to expel the amassed solids, flushing expelled particles from the sap with a concentrated arrangement of substitution tar. Thus, the creation of discharge limits the utilization of ion exchange in wastewater treatment. Distinctive kinds of solid and weak ion-exchange pitches are regularly utilized as a part of adsorption forms. Other than engineered saps, normal minerals, particularly zeolites, are broadly utilized for substantial metal expulsion from fluid arrangements because of their minimal effort and high wealth (Kurniawau et al., 2006).

1.3.3 Precipitation

In the precipitation process, synthetic substances respond with overwhelming metal particles to frame insoluble accelerates that are additionally isolated from the water by sedimentation or filtration. Metals can be expelled from wastewater by precipitation as hydroxide at hoisted pH or by sulfide precipitation. Precipitation is normally utilized for the evacuation of metal particles, phosphorus mixes, and radioactive components. The most ordinarily utilized precipitation procedure is hydroxide treatment because of its relative straightforwardness, minimal effort, and programmed pH control. $Ca(OH)_2$ and NaOH mixes are utilized as precipitants. The fundamental favorable circumstances incorporate the high level of metal expulsion even at low pH and the possibility of specific metal evacuation and recuperation. Metal sulfide slop additionally shows preferred thickening and dewatering qualities over the relating metal hydroxide muck. Restrictions of the procedure incorporate the development of lethal hydrogen sulfide vapor and the arrangement of sulfide colloidal precipitates (Fu and Wang, 2011).

1.3.4 Membrane Filtration

Membrane filtration, broadly utilized as a part of chemical and biotechnology firms, is now settled as a profitable method for separating and cleaning wastewater and modern process water. Even more as of late, tubular and winding layer plants have started to be utilized to channel contaminations from savoring water locales where traditional treatment turns out to be to be uneconomical. A membrane is a thin layer of semi-porous material that isolates substances when the main thrust is connected over the layer. Membrane forms are progressively utilized for evacuation of microbes, microorganisms, particulates, and characteristic natural material, which can give shading, tastes, and smells to water and responds with disinfectants to shape cleansing side effects (Bessbousse et al., 2008). As progressions are made in membrane generation and module outline, capital and working costs keep on declining. Regularly, the layer material is made from a manufactured polymer, albeit different structures, including fired and metallic "layers," might be accessible. All layers fabricated for drinking water are made of polymeric material since they are altogether more affordable than layers developed of different materials. Membrane filtration can be classified as ultrafiltration, microfiltration, and nanofiltration.

1.3.5 Electrochemical Processes

Electrochemical techniques rival other physic-compound innovations in offering answers for the necessities of numerous enterprises, for example, the decontamination of various kinds of wastewater. Electrochemical procedures can be proficiently connected for metal recuperation medications, with the fundamental preferred standpoint being dictated by the utilization of clean reagent-electrons. The guideline of the technique is the plating-out of metal particles on a cathode surface and their resulting recuperation in the essential state. The flexibility of electrochemical techniques (electrocoagulation, electroflotation, electrooxidation, and electrodeposition) can be found in their low ecological effect, simple use, and absence of hurtful or dangerous remains. Electrochemical advances have been explored as the emanating treatment forms for over a century. Central and building inquiries about them have set up the electrochemical testimony innovation in metal recuperation or overwhelming metal-emanating treatment. Electrocoagulation has been utilized in modern times and its predominant exhibitions have been shown in treating effluents containing suspended solids, oil and oil, and even natural or inorganic poisons that can be flocculated. Electroflotation is broadly utilized as a part of the mining enterprises and is finding expanding applications in wastewater treatment. The uniform and minor estimated bubbles created electrically give much preferable execution over either broken down air buoyancy, sedimentation, or considerably impeller buoyancy. This procedure is minimal and simple to encourage with programmed control. Regardless of the numerous focal points of electrochemical advances, it isn't relevant worldwide because of its high-power utilization and necessity of high capital speculation (Brillas and Sires, 2012).

Physico-chemical techniques used to diminish the level of various kinds of toxins in wastewater have their own points of interest and detriments. One of them leads to auxiliary water contamination, while others are exceptionally costly and not beneficial. Along these lines, later on, there is a need to grow new modest and earth-agreeable techniques for wastewater treatment.

1.4 BIOSORPTION PROCESS

As of late, microbial biomass has risen as a possibility for creating financial and eco-accommodating wastewater treatment processes; in this way, applying biotechnology in controlling and expelling metal contamination has been paid much consideration, and it step-by-step turns into a hotly debated issue in the field of metal contamination control as a result of its potential application (Saravanan et al., 2016). An elective procedure is biosorption, which uses different certain characteristic materials of organic birthplace, including microscopic organisms, organisms, yeast, green growth, and so forth (Michalak et al., 2013).

- Biosorption can be defined as the capacity of organic materials to remove overwhelming metals from wastewater through metabolically interceded or physico-chemical pathways of take-up.
- Biosorption, which is the capacity of certain microbial biomaterials to tie and focus overwhelming metals from even the most weakened watery arrangements, offers an actually doable and financially alluring option.
- Biosorption has been characterized as the property of certain biomolecules (or kinds of biomass) to tie and think chosen particles or different atoms from watery arrangements.

Biosorption – A Strategy for Wastewater Treatment

Green growth algae and microscopic organisms including bacteria, fungi, and yeast have ended up being potential metal biosorbents. Biosorption is viewed as a perfect elective technique for expelling contaminants from effluents; it involves a quick wonder of detached metal sequestration by the non-developing biomass/adsorbents. It has benefits over other conventional methods, a portion of which are recorded: ease; high proficiency; minimization of chemical as well as organic slop; no extra supplement prerequisite; recovery of biosorbent; and the likelihood of metal recuperation. The biosorption procedure includes a strong stage (sorbent or biosorbent; adsorbent; natural material) and a fluid stage (dissolvable, ordinarily water) containing broken down animal categories to be sorbent (adsorbate, metal). Because of the higher fondness of the adsorbent for the adsorbate species, the last is pulled in and bound along these distinctive systems. The procedure proceeds until the point when balance is set up between the measure of strong bound adsorbate species and its bit staying in the arrangement. The fondness between the adsorbent and the adsorbate decides its conveyance between the solid and fluid stages. The biosorption process, in which microorganisms are utilized to expel and recoup overwhelming metals from fluid arrangements, has been known for a couple of decades; however, it has risen as a minimal-effort promising innovation in the most recent decades. In this procedure, the take-up of overwhelming metals and radioactive mixes happens because of physico-chemical connections of metal particles with the cell mixes of organic species. Subsequently, the possibility of the utilization of a biomaterial for the take-up of substantial metals has been widely examined throughout the previous two decades. Biotechnological methodologies can prevail in those regions and are intended to cover such specialties (Gadd et al., 2012).

Microorganisms have advanced different measures to react to substantial metal pressure by means of procedures, for example, transport over the cell layer, biosorption to cell dividers and ensnarement in extracellular containers, precipitation, complexation, and oxidation-diminishment responses. They have demonstrated the ability to take up substantial metals from watery arrangements, particularly when the metal focuses in the profluent run from under 1 to around 20 mg/L. In addition, adaptability to deal with the scope of physico-chemical parameters in effluents, selectivity to expel just the coveted metals, and the cost-adequacy are a few included favorable circumstances of organic metal cleanup strategies. These variables have advanced broad research on the organic strategies for metal expulsion. This chapter surveys the utilization of various kinds of microorganisms, green growth organisms, and yeasts and their losses as biosorbents to expel substantial metals from wastewaters; these natural biosorbents have a very compelling and additionally dependable role in the expulsion of overwhelming metal particles from wastewater.

1.5 BIOSORPTION MECHANISM

Microorganisms are life forms that are equipped for enduring negative conditions, and these systems developed for a great many years. The capacity of microorganisms like bacteria, algae, plants, and fungal biomass to evacuate substantial metal particles and radionuclide, and/or to elevate their change to less dangerous structures, has pulled in the consideration of different natural researchers, designers, and biotechnologists for a long time. In this manner, different ideas for expulsion of overwhelming metals from squander streams and bioremediations of polluted conditions are being foreseen, some of which were conveyed at a pilot or mechanical level. There are numerous instruments included in biosorption; some are not completely comprehended. The biosorption system might be arranged by reliance on the cell metabolism, which is called metabolism dependent, or as indicated by the area where the metal expelled from the arrangement is discovered, which is called metabolism independent, like extracellular accumulation/precipitation, cell surface sorption/precipitation, and intracellular gathering (Kotrba, 2011).

Amid metabolism independent systems, metal take-up is by a physico-chemical association between the metal and the utilitarian gatherings introduced on the microbial cell surface. This depends on physical adsorption, ion exchange, and chemical sorption, which is not reliant on the cells' metabolism. Cell dividers of microbial biomass, mostly made out of polysaccharides, proteins, and lipids, have copious metal restricting gatherings, for example, carboxyl, sulfate, phosphate, and amino gatherings. This sort of biosorption, i.e. metabolism independent, is generally quick and can be reversible (Gadd and Fomina, 2011).

1.5.1 Bacteria

Bacterial biosorption is fundamentally utilized for the expulsion of toxins from effluents defiled with contaminations that are not biodegradable, similar to metal particles and dye materials. Notwithstanding, their confinement, screening, and gathering on a bigger scale might be entangled, yet at the same time stay one of the proficient methods for remediating contaminations. Diverse bacterial strains were utilized for the evacuation of various metal particles. Microorganisms have advanced various productive frameworks for detoxifying metal particles; they build up these opposition components generally for their own survival (Brierley, 1990).

1.5.2 Fungi

Fungi have been perceived as a promising material of low-cost adsorbents for expulsion of heavy metal ions from fluid waste streams. The ability of the numerous kinds of growths to deliver extracellular chemicals for the digestion of complex starches for the previous hydrolysis makes skilled the corruption of different degrees of contaminations. They likewise have the advantage of being moderately uncomplicated to develop in fermenters; in this way being fitting for vast scale creation. Another advantage is the simple partition of contagious biomass by filtration in light of its filamentous structure. In contrast with yeasts, filamentous growths are less touchy to varieties in supplements, air circulation, pH, and temperature, and have a lower nucleic content in the biomass (Dhankhar and Hooda, 2011).

1.5.3 Algae

Algae are productive and low-cost biosorbents, as the prerequisite of the supplement by green growth is less (Suganya et al., 2016; Gunasundari et al., 2017). In light of the factual investigation on green growth possibility in biosorption, it has been accounted for that green growth ingests around 15–85%, which is higher when contrasted with other microbial biosorbents. In all the sort of green growth, dark-colored green growth was known to have a high ingestion limit. Biosorption of metal particles happens on the cell surface by methods for ion exchange technique. Dark brown colored marine algae have the ability to ingest metals like nickel, lead, and cadmium through substance bunches on their surface, for example, carboxyl, sulfonate, amino, and also sulfhydryl (Davis et al., 2003).

1.6 FACTORS AFFECTING THE BIOSORPTION PROCESS

The process of biosorption is influenced by various parameters such as pH, temperature, biomass concentration and characteristics, initial metal ion concentration, type of biomass, etc. A few factors are discussed in this section.

1.6.1 pH

pH is the most vital parameter in the biosorption forms. Biosorption is like an ion exchange process, i.e. biomass can be considered as characteristic ion exchange materials, which chiefly contain feebly acidic and essential gatherings. In this way, pH of arrangement affects the idea of biomass-restricting destinations and metal solvency; it influences the arrangement chemistry of the metals, the action of the utilitarian gatherings in the biomass, and the opposition of metallic particles. Metal biosorption has as often as possible appeared to be emphatically pH subordinate in all frameworks analyzed, including microscopic organisms, cyanobacteria, green growth, and parasites. Rivalry amongst cations and protons for restricting locales implies that biosorption of metals like zinc, copper, cobalt, nickel, and cadmium is frequently diminished at low pH esteems. The overwhelming metal take-up for the majority of the biomass composes decays essentially when the pH of the metal arrangements is diminished from pH 6.0 to 2.5. At pH under 2, there is the least or irrelevant expulsion of metal particles from arrangements. The metal take-up increments when pH increments from 3.0 to 5.0. The ideal estimation of pH is imperative to get the most elevated metal sorption, and this limit will diminish with additional increment in pH esteem (Deng and Wang, 2012).

1.6.2 Temperature

As a variation to the bioaccumulation process, biosorption efficiency remains unaffected inside the range of 20–35°C; however, high temperatures of 50°C may grow biosorption; in any case, these high temperatures may hurt microbial living cells and after that reduce metal take-up. Adsorption responses are for the most part exothermic, and the degree of adsorption increments with diminishing temperature. For example, the most extreme biosorption limit with respect to lead and nickel by *S. cerevisiae* is 25°C, and it was found to diminish as the temperature was expanded to 40°C (White et al., 1997).

1.6.3 CHARACTERISTICS AND CONCENTRATION OF BIOMASS

The type of the biomass or determined product might be viewed as one of the critical elements, including the idea of its application, for example, uninhibitedly suspended cells, immobilized arrangements, living biofilms, and so on. Physical medicines, for example, bubbling, drying, autoclaving and mechanical interruption, will all influence restricting properties, while synthetic medications, for example, antacid treatment, frequently enhance the biosorption limit (Wang and Chen, 2006). Development and nourishment on the biomass can likewise affect biosorption because of changes in cell measure, divider arrangement, extracellular item development, and so on. The grouping of biomass in arrangement influences the particular take-up. At equilibrium concentration, the biomass adsorbs more metal particles at low cell densities than at high densities. So, electrostatics communication between the cells plays an imperative part in metal take-up. At low biomass fixation, the particular take-up of metals is expanded in light of the fact that an expansion in biosorbent focus prompts obstruction between the binding regions. High biomass fixation limits the entrance of metal particles to the binding destinations (Nuhoglu and Malkoc, 2005).

1.6.4 INITIAL METAL ION CONCENTRATION

The initial metal ion concentration gives an essential main impetus to defeat all mass exchange resistance of metal between the fluid and strong stages. Expanding measures of metal adsorbed by the biomass will be expanded with the underlying fixation of metals. The ideal level of metal evacuation can be taken at a low beginning metal focus. Along these lines, at a given concentration of biomass, the metal take up increments with increment in initial concentration (Zouboulis et al., 1997).

1.7 BIOSORBENTS

Any sort of natural material has a fondness for inorganic and natural contaminations, which means there is huge biosorption potential inside endless sorts of biomaterials. In the scan for exceptionally proficient and modest biosorbents and new trends for contamination control, component recuperation, and reusing, a wide range of microbial, plant and creature biomass, and determined items have gotten the examination in an assortment of structures and in connection to an assortment of substances. The sorts of substrates of natural starting points that have been researched for biosorbent preparation incorporate microbial biomass (microorganisms, archae, cyanobacteria, filamentous parasites and yeasts, microalgae), ocean growth (macro algae), wastes from industries (fermented food waste, sludge, and so on), agricultural wastes (organic product/vegetable wastes, rice straw, wheat grain, sugar beet mash, soybean frames, and so forth), characteristic build-ups (plant deposits, sawdust, tree coverings, weeds, etc.), and different materials. The biosorptive limit of a biosorbent generally relies upon test conditions and its ancient times and pretreatment (Park et al., 2010). When looking at biosorptive limits of biosorbents for a specific contaminant, the test information of every specialist ought to be deliberately considered in the light of these elements. A noteworthy test has been to choose the most encouraging sorts of biomass from an expansive pool of promptly accessible and economical biomaterials. In principle, for expansive scale modern uses, the biosorbent ought to be promptly accessible and could originate from

- Mechanical wastes, which could be accessible free or at low charge
- Life forms available freely on earth
- Living beings that can be developed effectively for biosorption purposes

A typical method of reasoning is that "waste" biomass will give a financial preferred standpoint. Despite the fact that no new biomass can be produced for investigation, one of the issues is that the local biomass organization may not change altogether between various types of similar families. Microorganisms can discharge numerous sorts of metal-restricting metabolites. Another part of the idea of biosorbents that ought to be featured here is that biomass utilized for biosorption might be alive or dead. While the utilization of dead biomass or determined items might be less demanding by decreasing many-sided quality, the impact of metabolic procedures on sorption is frequently neglected, especially where there is an organic contribution to the issue. The usage of dead biomass is apparently a favored choice for the larger part of metal removal. This can be an interesting alternative for the bioremediation processes of heavy metals, which can be explained by the following points:

- Lack of harmfulness constraints
- Nonappearance of necessities for development of media and supplements in the feed arrangement
- Simple absorbance and recuperation of biosorbed metals
- Simple recovery and reuse of biomass
- Probability of simple immobilization of dead cells
- Less demanding numerical displaying of metal take-up

In spite of the conspicuous focal points of utilizing dead biomass over living microorganisms, numerous characteristics of living microorganisms stay unexploited in a modern setting. Living microorganisms degenerate common poisons and can sorb, transport, complex, and change metals, metalloids, and radionuclides, and an extensive variety of techniques may add to the general removal process. They can be used for specific applications when refined biosorptive metal removal isn't conceivable and may be of a motivating force in systems where additional points of interest will occur due to the metabolic activity (Dhankhar and Hooda, 2011).

1.7.1 Techniques for Biosorbent Characterization

The biosorption process can be studied using various analytical techniques as follows:

- UV vis-spectrophotometer
- Atomic adsorption spectrophotometer (AAS)

- Fourier transform infrared spectroscopy (FTIR)
- Scanning and transmission electron microscopy with energy dispersive X-ray spectroscopy (SEM/TEM with EDX)
- Thermo gravimetric analysis (TGA)
- X-ray diffraction analysis (XRD)

These techniques are interconnected in one way or the other for determining the process of biosorption (Park et al., 2010; Ngwenya, 2007).

1.8 BIOSORPTION ISOTHERMS

Broad research exertion has been committed to a sound comprehension of biosorption isotherm, energy, and thermodynamics. Contrasted with biosorption isotherm, there is an absence of a hypothetical premise behind the detailed kinetics of biosorption information. In such a manner, pseudo first- and second-order kinetic conditions have been generally used to depict the time development of biosorption under non-equilibrium conditions.

The two most broadly utilized adsorption isotherms are the Langmuir and Freundlich isotherms. The Langmuir isotherm expects a surface with homogeneous restricting destinations, proportional sorption energies, and no collaboration between sorbed species. The Freundlich isotherm is an exact condition in view of an exponential appropriation of sorption destinations and energies. These linearized harmony adsorption isotherm models for single solute framework are given by the accompanying conditions:

1.8.1 LANGMUIR ISOTHERM

The Langmuir adsorption display is the most widely recognized one used to measure the sum of adsorbate adsorbed on an adsorbent as a component of incomplete weight at guaranteed temperature. The Langmuir isotherm depicts the reliance of the surface scope of an adsorbed gas on the weight of the gas over the surface at a settled temperature. There are numerous different kinds of isotherm which contrast in at least one of the suppositions made in determining the articulation for the surface scope; specifically, on how they treat the surface scope reliance of the enthalpy of adsorption. While the Langmuir isotherm is one of the least difficult, despite everything it gives a valuable knowledge into the weight reliance of the degree of surface adsorption.

The Langmuir adsorption isotherm depends on the accompanying presumptions:

(a) The adsorbed particle stays at the site of adsorption until the point that it is desorbed (i.e. the adsorption is restricted).
(b) At greatest adsorption, just a monolayer is framed; particles of adsorbate do not store on the other, as of now adsorbed, particles of adsorbate, just on the free surface of adsorbent.
(c) The surface of the adsorbents is uniform; that is, all the adsorption locales are proportional.
(d) There is no association between particles adsorbed on neighboring destinations.

(e) All adsorption happens through a similar system.
(f) Particles are adsorbed at the characteristic locales on the adsorbent surface.

Langmuir isotherm equation can be described as follows (Langmuir, 1918):

$$q_e = \frac{q_m K_L C_e}{1 + K_L C_e} \tag{1.1}$$

Where q_e is milligrams of metal gathered per gram of the biosorbent material; C_e is the metal remaining fixation in arrangement; q_m is the most extreme particular take-up relating to the site immersion; and K_L is the proportion of adsorption and desorption rates.

1.8.2 Freundlich Isotherm

Freundlich gave an exact articulation speaking to the isothermal variety of adsorption of an amount of gas adsorbed by unit mass of strong adsorbent with weight. This condition is known as the Freundlich adsorption isotherm or Freundlich adsorption condition.

The Freundlich isotherm model can be described as follows (Freundlich, 1960):

$$q_e = K_F C_e^{1/n} \tag{1.2}$$

K_F is the Freundlich steady [(mg/g)/(L/mg)1/n] identified with the holding vitality, n is the Freundlich steady used to quantify the deviation from linearity of adsorption. The outcome of n is given as takes after: $n = 1$ (direct); $n > 1$ (physical process); $n < 1$ (concoction process).

1.8.3 Constraint of Freundlich Adsorption Isotherm

Tentatively it was resolved that degree of adsorption differs straightforwardly with weight until immersion weight is reached. Past that, point rate of adsorption soaks even in the wake of applying higher weight. Therefore, the Freundlich adsorption isotherm fizzled at higher weight.

1.9 KINETICS STUDY

The procedure plan, task control, and adsorption energy are vital. The adsorption energy in a wastewater treatment is huge, as it gives important experiences into the response pathways and the system of an adsorption response. Likewise, the energy depicts the solute take-up, which thusly controls the living arrangement time of adsorbate at the strong arrangement interface.

Adsorption/biosorption is a multi-step process, including four back-to-back basic advances:

(a) Exchange of solute from the majority of answers for the fluid film encompassing the particles

(b) Transport of the solute from the limit fluid film to the surface of the particles (outer dissemination)
(c) Exchange of solute from the surface to the inside dynamic restricting locales (intra-molecule dissemination)
(d) Connection of the solute with the dynamic restricting locales

Pseudo First-Order Kinetic Model (Lagergren, 1898)

The Lagergren rate condition was the top-notch condition for the adsorption of fluid/strong framework in view of strong limit. The Lagergren rate condition is a standout amongst the most generally utilized sorption rate conditions for the adsorption of a solute from a fluid arrangement. It might be spoken to as:

$$q_t = q_e\left(1-\exp(-k_1 t)\right) \qquad (1.3)$$

Where t is the time (min) and k_1 is the pseudo first-order kinetic rate constant (1/min).

Pseudo Second-Order Kinetic Model (Ho and McKay 1999)

There are sure suspicions in portrayal of this active model.

(a) There is a monolayer of adsorbate on the surface of adsorbent.
(b) The vitality of adsorption for every adsorbent is the same and autonomous of surface scope.
(c) The adsorption happens just on confined destinations and includes no communications between adsorbed poisons.
(d) The rate of adsorption is relatively insignificant in examination with the underlying rate of adsorption.

The kinetic rate equation can be composed as takes after:

$$q_t = \frac{q_e^2 k_2 t}{1+q_e k_2 t} \qquad (1.4)$$

Where k_2 is the pseudo second-order kinetic rate constant (g/mg min)

1.10 CONCLUSION

Many years of biosorption have uncovered the unpredictability of the procedure, its reliance on physico-chemical and organic components, and vulnerability about the components included. Biosorption has not been financially effective, and its conventional heading as a minimal effort and ecologically neighborly contaminant treatment strategy ought to be re-examined. Endeavors to enhance biosorption (limit, selectivity, energy, re-use) by physico-chemical and biotic controls increment cost and may raise natural issues. Down to business market and cost method of reasoning ought to be considered in coordinating further research into elective applications, for example, organics expulsion, recuperation of pharmaceuticals, profitable metals and components, and the produce of advanced feed supplements and composts. Microbial biomass is one of the minimal efforts and proficient biosorbents of substantial metals expulsion from arrangements. The procedure of biosorption has numerous appealing highlights, including the evacuation of metal particles over the moderately wide scope of pH and temperature. Numerous analysts contemplated biosorption execution of various microbial biosorbents, which give respectable contentions to the usage of biosorption innovations for overwhelming metal expulsion from arrangements and furthermore to comprehend the system in charge of biosorption. Therefore, through unwavering exertion and research, overall full-scale on pilot and biosorption process, the circumstance is required to change sooner rather than later, with biosorption innovation winding up more valuable than as of now utilized physico-chemical advancements of substantial metal evacuation.

REFERENCES

Abbas, S. H., Ismail, I. M., Mostafa, T. M., and Sulaymon, A. H. 2014. Biosorption of heavy metals: A review. *Journal of Chemical Science and Technology* 3:74–102.

Aksu, Z. 2005. Application of biosorption for the removal of organic pollutants: A review. *Process Biochemistry* 40:997–1026.

Babel, S. and Kurniawan, T. A. 2003. Low-cost adsorbents for heavy metals uptake from contaminated water: A review. *Journal of Hazardous Material* 97:219–43.

Bessbousse, H., Rhlalou, T., Verchere, J. F., and Lebrun, L. 2008. Removal of heavy metal ions from aqueous solutions by filtration with a novel complexing membrane containing poly(ethyleneimine) in a poly(vinyl alcohol) matrix. *Journal of Membrane Science* 307:249–259.

Brierley, C. L. 1990. Metal immobilization using bacteria. In: H. J. Ehrlich and C. L. Brierley (Eds.), *Microbial Mineral Recovery*. McGraw-Hill, New York, 303–324.

Brillas, E. and Sires, I. 2012. Electrochemical remediation technologies for waters contaminated by pharmaceutical residues. In: E. Lichtfouse, J. Schwarzbauer, and D. Robert (Eds.), *Environmental Chemistry for a Sustainable World*. Springer, Dordrecht, 297–346.

Carolin, C. F., Kumar, P. S., Saravanan, A., Joshiba, G. J., Naushad, Mu. 2017. Efficient techniques for the removal of toxic heavy metals from aquatic environment: A review. *Journal of Environmental Chemical Engineering* 5:2782–2799.

Davis, T. A., Mucci, A., and Volesky, B. 2003. A review of the biochemistry of heavy metal biosorption by brown algae. *Water Research* 37:4311–4330.

Deng, X. and Wang, P. 2012. Isolation of marine bacteria highly resistant to mercury and their bioaccumulation process. *Bioresource Technology* 121:342–347.

Dhankhar, R. and Hooda, A. 2011. Fungal biosorption – An alternative to meet the challenges of heavy metal pollution in aqueous solutions. *Environmental Technology* 32:467–491.

Freundlich, H. M. F. 1960. Over the adsorption in solution. *Journal of Physical Chemistry* 57:385–470.

Fu, F. and Wang, Q. 2011. Removal of heavy metal ions from wastewaters: A review. *Journal of Environmental Management* 92:407–418.

Gadd, G. M. and Fomina, M. 2011. Uranium and fungi. *Geomicrobiology Journal* 28:471–482.

Gadd, G. M., Rhee, Y. J., Stephenson, K., and Wei, Z. 2012. Geomycology: Metals, actinides and biominerals. *Environmental Microbiology Reports* 4:270–296.

Gunasundari, E., Kumar, P. S., Christopher, F. C., Arumugam, T., Saravanan, A. 2017. Green synthesis of metal nanoparticles loaded ultrasonic-assisted Spirulina platensis using algal extract and their antimicrobial activity. *IET Nanobiotechnology* 11:754–758.

Ho, Y. S. and McKay, G. 1999. Pseudo-second order model for sorption processes. *Process Biochemistry* 34:451–465.

Kimura, M., Matsui, Y., Kondo, K., Ishikawa, T. B., Matsushita, T., and Shirasaki, N. 2013. Minimizing residual aluminium concentration in treated water by tailoring properties of polyaluminum coagulants. *Water Research* 47:2075–2084.

Kotrba, P. 2011. Microbial biosorption of metals-general introduction. In: P. Kotrba, M. Mackova, and T. Macek (Eds.), *Microbial Biosorption of Metals*. Springer Netherlands, Dordrecht, 1–6.

Krstic, V., Urosevic, T., and B. Pesovski. 2018. A review on adsorbents for treatment of water and wastewaters containing copper ions. *Chemical Engineering Science* 192:273–87.

Kurniawan, T. A., Chana, G. Y. S., Lo, W. H., and Babel, S. 2006. Physico-chemical treatment techniques for wastewater laden with heavy metals. *Chemical Engineering Journal* 118:83–98.

Lagergren, S. 1898. About the theory of so-called adsorption of soluble substances. *Kungl Svenska vetenskapsakademiens handlingar* 24:1–39.

Langmuir, I. 1918. The adsorption of gases on plane surfaces of glass, mica and platinum. *Journal of the American Chemical Society* 40:1361–1403.

Michalak, I., Chojnacka, K., and Witek-Krowiak, A. 2013. State of the art for the biosorption process: A review. *Applied Biochemistry and Biotechnology* 170:1389–1416.

Ngwenya, B. T. 2007. Enhanced adsorption of zinc is associated with aging and lysis of bacterial cells in batch incubations. *Chemosphere* 67:1982–1992.

Nuhoglu, Y. and Malkoc, E. 2005. Investigations of nickel (II) removal from aqueous solutions using tea factory waste. *Journal of Hazardous Materials* 127:120.

Park, D., Yun, Y-S., and Park, J-M. 2010. The past, present, and future trends of biosorption. *Biotechnology and Bioprocess Engineering* 15:86–102.

Saravanan, A., Kumar, P. S., Preetha, B., 2016. Optimization of process parameters for the removal of chromium(VI) and nickel(II) from aqueous solutions by mixed biosorbents (custard apple seeds and Aspergillus niger) using response surface methodology. *Desalination and Water Treatment* 57:14530–14543.

Suganya, S., Saravanan, A., Kumar, P. S., Yashwanthraj, M., Rajan, P. S., Kayalvizhi, K., 2016. Sequestration of Pb(II) and Ni(II) ions from aqueous solution using microalga Rhizoclonium hookeri: adsorption thermodynamics, kinetics, and equilibrium studies. *Journal of Water Reuse and Desalination* 7:214–227.

Thekkudan, V. N., Vaidyanathan, V. K., Ponnusamy, K. S., Charles, C., Sundar, S., Vishnu, D., Anbalagan, S., Kumar, S., Vaithyanathan, V. K., Subramanian, S. 2017. Review on nanoadsorbents: a solution for heavy metal removal from wastewater. *IET Nanobiotechnolgy* 11:213–224.

Wang, J. L. and Chen, C. 2006. Biosorption of heavy metals by Saccharomyces cerevisiae: A review. *Biotechnological Advances* 24:427–451.

Wang, J. L. and Chen, C. 2009. Biosorbents for heavy metals removal and their future: A review. *Biotechnology Advances* 27: 195–226.

White, C., Sayer, J. A., and Gadd, G. M. 1997. Microbial solubilisation and immobilization of toxic metals: Key biogeochemical processes for treatment of contaminant. *FEMS Microbiological Review* 20:503–516.

Wiesmann, U. 2007. Historical development of wastewater collection and treatment. In: S. Choi and E. M. Dombrowski (Eds.), *Fundamentals of Biological Wastewater Treatment*. Wiley-VCH Verlag GmbH & Co. KGaA, Weinheim, 1–23.

Xu, W., Gao, B., Wang, Y., Yue, Q., and Ren, H. 2012. Effect of second coagulant addition on coagulation efficiency, floc properties and residual Al for humic acid treatment by Al13 polymer and poly-aluminum chloride (PACl). *Journal of Hazardous Materials* 215–216:129–137.

Zouboulis, A. L., Matis, K. A., and Hancock, I. C. 1997. Biosorption of metals from dilute aqueous solutions. *Separation and Purification Methodology* 26:255–295.

2 Recent Advancements and Perspectives on Biological Degradation of Azo Dye

Rajmohan K.S., Ramya C. and Murali Mohan Seepana

CONTENTS

2.1 Introduction 18
2.2 Industrial Application of Azo Dye 18
2.3 Environmental Concerns Due to Azo Dye 19
2.4 Technologies Available for Degradation of Azo Dye 21
    2.4.1 Physico-Chemical Degradation 22
        2.4.1.1 Coagulation 22
    2.4.2 Advanced Oxidation Processes (AOPs) and Ozonation 22
    2.4.3 Biosorption 22
    2.4.4 Enzymatic Degradation 22
    2.4.5 Enzymatic Methods 23
2.5 Biological Degradation and Its Mechanism 24
    2.5.1 Biodegradation by Plants 24
    2.5.2 Biodegradation by Microbes 25
        2.5.2.1 Factors that Control Microbial Dye Decoloration 25
        2.5.2.2 Limitation of Microbial Dye Degradation 25
    2.5.3 Bioaugmentation of Microbes for Degradation 25
        2.5.3.1 Degradation of Effluents Using Genetic Engineering 27
    2.5.4 Role of Nanotechnology in Biological Remediation 27
2.6 Reactors for Biological Degradation of Azo Dye 28
    2.6.1 Continuously Stirred Anaerobic Digester 29
    2.6.2 Up-Flow Anaerobic Sludge Blanket Reactor 29
    2.6.3 Fluidized and Expanded Bed Reactors 30
    2.6.4 Anaerobic Filters (AF) 30
    2.6.5 Microbial Fuel Cells 30
2.7 Recent Advancements and Future Perspectives 31
2.8 Conclusion 31
References 31

2.1 INTRODUCTION

Dyes are the most common coloring chemical compounds employed in textile industries. Both organic as well as synthetic chemicals that can give color are known as dyes. Dyes are classified by chromophore groups in their chemical structures. Among them, azo, anthraquinone, phthalocyanine and triarylmethane are main groups of dyes. Azo dyes are widely used in textile, paper, food and the pharmaceutical industry. Azo dyes present in textile effluent are a major environmental concern. Their toxicity is due to the presence of a nitro group which is mutagenic in nature, and after breakdown they generate toxic products such as 1, 4-phenylenediamine, o-tolidine. The European Commission (EU) banned azo dye usage in 2002 due to the carcinogenic effect in order to restrict the marketing of the products that contain certain azo dyes, which are not environmentally safe (Miralles-Cueves et al., 2017). The discharge of effluent wastes from the textile industry was treated to remove toxic dyes by physical and chemical purification methods, including the advanced oxidation processes (i.e. application of ozone, hydrogen peroxide and ultraviolet). However, these processes involve high expenditure, which limits the use of that operation in the large-scale treatment profile of wastewater from the industry.

In recent decades, biological degradation of dye has gained interest mainly for its minimum energy consumption, effectiveness, safety and eco-friendliness. The use of organisms results in partial or complete bioconversion of organic pollutants to nontoxic substances. This is due to the fact that certain organisms need oxygen to thrive in the environment. Many bacterial, fungal and algal species have the capability to absorb the azo dyes on the cell wall owing to the presence of functional groups that have the potential to bind with the dye molecule. The earliest method of color removal by bacterial cells was by adsorption of the dye onto the biomass, which is like other physical adsorption mechanisms. Color removal using adsorption is not suitable for long-term treatment because during adsorption the dye gets concentrated onto the biomass, which would become saturated with time, and the dye-adsorbent composition should also be disposed of (Iyer et al., 2016).

Bacteria are the most frequently used microorganisms for the removal of dyes from textile effluents because they are easy to cultivate, adapted to survive in extreme environmental conditions both aerobic and anaerobic and decolorize the azo dyes at a faster rate. For mineralization of azo dyes, treatment systems having mixed microbial populations or certain species consortium, which are found to be more effective due to concerted metabolic activities of the microbial community.

In this chapter, biodegradation of azo dye(s) using microorganisms has been discussed. Moreover, the development of innovative nanotechnology coupled with the conventional biological process has been analyzed for the treatment of textile wastewater. Furthermore, the limitation and future scope of degradation of azo dyes were also examined.

2.2 INDUSTRIAL APPLICATION OF AZO DYE

Azo dyes are the synthetic coloring compounds containing electron-deficient xenobiotic additives and azo linkage N = N- that are linked to an aromatic ring.

These dyes have their versatile uses among diverse sectors such as textiles, prescription drugs, food, cosmetics and leather-based industries (Chang and Chia-Yu, 2001). Azo dyes are cheap, effortlessly available and offer strong shades, especially on cotton textile fabrics. The tendency to generate an electron deficiency xenobiotic aspect in these dyes makes the degradation procedure complex in diverse ecosystems (Anburaj et al., 2011).

Due to advancements in chemical technology, there are many kinds of dyes available in the market. Factors such as industrial application, structure, complex reaction and colors determine the major classification. The majority of the synthetic dyes fall under three major classifications such as anionic, cationic and non-ionic, depending on the color and application. Further, the dyes are classified into (a) acid dye, (b) direct dyes, (c) reactive dyes (basic dyes), (d) vat dyes, (e) sulfur dyes and (f) disperse dyes depending on the industrial application. The classification of dyes is illustrated in Figure 2.1. Upon coloring the textile fabrics, the effluents (unbound dyes) of these dyes are released to bodies of water, causing severe concerns to aquatic ecosystems and mankind (Robinson et al., 2001).

2.3 ENVIRONMENTAL CONCERNS DUE TO AZO DYE

Azo dyes constitute the major group of synthetic dyes, commonly used in the textile industries (Arica and Bayramoğlu, 2007). Approximately 1.3 tons of textile sectors have their greatest demand for cotton and polyester fabric materials in order to meet growing needs. The annual effluent disposal of these dyes is around 4,500,000 tons of color and/or unbound dye products, which become a concern due to their toxicity. The presence of chemical compounds gives effluents an intense color, leading to environmental and aesthetic problems. However, all of the dyes added to the fabric do not bind completely to the textile clothing materials (Asad et al., 2007). The unbound colorants after the bleaching process are released as industrial wastes.

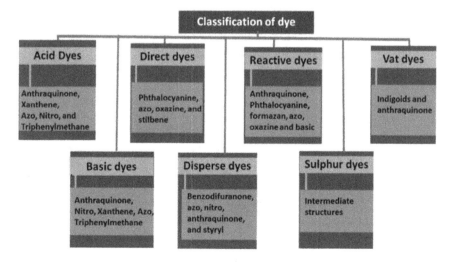

FIGURE 2.1 Classification of dyes (Yagub et al., 2014).

Yu et al.'s (2005) research study survey found that 2.81 tons of unbound dyes are released into the environment without any declarification process. These industrial effluents are harmful and toxic to flora and fauna due to the amine and heavy metal content (Weisburger, 2002).

As mentioned in Figure 2.2, released dyes in the environment cause serious health problems to mankind and animals (He et al., 2004). E.g., Metanil Yellow effluent causes serious hepatotoxic problems in albino rats (Singh et al., 1987). The unbound and toxic dyes have a greater impact in the floral water ecosystem. The presence of aromatics and heavy metals as effluents affects the sunlight penetration into the water, thereby restricting the exchange of gases to the environment (Mishra and Tripathy, 1993). When dyes are released as effluents in the water system, the penetration of sunlight into deeper layers is greatly reduced, which disturbs photosynthetic activity, leading to water quality deterioration and reduced gas solubility, ultimately causing acute toxic effects on aquatic flora and fauna, as well as human beings. The amine group of the synthetic dyes makes the effluent highly toxic (Robinson et al., 2001).

Exposure of these toxic effluents in the water bodies and environment causes genetic abnormalities and life-threatening diseases. Therefore, there is a need to study various technologies to remove the unbound dyes before releasing them into bodies of water. Prolonged effluent discharge causes long-term effects in color, oxygen level, BOD and COD levels in the aquatic system (Saratale et al., 2011). Suspended particles such as chlorine and heavy metals can shock and kill

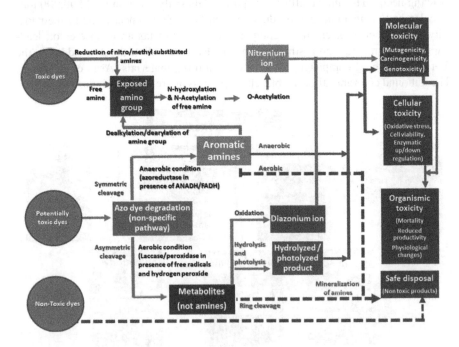

FIGURE 2.2 Effects of color dyes and consequences to the ecosystem (Rawat Deepak et al., 2016).

fish gills. The toxicity inhibits the growth of algae and photosynthetic processes (Holkar et al., 2016).

2.4 TECHNOLOGIES AVAILABLE FOR DEGRADATION OF AZO DYE

Several techniques available for wastewater treatment including physicochemical, chemical, biological and electrochemical methods are reported in the literature (Holkar et al., 2016; Rajmohan et al., 2014, 2016, 2017). Various technologies available for degradation of azo dye are illustrated in Figure 2.3. Each technique has its advantages and challenges. A comparison of merits and demerits of various dye degradation technologies is listed in Table 8.1. A brief description of the most commonly used techniques follows.

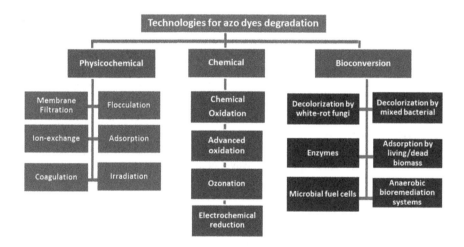

FIGURE 2.3 Treatment methods for the degradation of dyes in textile wastewater.

TABLE 8.1
Advantages and Disadvantages of Dye Removal Methods (Yagub et al., 2014)

Methods	Merits	Challenges
Oxidative process	Simple to apply	Requirement for activation of H_2O_2 agent
Ozonation	No increase in sludge	Short life
Electrochemical reduction	No toxic chemicals	High energy requirement
White-rot fungi	Easy dye degradation	Unreliable
Adsorption by microbial biomass	Specific adsorption	Unreliable
Adsorption	Variety of dye removal	Expensive
Membrane filtration	Removes all types of dyes	Concentrated sludge

2.4.1 Physico-Chemical Degradation

2.4.1.1 Coagulation

Physical methods based on flocculation are useful for the decolorization of wastewater containing dispersed dyes. Adsorption techniques have increased the trend in the decoloration of azo dyes. Many factors such as the ability to bind to the immobilized site, absorbent regeneration and stability have to be taken into account for dye degradation (Jadhav and Srivastava, 2013). Activated carbon is an effective absorbent for decoloration. The manufacturing cost, sludge generation and disposal of sludge waste hinder the usage (Galan et al., 2013).

2.4.2 Advanced Oxidation Processes (AOPs) and Ozonation

Advanced oxidation processes (AOPs) are processes containing hydroxyl radicals, which are potent key agents for oxidation. This process involves light-enhanced oxidation (use of the sunlight for the activation of the semiconductor catalyst) and chemical reactions between H_2O_2 and Fe^{3+} ions. The ferric effluent produced due to the combined flocculation of the reagent and the color molecules is one of the main drawbacks of the Fenton method. Oxidizing agents have a low degradation rate as equated with AOP processes due to lower hydroxyl radical production.

An important benefit of the ozonation process is the fact that ozone has been used in gassing and therefore does not raise wastewater volumes or generate sludge. However, the main disadvantage of ozone is that even biodegradable dyes in wastewater can produce toxic sub-products (Hayat et al., 2015).

2.4.3 Biosorption

Biosorption involves the interaction of ion exchange, adsorption and electrostatic methods for the discoloration of the chemical dyes. For instance, in the degradation of Disperse Red 1 dye, the amine and lipids are the functional regions in the electrostatic adsorption. Autoclaving at 121°C helps to identify the binding region and break the fungal anatomy resulting in flexible degradation. The binding interaction between effluent blue dye and the biomass is achieved by acid pre-treatment (Arica and Bayramoğlu, 2007).

2.4.4 Enzymatic Degradation

Many expensive physiochemical methods such as biosorption, chemical treatment, immobilization and ion extraction produce large amounts of sewage upon treatment. These processes need environmentally safe disposal to minimize the toxic effects. Decoloration of synthetic dyes using enzymes such as azo reductases and laccases has increased potential for the various industrial dyes (Rodriguez et al., 1999; Reyes et al., 1999). Added to laccases, peroxidases such as MnP, lignin and PPO have proven degradation among the various textile effluents. A simple direct enzymatic degradation process is shown in Figure 2.4.

Biological Degradation of Azo Dye

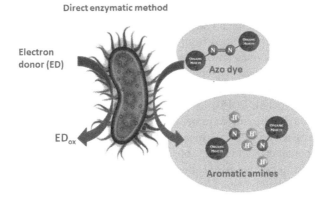

FIGURE 2.4 Enzymatic method of degradation (Singh et al., 2015).

2.4.5 ENZYMATIC METHODS

Azoreductases (EC 1.7.1.6) enzymes decolorize the unbleached heavy aromatic amines from the industrial effluents with reducing agents such as NADH, NADPH and FADH2 (Pandey and Dubey, 2012). NADH will disintegrate the azo linkage bond at the nuclear region of the bacterial cell membrane and the reducing agents act as an electron donor (Zimmermann et al., 1982). Figure 2.5 shows the enzymatic degradation of colored dyes in the presence of reducing agents (Keck et al., 1997). Decoloration is often caused by the breakdown of azo bonds, and the entire process needs the anaerobic environment for the cleavage-redox mechanism.

Laccases glycoprotein enzymes have the capacity to discolor synthetic dye effluents (Chivukula and Renganathan, 1995). These are a multimeric glycoprotein with

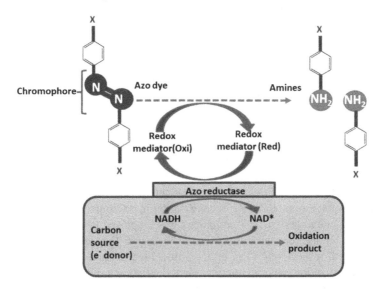

FIGURE 2.5 Discoloration process of synthetic dyes.

a copper molecule that is involved in the oxidation process. A large number of laccases enzymes are from fungal and plant origins, and a limited number of enzymes have been expressed and studied from the bacterial host (Claus, 2003; Gianfreda et al., 1999). Various enzymes have been tested for the effective phenol and aromatic effluent discoloration (Ezeronye et al., 1999). The synthesis of oxidases such as peroxidases and phenol oxidases determines the fungal ability to discolor the effluent toxic dyes (Haddaji et al., 2015).

Polyphenol or PPO is the tetramer enzyme (EC 1.14.18.1) containing four copper atmospheres bound to two compounds, namely aromatic ring and oxygen. The enzyme can operate on a wide range of substrates and can remove organic pollutants that are very low at contaminated sites (Babuponnusami et al., 2014). Textile effluents containing high polyphenol activity have been found to degrade by ammonium sulfate precipitation of potato.

2.5 BIOLOGICAL DEGRADATION AND ITS MECHANISM

Effluent treatment plants (ETPs) using biological methods are more likely to be based on lower effluent waste production and environmentally friendly and complete discoloration of dyes. Bio-based methods have benefits, such as (a) being green, (b) competitive costs, (c) reduced sludge production, (d) non-toxic metabolites or full mineralization and (e) lower water consumption (higher levels or lower requirements for dilute solutes) than physical/oxidant methods (Yagub et al., 2014).

2.5.1 BIODEGRADATION BY PLANTS

Phytoremediation uses plants to protect the environment by removal of pollutants including azo dye, and it is one of the most effective, environmentally friendly strategies for removal of dye waste. Adaptation in genetic levels is the basic strategy behind plants that are able to manage the contaminants from the polluted site. Fast-growing plants with long fibrous roots are apt for phytoremediation. *Eichhornia crassipes*, *Scirpus grossus* and the aquatic plant *Spirodela polyrhiza* are a few commonly used plants for dye degradation. A consortium of *Gaillardia x grandiflora* and *Petunia grandiflora* plants has displayed effectiveness towards dye degradation (Watharkar and Jadhav, 2014). A combined technique involving plant-associated microorganisms with *Sesbania cannabina Pers.* and *Medicago sativa L.* has the potential for azo dye decolorization. For instance, *G. pulchella (sweet)* and *M. verbena (Tronc)* have been used to treat sulfonated azo dye Green HE4B, and three non-toxic metabolites were formed (Kabra et al., 2011). Moreover, *V. radiata* (green gram), *P. vulgaris* (kidney beans) and *T. foenum-graecum* (fenugreek) were tested towards Indigo blue dye decolorization, and 94% decolorization in 48 hours was reported (Ebency et al., 2013). In another study, *E. crassipes* was employed towards Methylene Blue and Methyl Orange decolorization, and 98% and 67% decolorization efficiency were obtained, respectively (Tan et al., 2016). Five methods of phytoremediation are phytostabilization, phytoextraction, phytodegradation, phototransformation and rhizofiltration, reported elsewhere (Vidali, 2001). However, phytoremediation is an unfamiliar research area mainly due to space constraints and complexity.

2.5.2 Biodegradation by Microbes

Microorganisms are a useful alternative to the development of bio-remediation processes for processing textile wastewater for discoloration of azo dyes (Solis et al., 2012). The azo bonding linkage and sulfonic group in enzyme degradation, coupled with other electron withdrawal groups, make the discoloration susceptible to decay by microorganisms more efficient.

Azo reductases only catalyze this reaction in the presence of reductional equivalents, like FADH, NADH and NADPH (Chacko, 2011; Santos et al., 2007). Azo dyes can be degraded completely by using microorganisms which excrete several oxidases, reductases and detoxify contaminated textile water. Most azo dyes have substituent groups for sulfonate and have a high molecular weight that will not pass through cell membranes. This makes the discoloration difficult (Yang et al., 2005).

2.5.2.1 Factors that Control Microbial Dye Decoloration

Bioremediation isolated from industrial textile sewage treatment plant buckets is considered useful because they have been adapted to grow microbes under extremely severe conditions (Waghmode et al., 2011). Factors such as dyestuffs, high salinity, pH fluctuations and aromatic compounds are sensitive to microorganisms (Dua et al., 2002; Yang et al., 2009). Different microorganisms accept carbon sources differently and have a major effect on the extent of coloration. However, high carbon levels lead to low bleaching because microorganisms preferably use the carbon source or the coloration (Waghmode et al., 2011).

2.5.2.2 Limitation of Microbial Dye Degradation

The biggest drawback of degradation of microbial dye is the failure to assess the ecotoxicity of degraded products while using color degrading microbes (Rawat Deepak et al., 2016). Azo compounds, after decolorization and inadequate toxicity assessment, are likely to form more dangerous products. A global priority to achieve cost-efficient techniques, energy needs, reduced environmental efficiency and toxic sludge production remains debatable in dye detoxification efficacy (Esther et al., 2004). Figure 2.6 shows the degradation pathway using microbial sources.

2.5.3 Bioaugmentation of Microbes for Degradation

Yeast and filamentous fungi have a greater ability to degrade heavy mineralization metals such as cadmium and lead in textile effluents (Ertuğrul et al., 2008; Fairhead and Thöny-Meyer, 2012) at a very low cost. Filamentous fungi sustain growth at different levels of carbon and nitrogen sources and in lower PH level (40). E.g., degradation of Orange G and Congo red has been treated by white rot fungus *Thelephora sp.* and tested in the modified C-limited medium and low PH (Selvam et al., 2003).

Bacterial activity is inhibited by the aromatic amines in the degradation process (Qu et al., 2010). Catalytic treatments are used by the fungi to decolorize the dye compounds. Identifying the correct functional filamentous fungi can help in the better degradation of effluent wastes (Gomi et al., 2011).

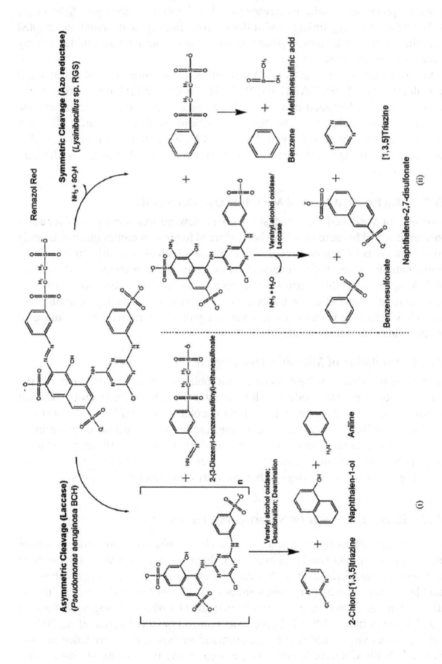

FIGURE 2.6 Microbial degradation of azo dye and its degradation pathways (Reproduced with permission from Deepak Rawat et al., 2016).

2.5.3.1 Degradation of Effluents Using Genetic Engineering

The bioremediation methodology in the treatment of effluents is limited by many factors such as PH level, salt, temperatures, carbon source etc.; the effluent compounds can lead to deactivation of enzymes and fungal cells stability. To maintain high enzyme stability and activity, various biotechnology methodologies have been adopted. Techniques such as recombinant cloning, mutation-directed evolution and protein engineering help to clone the possible gene in any microbial host cells such as fungal or bacterial host cells (Jin et al., 2007). This process involves the selection of enzyme screening with possible gene recombination to understand biological pathways for the effective degradation process. However, the cost and knowledge are still in debate, as the process is iterative.

2.5.4 ROLE OF NANOTECHNOLOGY IN BIOLOGICAL REMEDIATION

Due to the increased popularity of water treatment in recent years, several membrane technology methods have been evolved for the reduction and safe refusal of textile dyes. The nanofiltration (NF) membrane process works at a relatively low-pressure range, from 500–1,000 kPa with nano-filter carbon membranes (Dasgupta et al., 2015). There are many advantages to consider in the membrane process for sludge treatment. This includes the use of high solvent penetration into the membrane, the retention volume, higher molecular weight cut off range, 150Da modular structure, easy scale-up, easy chemical cleaning, membrane ability to resist high temperatures (approximately 70°C) and better energy consumption (Schäfer et al., 2005; Fairhead and Thöny-Meyer, 2012).

Reverse osmosis (RO) is another membrane permeability process for the ion removal from the textile effluent discharge; the treated sludge product has a low total salt content, and the color of the dyes has been greatly reduced (Kołtuniewicz and Drioli, 2008). The use of dense polymer membranes, the higher salt concentration in the effluent waste and high osmotic pressure significantly affect the membrane performance. For the effective degradation process, it is suggested that the transmembrane pressure should be greater than 2,000 kPa (Schäfer et al., 2005).

Microfiltration is used to remove the particulate suspension and colloidal coloring from the synthetic dyes and discharge from industrial effluents, as illustrated in Figure 2.7 (Dasgupta et al., 2015). Due to its close similarity to conventional crude filtration processes, microfiltration has limited application in textile effluents (Mulder, 1996). However, microfiltration membranes allow unused auxiliary chemicals, dissolved organic pollutants and other soluble contaminants to escape from the permeate. It is mainly used for the degradation and discoloration of textile industrial effluents.

Ultrafiltration is the membrane process that is used primarily to separate macromolecules and colloidal substances from solutions; UF membranes can retain the filtered solutes of several thousand Dalton molecular weights (Mulder, 1996). However, the limitation in the molecular weight of textile dye effluents makes the ultrafiltration process hinder the applications in diverse fields. Thus, dye refusal

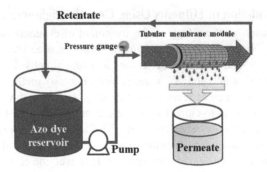

FIGURE 2.7 Experimental set-up for microfiltration.

due to ultrafiltration does not exceed 90%, although hydrophobic ultrafiltration membranes like polyether sulfones and poly(vinylidene fluoride) (PVDF) UF membranes are shown to have higher percent retention rates of dye or COD removal (Schäfer et al., 2005).

2.6 REACTORS FOR BIOLOGICAL DEGRADATION OF AZO DYE

Bioreactors used for azo dye removal can be classified as anaerobic reactor types, aerobic reactor types and integrated aerobic-anaerobic reactors (Van der Zee and Villaverde, 2005). Examples of various bioreactors reported in the literature are summarized in Figure 2.8.

Among all of the above, the most commonly used bioreactors are completely mixed anaerobic digester, UASB reactor, AFB or EGSB reactor and up-flow anaerobic filters, as shown in Figure 2.9. A brief description of the same is explained in the following section.

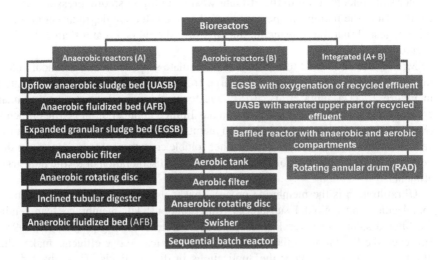

FIGURE 2.8 Classification of bioreactors for azo dye removal.

Biological Degradation of Azo Dye

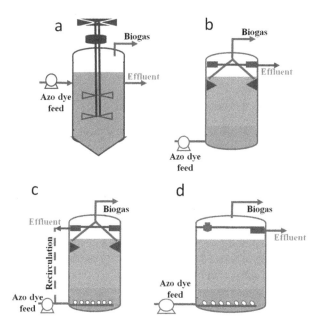

FIGURE 2.9 Most commonly used anaerobic reactor types: (a) completely mixed anaerobic digester, (b) UASB reactor, (c) AFB or EGSB reactor, (d) up-flow anaerobic filters.

2.6.1 Continuously Stirred Anaerobic Digester

This is a simple treatment system with the same solids retention time and hydraulic retention time of 20–35 days for operation and stability. These digesters do not have to be recycled and are most suitable for high solid concentration waste. A limitation of the completely mixed anaerobic digester is that more volumetric loading rate is achieved with chemical oxygen demand as high as 8,000 to 50,000 mg/L. Nevertheless, most of the waste streams are dilute. Consequently, this system reduces the economic merits of anaerobic treatment with lower COD loading per unit volume (Ersahin et al., 2011).

2.6.2 Up-Flow Anaerobic Sludge Blanket Reactor

Up-flow anaerobic sludge blanket reactor (UASB) is the most important subcategory of anaerobic treatment process technology, and it was invented by Lettinga et al. in 1980. This technology has a large window of application in treating industrial wastewater and municipal wastewater. The chemical industries, especially paper and food processing industries, are using this technology. Wastewater flows into the reactor at the bottom and moves up through sludge blanket, is distributed around a funnel with an increase in solid retention and a decrease in up-flow, and separates the solids from the wastewater. A population of bacteria which naturally forms after several weeks from granules is responsible for substrate methane fermentation. Low retention times, a good settle ability, higher biomass concentrations (30–80 g/L), elimination of the packing material cost, operation at higher loading rates and excellent solids/liquid separation eases the operation of UASB systems (Speece, 1996). The high solid content in the wastewater inhibits the granular sludge development, and this is the only limitation of this technology.

2.6.3 FLUIDIZED AND EXPANDED BED REACTORS

The bacteria are attached to small media like granular or sand in this anaerobic fluidized bed (AFB) reactor. Less clogging, good mass transfer, high specific surface area, short-circuiting and larger pore aids make these fluid bed reactors more efficient (Ersahin et al., 2011). Nevertheless, challenges in developing firmly attached biofilm comprising the correct methanogens, negative effects of the dilution at the inlet, the risk of detachment microorganisms, high recycle and high energy costs still need to be addressed (Ismail and Khudhair, 2015). The expanded granular sludge bed (EGSB) system is a modified AFB reactor with a different fluid upward velocity. This reactor makes use of lesser fluidization velocity of the bed that leads to partial fluidization of the bed (Olvera et al., 2012).

2.6.4 ANAEROBIC FILTERS (AF)

The anaerobic filter is generally used in the chemical industries, food-processing, beverage and pharmaceutical industries. This is mainly for its high capability towards biosolids retention (Ersahin et al., 2011). However, clogging due to precipitated minerals, influent suspended solids and biosolids is a major challenge to be addressed. It has been observed that down-flow packed bed (DFPB) is more efficient than the up-flow packed bed. In DFPB, Prevention of methanogens, stripping of sulfide and solids removal can be performed easily at the lower levels, upper part and top of the reactor, respectively. However, DFPB poses a threat of higher biosolids loss to the effluent, which is still considered a major hindrance (Rittmann and McCarty, 2001).

2.6.5 MICROBIAL FUEL CELLS

The microbial fuel cell (MFC) is a technology using microorganisms to catalyze the conversion of organic matters into electricity (Uma Maheswari et al., 2016). DC power can be harvested from wastewater, and simultaneously pollutants including an azo dye in wastewater can be significantly treated with microorganisms. Thus, power production, together with the electrochemical reduction of azo dyes, can be carried out. The energy requirement for the electrochemical reduction of azo dye is borne by the energy produced by the MFC. Low cost and easy handling make the MFC a promising technique for azo dye removal. Liu et al. (2009) and Das and Mishra (2019) have demonstrated the decolorization of Remazol Navy Blue in an integrated microbial fuel cell-aerobic system using bacterial consortium to lesser toxic intermediates, which were further degraded into simpler compounds in the successive aerobic treatment stage. Kong et al. (2018) have demonstrated a novel, single-chamber bioelectrochemical system (BES) to decolorize azo dye Acid Orange 7, and decolorization efficiency of 80% was achieved with a 3- module MFC.

2.7 RECENT ADVANCEMENTS AND FUTURE PERSPECTIVES

Textile effluents, when released into water bodies, cause a high toxic impact on the environment, affecting flora and fauna. Though the different kinds of degradation methods have been discussed earlier, still there is ongoing debate to manage the process to be cost-effective and environmentally friendly. Current research has been focused on adopting biological methods and physiochemical processes to detoxify the released effluents. The use of absorbents such as activated carbon and peat has demonstrated the ability to decolorize the released industrial wastes (Ayati et al., 2016).

Due to the limiting factors, such as cost and longer regeneration time, the use of absorbent has been minimized. The advent of recombinant biological techniques, enzymes and use of microbes such as fungi and bacteria has become a proven technique for bioremediation of these dyes. However, extensive research has to be done in choosing the correct biological system, proper shear process and genetic manipulation in the host cells (Zhong, 2010). Development in the bioreactor plants under limiting aerobic and anaerobic conditions has been developed. Immobilization of the selected enzymes on the gel beads has improved the catalytic activity and stability of the enzymes (Jiang et al., 2014).

Immobilization techniques involve crosslinking of enzymes, enzymes encapsulated into a polymer substrate and binding into polymers. These immobilization processes are environmentally friendly, and the coupling process can be easily achieved using biocatalysts (Patel and Gupte, 2015). The use of microbial fluid cells has developed a recent trend in the bioremediation technique. The microorganisms are allowed to oxidize the substrate at the anode end, and the electrons are further transmitted to the cathode, resulting in the generation of electric current (Zhang et al., 2016). This process needs an ion exchange membrane for the breakdown of the azo chemical bond.

2.8 CONCLUSION

Although all of the bioremediation processes have been proposed to achieve the effective decoloration process, the cost and the removal of post-treatment sludge is still a big debate. Sewage treatment plants after post-treatment release huge amounts of inorganic sludge, which affects the stability of water and dissolved oxygen levels. The regrowth of microbial flora is affected due to the post-treatment waste sludge. Advancements in modern biotechnology and bioreactors aim to design the desired gene of interest to achieve the dye bioremediation process in the aerobic and anaerobic environment. However, all these techniques need proper operational parameters such as stability and homogenization to remove recalcitrant dyes to make them environmentally friendly.

REFERENCES

Anburaj, J., T. Kuberan, C. Sundaravadivelan, and P. Kumar. "Biodegradation of azo dye by Listeria sp." *International Journal of Environmental Sciences* 1(7) (2011): 1760.

Arica, M. Y., G. Bayramoğlu, and G. Bayramoğlu. "Biosorption of Reactive Red-120 dye from aqueous solution by native and modified fungus biomass preparations of Lentinussajorcaju." *Journal of Hazardous Materials* 149(2) (2007): 499–507.

Asad, S., M. A. Amoozegar, A. A. Pourbabaee, M. N. Sarbolouki, and S. M. M. Dastgheib. "Decolorization of textile azo dyes by newly isolated halophilic and halotolerant bacteria." *Bioresource Technology* 98(11) (2007): 2082–2088.

Ayati, A., M. Niknam Shahrak, B. Tanhaei, and M. Sillanpää. "Emerging adsorptive removal of azo dye by metal–organic frameworks." *Chemosphere* 160 (2016): 30–44.

Babuponnusami, A., and K. Muthukumar. "A review on Fenton and improvements to the Fenton process for wastewater treatment." *Journal of Environmental Chemical Engineering* 2(1) (2014): 557–572.

Chacko, J. T., and K. Subramaniam. "Enzymatic degradation of azo dyes-a review." *International Journal of Environmental Sciences* 1(6) (2011): 1250.

Chang, J.-S., and C.-Y. Lin. "Decolorization kinetics of a recombinant Escherichia coli strain harboring azo-dye-decolorizing determinants from Rhodococcus sp." *Biotechnology Letters* 23(8) (2001): 631–636.

Chivukula, M., and V. Renganathan. "Phenolic azo dye oxidation by laccase from Pyricularia oryzae." *Applied and Environmental Microbiology* 61(12) (1995): 4374–4377.

Claus, H. "Laccases and their occurrence in prokaryotes." *Archives of Microbiology* 179(3) (2003): 145–150.

Das, A., and S. Mishra. "Complete biodegradation of azo dye in an integrated microbial fuel cell-aerobic system using novel bacterial consortium." *International Journal of Environmental Science and Technology* (16) (2019): 1069–1078.

Dasgupta, J., J. Sikder, S. Chakraborty, S. Curcio, and E. Drioli. "Remediation of textile effluents by membrane-based treatment techniques: A state of the art review." *Journal of Environmental Management* 147 (2015): 55–72.

del Real Olvera, J., and A. Lopez-Lopez. "Biogas production from anaerobic treatment of agro-industrial wastewater." In: *Biogas*. InTech, London, 2012: 91–112.

Dos Santos, A. B., F. J. Cervantes, and J. B. Van Lier. "Review paper on current technologies for decolourisation of textile wastewaters: Perspectives for anaerobic biotechnology." *Bioresource Technology* 98(12) (2007): 2369–2385.

Dua, M., A. Singh, N. Sethunathan, and A. Johri. "Biotechnology and bioremediation: Successes and limitations." *Applied Microbiology and Biotechnology* 59(2–3) (2002): 143–152.

Ebency, C., I. Lebanon, S. Rajan, A. G. Murugesan, R. Rajesh, and B. Elayarajah. "Biodegradation of textile azo dyes and its bioremediation potential using seed germination efficiency." *International Journal of Current Microbiology and Applied Science* 2(10) (2013): 496–505.

Ersahin, M. E., H. Ozgun, R. K. Dereli, and I. Ozturk. "Anaerobic treatment of industrial effluents: An overview of applications." In *Wastewater-Treatment and Reutilization*. InTech, London, 2011: 9–13.

Ertuğrul, S., M. Bakır, and G. Dönmez. "Treatment of dye-rich wastewater by an immobilized thermophiliccyanobacterial strain: Phormidium sp." *Ecological Engineering* 32(3) (2008): 244–248.

Esther, F., C. Tibor, and O. Gyula. "Removal of synthetic dyes from wastewaters: a review." *Environment international* 30(7) (2004): 953–971.

Ezeronye, O. U., and P. O. Okerentugba. "Performance and efficiency of a yeast biofilter for the treatment of a Nigerian fertilizer plant effluent." *World Journal of Microbiology and Biotechnology* 15(4) (1999): 515–516.

Fairhead, M., and L. Thöny-Meyer. "Bacterial tyrosinases: Old enzymes with new relevance to biotechnology." *New Biotechnology* 29(2) (2012): 183–191.

Galán, J., A. Rodríguez, J. M. Gómez, S. J. Allen, and G. M. Walker. "Reactive dye adsorption onto a novel mesoporous carbon." *Chemical Engineering Journal* 219 (2013): 62–68.

Gianfreda, L., F. Xu, and J.-M. Bollag. "Laccases: A useful group of oxidoreductive enzymes." *Bioremediation Journal* 3(1) (1999): 1–26.

Gomi, N., S. Yoshida, K. Matsumoto, M. Okudomi, H. Konno, T. Hisabori, and Y. Sugano. "Degradation of the synthetic dye amaranth by the fungus Bjerkanderaadusta Dec 1: Inference of the degradation pathway from an analysis of decolorized products." *Biodegradation* 22(6) (2011): 1239–1245.

Haddaji, D., O. Saadani L. Bousselmi, I. Nouairi, and Z. Ghrabi-Gammar. "Enzymatic degradation of azo dyes using three macrophyte species: Arundodonax, Typhaangustifolia and Phragmitesaustralis." *Desalination and Water Treatment* 53(4) (2015): 1129–1138.

Hayat, H., Q. Mahmood, A. Pervez, Z. A. Bhatti, and S. A. Baig. "Comparative decolorization of dyes in textile wastewater using biological and chemical treatment." *Separation and Purification Technology* 154 (2015): 149–153.

He, F., W. Hu, and Y. Li. "Biodegradation mechanisms and kinetics of azo dye 4BS by a microbial consortium." *Chemosphere* 57(4) (2004): 293–301.

Holkar, C. R., A. J. Jadhav, D. V. Pinjari, N. M. Mahamuni, and A. B. Pandit. "A critical review on textile wastewater treatments: Possible approaches." *Journal of Environmental Management* 182 (2016): 351–366.

Ismail, Z. Z., and H. A. Khudhair. "Recycling of immobilized cells for aerobic biodegradation of phenol in a fluidized bed bioreactor." *Systemics, Cybernetics and Informatics* 13(5) (2015): 81–86.

Iyer, P. B., B. Atchaya, K. Sujatha, and K. Rajmohan. "Comparison of synthetic dyes decolourisation by Ganoderma sp. using immobilized enzyme." *Journal of Environmental Biology* 37(6) (2016): 1507.

Jadhav, A. J., and V. C. Srivastava. "Adsorbed solution theory based modeling of binary adsorption of nitrobenzene, aniline and phenol onto granulated activated carbon." *Chemical Engineering Journal* 229 (2013): 450–459.

Jiang, Y., W. Tang, J. Gao, L. Zhou, and Y. He. "Immobilization of horseradish peroxidase in phospholipid-templated titania and its applications in phenolic compounds and dye removal." *Enzyme and Microbial Technology* 55 (2014): 1–6.

Jin, G.-Q., L. Xian-Chun, Z.-H. Xu, and W.-Y. Tao. "Decolorization of a dye industry effluent by Aspergillus fumigatus XC6." *Applied Microbiology and Biotechnology* 74(1) (2007): 239–243.

Kabra, A. N., R. V. Khandare, M. B. Kurade, and S. P. Govindwar. "Phytoremediation of a sulphonated azo dye Green HE4B by Glandularia pulchella (Sweet) Tronc. (Moss verbena)." *Environmental Science and Pollution Research International* 18(8) (2011): 1360–1373.

Keck, A., J. Klein, M. Kudlich, A. Stolz, H.-J. Knackmuss, and R. Mattes. "Reduction of azo dyes by redox mediators originating in the naphthalene sulfonic acid degradation pathway of Sphingomonas sp. strain BN6." *Applied and Environmental Microbiology* 63(9) (1997): 3684–3690.

Kołtuniewicz, A., and E. Drioli. *Membranes in Clean Technologies*. Wiley-VCH, Weinheim, 2008.

Kong, F., H.-Y. Ren, S. G. Pavlostathis, A. Wang, J. Nan, and N.-Q. Ren. "Enhanced azo dye decolorization and microbial community analysis in a stacked bioelectrochemical system." *Chemical Engineering Journal* 354 (2018): 351–362.

Lettinga, G. A. F. M., A. F. M. Van Velsen, S. W. de Hobma, W. De Zeeuw, and A. Klapwijk. "Use of the upflow sludge blanket (USB) reactor concept for biological wastewater treatment, especially for anaerobic treatment." *Biotechnology and Bioengineering* 22(4) (1980): 699–734.

Liu, L., F.-B. Li, C.-H. Feng, and X-Z. Li. "Microbial fuel cell with an azo-dye-feeding cathode." *Applied Microbiology and Biotechnology* 85(1) (2009): 175.

Miralles-Cuevas, S., I. Oller, A. Agüera, M. Llorca, J. A. Sánchez Pérez, and S. Malato. "Combination of nanofiltration and ozonation for the remediation of real municipal wastewater effluents: Acute and chronic toxicity assessment." *Journal of Hazardous Materials* 323(A) (2017): 442–451.

Mishra, G., and M. Tripathy. "A critical review of the treatments for decolourization of textile effluent." *Colourage* 40 (1993): 35–35.

Mulder, M. "Preparation of synthetic membranes." In: *Basic Principles of Membrane Technology*. Springer, Dordrecht, 1996: 71–156.

Pandey, A. K., and V. Dubey. "Biodegradation of azo dye Reactive Red BL by Alcaligenes sp. AA09." *International Journal of Engineering Science* 1(12) (2012): 51–60.

Patel, Y., and A. Gupte. "Biological treatment of textile dyes by agar-agar immobilized consortium in a packed bed reactor." *Water Environment Research: A Research Publication of the Water Environment Federation* 87(3) (2015): 242–251.

Qu, Y., S. Shi, F. Ma, and B. Yan. "Decolorization of reactive dark blue KR by the synergism of fungus and bacterium using response surface methodology." *Bioresource Technology* 101(21) (2010): 8016–8023.

Rajmohan, K. S., M. Gopinath, and R. Chetty. "Review on challenges and opportunities in the removal of nitrate from wastewater using electrochemical method." *Journal of Environmental Biology* 37(6) (2016): 1519.

Rajmohan, K. S., and R. Chetty. "Enhanced nitrate reduction with copper phthalocyanine-coated carbon nanotubes in a solid polymer electrolyte reactor." *Journal of Applied Electrochemistry* 47(1) (2017): 63–74.

Rajmohan, K. S., and R. Chetty. "Nitrate reduction at electrodeposited copper on copper cathode." *ECS Transactions* 59(1) (2014): 397–407.

Rawat, D., V. Mishra, and R. S. Sharma. "Detoxification of azo dyes in the context of environmental processes." *Chemosphere* 155 (2016): 591–605.

Reyes, P., M. A. Pickard, and R. Vazquez-Duhalt. "Hydroxybenzotriazole increases the range of textile dyes decolorized by immobilized laccase." *Biotechnology Letters* 21(10) (1999): 875–880.

Rittmann, B. E., and P. L. McCarty. *Environmental Biotechnology: Principles and Applications*. McGraw-Hill, New York, 2001.

Robinson, T., G. McMullan, R. Marchant, and P. Nigam. "Remediation of dyes in textile effluent: A critical review on current treatment technologies with a proposed alternative." *Bioresource Technology* 77(3) (2001): 247–255.

Rodriguez, E., M. A. Pickard, and R. Vazquez-Duhalt. "Industrial dye decolorization by laccases from ligninolytic fungi." *Current Microbiology* 38(1) (1999): 27–32.

Saratale, R. G., G. D. Saratale, J.-S. Chang, and S. P. Govindwar. "Bacterial decolorization and degradation of azo dyes: A review." *Journal of the Taiwan Institute of Chemical Engineers* 42(1) (2011): 138–157.

Schäfer, A. I., A. G. Fane, and Th. D. Waite, eds. *Nanofiltration: Principles and Applications*. Elsevier, The Netherlands, 2005.

Selvam, K., K. Swaminathan, and K.-S. Chae. "Decolourization of azo dyes and a dye industry effluent by a white rot fungus Thelephora sp." *Bioresource Technology* 88(2) (2003): 115–119.

Singh, R. L., P. K. Singh, and R. P. Singh. "Enzymatic decolorization and degradation of azo dyes—A review." *International Biodeterioration and Biodegradation* 104 (2015): 21–31.

Singh, R. L., S. K. Khanna, and G. B. Singh. "Safety evaluation studies on metanil yellow, (I) Acute and subchronic exposure response." *Bev Food World* 74 (1987): 9–13.

Solís, M., A. Solís, H. Inés Pérez, N. Manjarrez, and M. Flores. "Microbial decolouration of azo dyes: A review." *Process Biochemistry* 47(12) (2012): 1723–1748.

Speece, R. E. "Anaerobic biotechnology for industrial wastewaters." In: *Anaerobic Biotechnology for Industrial Wastewaters* (1996).

Tan, K. A., N. Morad, and J. Q. Ooi. "Phytoremediation of methylene blue and methyl orange using Eichhornia crassipes." *International Journal of Environmental Science and Development* 7(10) (2016): 724.

Uma M. R., C. Mohanapriya, P. Vijay, K. S. Rajmohan, and M. Gopinath. "Bioelectricity production and desalination of Halomonas sp.—the preliminary integrity approach." *Biofuels* (2016): 1–9.

Van der Zee, F. P., and S. Villaverde. "Combined anaerobic–aerobic treatment of azo dyes—A short review of bioreactor studies." *Water Research* 39(8) (2005): 1425–1440.

Vidali, M. "Bioremediation. An overview." *Pure and Applied Chemistry* 73(7) (2001): 1163–1172.

Waghmode, T. R., M. B. Kurade, and S. P. Govindwar. "Time dependent degradation of mixture of structurally different azo and non azo dyes by using Galactomycesgeotrichum MTCC 1360." *International Biodeterioration and Biodegradation* 65(3) (2011): 479–486.

Watharkar, A. D., and J. P. Jadhav. "Detoxification and decolorization of a simulated textile dye mixture by phytoremediation using Petunia grandiflora and, Gaillardia grandiflora: A plant–plant consortial strategy." *Ecotoxicology and Environmental Safety* 103 (2014): 1–8.

Weisburger, J. H. "Comments on the history and importance of aromatic and heterocyclic amines in public health." *Mutation Research/Fundamental and Molecular Mechanisms of Mutagenesis* 506 (2002): 9–20.

Yagub, M. T., T. K. Sen, S. Afroze, and H. M. Ang. "Dye and its removal from aqueous solution by adsorption: A review." *Advances in Colloid and Interface Science* 209 (2014): 172–184.

Yang, Q., M. Yang, A. Yediler, and A. Kettrup. "Decolorization of an azo dye, Reactive Black 5 and MnP production by yeast isolate: Debaryomycespolymorphus." *Biochemical Engineering Journal* 24(3) (2005): 249–253.

Yang, Q., C. Li, H. Li, Y. Li, and Y. Ning. "Degradation of synthetic reactive azo dyes and treatment of textile wastewater by a fungi consortium reactor." *Biochemical Engineering Journal* 43(3) (2009): 225–230.

Yu, Z., and X. Wen. "Screening and identification of yeasts for decolorizing synthetic dyes in industrial wastewater." *International Biodeterioration and Biodegradation* 56(2) (2005): 109–114.

Zhang, Q., J. Hu, and D.-J. Lee. "Microbial fuel cells as pollutant treatment units: Research updates." *Bioresource Technology* 217 (2016): 121–128.

Zhong, J.-J. "Recent advances in bioreactor engineering." *Korean Journal of Chemical Engineering* 27(4) (2010): 1035–1041.

Zimmermann, T., H. G. Kulla, and T. Leisinger. "Properties of purified Orange II azoreductase, the enzyme initiating azo dye degradation by Pseudomonas KF46." *European Journal of Biochemistry* 129(1) (1982): 197–203.

3 Performance Analysis of Anaerobic Digestion of Textile Dyeing Industry Effluent in a Modified Sequential Batch Reactor

S. Venkatesh Babu and M. Rajasimman

CONTENTS

3.1 Introduction 37
3.2 Materials and Methods 38
    3.2.1 Sorbent and Support Media 38
    3.2.2 Anaerobic Sequential Batch Reactor (ASBR) 39
    3.2.3 Groundnut Shell Powder..39
    3.2.4 Experimental Procedure for Optimization Studies in ASBRs 39
    3.2.5 Screening of Sorbent 41
3.3 Results and Discussion 41
    3.3.1 Start-Up of ASBR for Textile Dyeing Effluent Treatment..................41
    3.3.2 Optimization of Process Variables in ASBR..44
    3.3.3 Continuous Study in ASBR 46
    3.3.4 SVI and Mixed Liquor Volatile Suspended Solid Concentration in ASBR..49
    3.3.5 Effect of HRT and Substrate Concentration on Volatile Fatty Acid 49
    3.3.6 Gas Production and F/M Ratio in ASBR...50
3.4 Conclusions...51
References..51

3.1 INTRODUCTION

A sequencing batch reactor (SBR) is an activated sludge process designed to operate under non-steady state conditions in the presence/absence of air. The conventional sequential batch reactor operates in a batch mode with aeration and sludge settlement both occurring in the same tank. In addition, the SBR system can be designed with the ability to treat a wide range of influent volumes, whereas the continuous system is based upon a fixed influent flow rate. Thus, there is a degree of flexibility associated with working in a time rather than in a space sequence. ASBR is similar to SBR, but during the reaction step, air is not supplied into the reactor. ASBR is an excellent reactor to treat a variety of wastewaters; they could be applied to treat domestic wastewater, landfill leachate, industrial wastewater, biological phosphorus and nitrogen removal, etc. The advantages of ASBR are: (i) flexible, (ii) easy to operate, (iii) less cost (construction and maintenance) and (iv) removal of nitrogen and phosphorus. ASBR is widely employed in wastewater treatment (Hudson et al., 2001; Mohan et al., 2005; Tsang et al., 2007; Oliveira et al., 2008; Neczaj et al., 2008; Rezaee et al., 2008; El-Gohary and Tawfik, 2009; Papadimitriou et al., 2009; Chiavola et al., 2010; Durai et al., 2011; Sathian et al., 2014; Rajasimman et al., 2017).

Textile dyeing wastewater is one of the high strength wastewaters. It has biodegradable substrates and inhibitory constituents. Presence of these inhibitory components affects the growth of microorganisms in the biological process. Hence, treatment of textile dyeing wastewater by biological methods alone will not give expected results. To enhance the performance of the ASBR system in treating the textile dye wastewater, adsorbents have been added. The sorbent addition process involves simultaneous biodegradation and adsorption processes. The advantages of adding sorbent to the biological process are: (i) improvement of the removal of organic matter; (ii) improvement of the stability to shock loads and toxic upsets; (iii) enhancement of the removal of toxic substances and priority pollutants; (iv) effective color removal; (v) improvement of sludge settling and dewatering; (vi) suppression of stripping of volatile organics; and (vii) less tendency to foam in aerator. Though several new technologies are promising in terms of cost and performance, they all suffer limitations which require further research and/or need broader validation.

To develop effective treatment technologies for textile dyeing effluents, no single solution has been satisfactory for remediating a broad diversity of textile wastes. So far, many researchers focused on various biological, chemical and physical techniques for treating synthetic and real dye industry effluent. All these methods have their own merits and demerits. Combinations of chemical and biological or physical and biological treatment have also proven to be effective. Hence this study is focused on the biological and combination of physical and biological processes for the treatment of textile effluent. The main objective of this study is to examine the performance of the anaerobic sequential batch reactor for the treatment of textile dyeing industry effluent. The specific objectives are to optimize the process parameters of the ASBR such as cycle time, sorbent dosage and biomass support using RSM to study the performance of ASBR during the treatment of textile dye industry wastewater by varying the parameters such as hydraulic retention time and initial substrate concentration.

3.2 MATERIALS AND METHODS

3.2.1 SORBENT AND SUPPORT MEDIA

In the present study, various sorbents are screened for color removal in textile dye industry wastewater. The sorbents used are cheap materials and algae viz. *Sargassum tenerrimum* (A), *Hydrilla verticillata* (B), groundnut shell powder (C), neem sawdust (D), wheat bran (E), *Hypnea valentiae* (F), paddy straw (G), tamarind seed (H), tea plant leaves (I), *Turbinaria ornata* (J), pressmud (K), *Turninaria conoides* (L), sugarcane bagasse (M), coconut shell (N), bamboo waste (O), *Scenedesmus obliquus* (P), guava leaf (Q) and tamarind hull powder (R). The selected sorbent is used in the ASBR.

3.2.2 Anaerobic Sequential Batch Reactor (ASBR)

ASBR is constructed using 12 mm-thick Plexiglas that is cylindrical in shape and used for conducting the research work. The total volume of each of the reactors is 1.5 L, of which 1 L is used as working volume and the remaining 0.5 L for head space. Tubes are inserted into the reactors to ensure the filling and withdrawal of the effluent using peristaltic pumps. The mixing inside the reactors is achieved with a mechanical stirrer at the speed of 10 rpm at frequent intervals during the reaction time. The ASBR is filled with the textile dye industry effluent and inoculated with the mixed culture obtained from textile dye wastewater treatment ponds. The pH and temperature are maintained at the optimum value of 7.2 and $32 \pm 1°C$ respectively (Venkatesh Babu, 2016). The setup is left for 80 days in order to acclimatize the microorganisms. The textile dye industry wastewater is pumped into the reactor regularly and the COD reduction, MLVSS concentration and biogas production are monitored regularly in the ASBR. In ASBR, the mixed anaerobes are added along with the selected sorbent (ground nut shell powder), and the same procedure is repeated. The temperature and pH are maintained at the optimum conditions.

3.2.3 Groundnut Shell Powder

Groundnut shell is separated and soaked in water to remove the dirt particles. Then it is dried in an oven at 70°C for an hour. After that, it is crushed and sieved to different mesh sizes and used for the studies.

3.2.4 Experimental Procedure for Optimization Studies in ASBRs

After the start-up period, experiments are performed based on the central composite design (CCD) given in Table 3.1, in ASBR. This design is widely used to study the effects of the variables towards their responses and subsequently in the optimization studies. This method is suitable for fitting a quadratic surface, and it helps to optimize the effective parameters with a minimum number of experiments, as well as to analyze the interaction between the parameters. In order to determine the existence of a relationship between the factors and the response variables, the data collected are analyzed in a statistical manner, using regression. The parameters cycle time (12, 24 and 36 h) and sorbent dosage (5, 10 and 15 g/L) are optimized. The hydraulic retention time (HRT) in the reactor is maintained as eight days. During the fill period, textile dye wastewater is fed into the reactor. The reaction is carried out for a specified time (10 h, 20 h and 32 h for the cycle time; 12, 24 and 36 h respectively). Then the bio-sludge is allowed to settle, and after that the supernatant is removed during the withdrawal time. The same procedure is repeated several times.

TABLE 3.1
CCD-based Design and Results for the Optimization of Variables in ASBR

Run. No	A, h	B, g/L	Decolorization, %		COD reduction, %		SVI, mL/g	
			Experimental	Predicted	Experimental	Predicted	Experimental	Predicted
1	−1	1	63.3	63.95	67.2	68.31	105.0	103.71
2	−1	−1	60.1	59.32	65.1	64.50	107.1	108.66
3	0	−1	64.6	67.17	68.3	71.16	105.1	100.40
4	0	0	71.2	70.79	74.3	74.12	93.4	94.44
5	0	1	71.3	71.00	75.1	74.52	95.2	96.20
6	0	0	71.1	70.79	75.0	74.12	93.6	94.44
7	1	−1	68.2	66.42	72.6	70.35	100.2	103.34
8	0	0	71.4	70.79	74.8	74.12	93.6	94.44
9	1	0	67.5	69.63	70.3	73.09	101.2	97.76
10	0	0	71.3	70.79	74.6	74.12	93.8	94.44
11	1	1	69.8	69.45	73.8	73.26	99.6	99.89
12	−1	0	63.2	63.33	68.2	67.69	102.6	102.33
13	0	0	71.2	70.79	74.2	74.12	94.1	94.44

The excess bio-sludge is removed from the bottom of the ASBR to maintain the sludge concentration. Decolorization, sludge volume index (SVI) and COD are analyzed for each condition as per standard methods of analysis, APHA (1992). At the optimized conditions, ASBR performances are studied in terms of decolorization and COD reduction.

The regression and graphical analysis with statistical significance are carried out using Design-Expert software (version 7.1.5, Stat-Ease, Inc., Minneapolis, USA). In order to visualize the relationship between the experimental variables and responses, the three D plots are generated from the models. The optimum values of the process variables are found using the polynomial equation. The adequacy of the model is justified through analysis of variance (ANOVA). Lack-of-fit is a special diagnostic test for adequacy of a model that compares the pure error, based on the replicate measurements to the other lack-of-fit, based on the model performance (Noordin et al., 2004). F-value, calculated as the ratio between the lack-of-fit mean square and the pure error mean square, is the statistic parameter used to determine whether the lack-of-fit is significant or not, at a significance level.

At the optimized conditions, experiments are carried out to compare their performance at various organic loading rates (OLR). OLR is varied by changing the influent wastewater concentration (880, 1720 and 2600 mg/L) and HRT (8, 6 and 4 d). During the 75 days of operation, decolorization, COD reduction, mixed liquor suspended solids (MLSS), sludge volume index (SVI) and gas production are measured regularly at frequent intervals. Volatile fatty acids (VFA) and food to microorganism (F/M) ratio are measured at the end of operation. All the analyses are carried out as per standard methods of analysis, APHA (1992). Dye concentration is measured in bio-spectrophotometer (Model: BL-200, ELICO, India) at a wavelength of 395 nm. COD of the sample is analyzed using the procedure given in APHA. The performance of the reactor is analyzed in terms of percentage COD reduction and the percentage color removal.

3.2.5 SCREENING OF SORBENT

In this work, various agro wastes and algae are utilized as sorbent for the decolorization of textile dye wastewater in ASBRs. From the results, it is found that the ground nut shell (C) showed better ability in decolorizing the textile dye industry effluent. Hence, it is selected as a sorbent for decolorization process in ASBR.

3.3 RESULTS AND DISCUSSION

3.3.1 START-UP OF ASBR FOR TEXTILE DYEING EFFLUENT TREATMENT

In general, the start-up period of an anaerobic reactor is comparatively more when compared to the aerobic reactor due to the slow growth rate of anaerobic microorganisms. This is a vital concern for the selection of anaerobic reactors for wastewater treatment. After the successful start-up, an anaerobic reactor can be operated for a long period without much deterioration in biomass properties (Wei, 2007). Hence, monitoring the start-up phase of the anaerobic reactor is essential for any wastewater treatment.

The problems faced in anaerobic systems during the short HRT are given by Ndon and Dague (1997):

- High organic loading resulting in the domination of microorganisms, which are more dispersed
- Microorganisms have a longer settling velocity, which results in poor solid-liquid separation
- High hydraulic loading, which caused higher biomass loss in the effluent

Hence, a long hydraulic residential time is desired for the start-up of anaerobic reactors. Long HRT increases the sludge retention time of the system by preventing biomass washout and retaining the biomass in the reactor. In this work, the start-up phase of the ASBR denotes the period of operation ASBR from day one to until the performance of the ASBR is relatively stable. The performance of ASBR is analyzed in terms of MLVSS concentration, COD reduction and biogas production. The ASBR is operated continuously until the performance of the reactors stabilized is assumed as stable when the COD reduction is found to be almost constant. The start-up period of ASBR is found to be 75 to 80 days.

The biomass concentration in the ASBR is measured as MLVSS concentrations. The amount of biomass present in the reactor is associated with the food to microorganism (F/M) ratio. Also, the quality of the effluent is highly dependent on the extent of solid-liquid separation. Thus, the ability of ASBR to retain biomass will ultimately affect the effluent quality. Figure 3.1a shows the variation of the MLVSS concentrations in ASBR during the start-up period. From the figure it is observed that throughout the start-up, MLVSS is found to be almost stable (between 6100–6300 mg/L). The percentage reduction in COD in the ASBR during start-up is depicted in Figure 3.1b. It shows that the organic matter removal is more or less stable irrespective of fluctuations in the influent concentration of wastewater. During the first ten days of operation, the COD reduction is found to be less than 5%. After that, the removal efficiency increases when the COD concentration of the feed wastewater increased and remains stable.

After 80 days of operation of ASBR, it was observed that the COD reduction and biogas yield are stable (Figure 3.1c). The organic matter removal is the limiting factor for the successful start-up of ASBR. This is due to the slow growth of the anaerobic microorganisms group. It is known that both the growth yields and decay constants of heterotrophic bacteria are higher than that of methanogens and fermenters (Metcalf and Eddy, 2003). Thus, hydrolysis of complex organics into soluble compounds is much faster than the substrate utilization rate of fermentative or methanogenic microorganisms. As at the initial start-up period, heterotrophs in the seed sludge dominated the microbial population in the ASBR and hydrolysis of organics contributed to a higher amount of COD in the effluent. This is the reason for the low COD reduction efficiencies during the start-up period of ASBR.

Methanogenic bacteria are more vulnerable than acidogenic bacteria when subjected to a change in environmental conditions and food. Starvation causes inhibitions, which will decrease the rate of destruction of VFAs (Pavlostathis and Giraldo-Gomez, 1991; Wu et al., 1995). Also, acidogenic bacteria has a higher growth yield than methanogenic bacteria. This may lead to an accumulation of VFAs

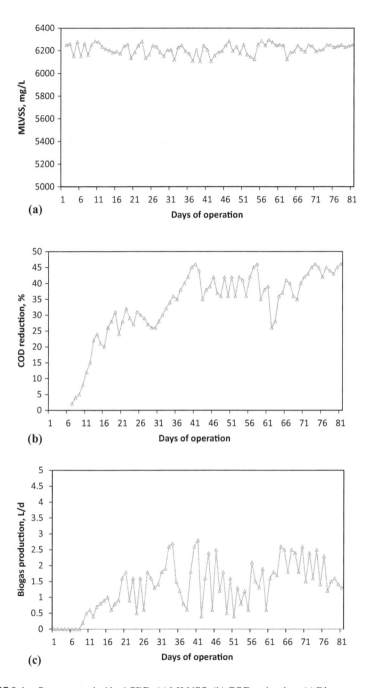

FIGURE 3.1 Start-up period in ASBR: (a) MLVSS, (b) COD reduction, (c) Biogas production.

(Sawyer et al., 1994). All the above points may inhibit the growth of methanogens and decrease their growth yield further. Hence, a potential challenge to improve the COD reduction efficiencies as well as the overall performance of the ASBR is the cultivation of methanogens.

During the start-up period, the biogas production is observed only after 8 d with significant fluctuations in the quantity of biogas produced. Whenever there is a fluctuation in temperature and pH in the reactor, a sudden drop in biogas production rate is observed. When the pH drops below 6.8 and temperature below 30°C, there is a decrease in the amount of biogas produced. During the start-up period, it is difficult to maintain the pH and temperature level in the reactor due to the varying influent quality, thus resulting in fluctuating biogas production rate.

3.3.2 Optimization of Process Variables in ASBR

The process variables, sorbent dosage and cycle time have been optimized using RSM in the ASBR. The mixed anaerobes obtained from textile dye wastewater treatment ponds are utilized for the treatment, along with the addition of sorbent. Based on the CCD design, experiments are performed, and the results, along with the predicted values, are given in Table 3.1. A second order equation relating the response and variables is:

$$\text{Decolorization}, \% = 70.79 + 4.45A + 2.71B - 0.80AB - 8.60A^2 - 3.40B^2 \quad (3.1)$$

$$\text{COD reduction}, \% = 74.12 + 3.82A + 2.38B - 0.45AB - 7.47A^2 - 2.57B^2 \quad (3.2)$$

$$\text{SVI, mL/g} = 94.44 - 3.23A - 2.97B + 0.75AB + 11.21A^2 + 7.71B^2 \quad (3.3)$$

The analysis of variance (ANOVA) results are tabulated in Table 3.2. The F value and P were used to check the applicability of the model. From the results, it is found that the model F-value for decolorization, COD reduction and SVI are significant. A low P value, less than 0.05, specifies that the model terms A, B, A^2 are found to be significant model terms for decolorization. The linear and square effects of cycle time are significant for COD reduction and square effect of cycle time is significant for SVI.

The predicted R^2 value for all the responses is in good agreement with the adjusted R^2 values. The signal to noise ratio represented by adequate precision is found to be more than 4 for all the responses. This ratio value greater than 4 is desirable. The coefficient of determination, R^2 value, greater than 0.90 for the three responses, indicates the suitability of the model. The coefficient of variation (CV) indicates the degree of precision with which the treatments are compared. Usually, the higher the value of the CV, the lower the reliability of the experiment. Here, a lower value of CV (<4) for the three responses indicates greater reliability of the experiments performed.

The response surface and contour plots give accurate geometrical representation and provide useful information about the behavior of the system within the experimental design. The optimization of process variables is targeted to give maximum percentage decolorization and COD reduction. The 3D plots for the treatment of textile dye wastewater in ASBR are shown in Figures 3.2 through 3.4. Figure 3.2 shows the 3D plot for the decolorization of textile dye wastewater in ASBR.

TABLE 3.2
ANOVA for the Treatment of Textile Dye Wastewater in ASBR

	Decolorization		COD Reduction		SVI	
Source	F	P	F	P	F	P
Model	169.38	0.0015	123.14	0.0124	261.02	0.0118
A	59.53	0.0016	43.74	0.01	31.28	0.0795
B	22.04	0.0188	17	0.0654	26.46	0.1013
AB	0.64	0.6203	0.2	0.8186	0.56	0.7913
A^2	51.11	0.0024	38.52	0.0134	86.77	0.0112
B^2	8	0.1096	4.56	0.2958	41.05	0.0512
Std. Dev.		1.54		1.89		2.73
Mean		68.02		71.81		98.81
C.V. %		2.27		2.63		2.76
R-Squared		0.9103		0.9001		0.9010
Adj R-Squared		0.8463		0.8508		0.8480
Pred R-Squared		0.7247		0.7401		0.7572
Adeq Precision		11.142		7.811		7.672

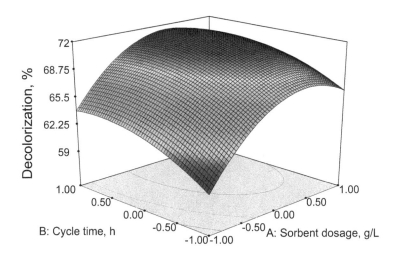

FIGURE 3.2 3D plot of the combined effects of sorbent dosage and cycle time on decolorization of textile dye wastewater in ASBR.

Figure 3.2 shows the decolorization as a function of cycle time and sorbent dosage. From the figure it is observed that, as the sorbent dosage increases, the decolorization increases. Increase in sorbent dosage up to 11.2 g/L increases %decolorization, and further increase in sorbent dosage decreases the response. The increase in the %decolorization of textile dye wastewater with the increase in sorbent dosage is due to the availability of more surface area. The decrease in %decolorization may be due to the solute transfer rate onto the sorbent surface (Nandi et al., 2008). In addition, the amount of dye compounds sorbed onto unit weight of sorbent gets split with increasing sorbent dosage (Santhi et al., 2009). Increase in the cycle period did not have significant effects on the decolorization. A similar profile is obtained for COD reduction and is shown in Figure 3.3.

Sludge volume index is used to represent the sludge settling properties and sludge bulking. Usually, SVI varies from 30 to 400 mL/g. SVI value less than 150 mL/g indicates the good settling property of the sludge. Above this value, it is termed as sludge bulking (Palm and Jenkins, 1980). A proper SVI value of below 100 mL/g is of major importance in the activated sludge systems (Tyagi, 1990).The SVI values in this study (93–107 mL/g) are found to be low when compared to the results reported by other authors (Palm and Jenkins, 1980; Janczukowicz et al., 2001; Zinatizadeh et al., 2011). It shows that the ASBR can be applied to textile dye wastewater. Increase in the sorbent dosage (up to 11.2 g/L) resulted in decrease in SVI, and the variable cycle time has no effect on SVI value (Figure 3.4). The polynomial regression equation obtained is solved using the sequential quadratic programming in MATLAB 7. The optimum values for the maximum percentage of decolorization and COD reduction are: cycle time – 24 h, sorbent dosage – 11.2 g/L.

3.3.3 Continuous Study in ASBR

The textile dyeing wastewater was treated in ASBR using anaerobes at various HRTs of 8, 6 and 4 days. The average initial substrate concentration was maintained at 880, 1720 and 2600 mg COD/L. The results attained in ASBR were shown in Figure 3.5. Initially, the HRT of the ASBR is fixed as 8 days, and this leads to an organic loading of 0.110 kg COD/m^{3}d. At this loading rate, the reactor reached steady state within 7–10 d. When the organic loading rate is further increased, it was found that the

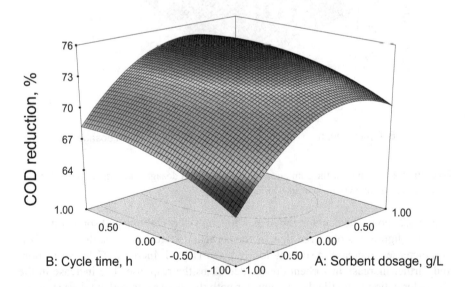

FIGURE 3.3 3D plot of the combined effects of sorbent dosage and cycle time on COD reduction of textile dye wastewater in ASBR.

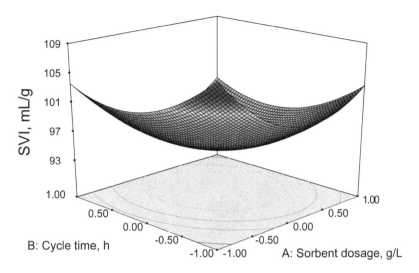

FIGURE 3.4 3D plot of the combined effects of sorbent dosage and cycle time on SVI of textile dye wastewater in ASBR.

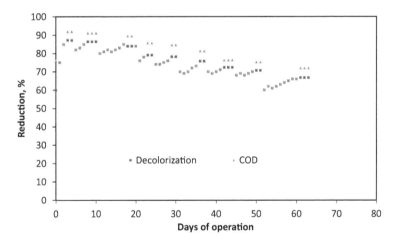

FIGURE 3.5 Decolorization and COD reduction of textile dyeing effluent using anaerobes in ASBR.

effluent qualities are almost stable. The reduction in COD and color of the system decreased with the increase in organic loading or decrease in HRT. A substantial amount of reduction in color and COD happens when the HRT was reduced from 6 days to 4 days. It was also found that there is no major increase in the removal above 6 days. Similar trends were observed for the influent initial substrate concentration of 1720 mg COD/L and 2600 mg COD/L.

The results obtained at the end of operation in the ASBR are presented in Table 3.3. From the results, it is found that increase in OLR leads to decrease in decolorization, and COD reduction and significant percentage degradation is observed under

TABLE 3.3
Performance of ASBR at Various Operating Conditions

HRT, d	SC, mgCOD/L	OLR, KgCOD/m³d	Decolorization, %	COD red, %	F/M	SVI, mL/g	MLVSS, mg/L	VFA, mg/L
8	880	0.110	87.2	91.9	0.14	94	6485	48
6		0.147	86.4	91.1	0.14	93	6420	49
4		0.220	84.0	89.5	0.14	94	6325	52
8	1720	0.215	79.1	85.7	0.27	95	6420	53
6		0.287	78.2	84.6	0.27	96	6355	54
4		0.430	75.8	81.3	0.27	95	6140	57
8	2600	0.325	72.3	76.4	0.41	95	6405	59
6		0.433	70.7	75.3	0.41	98	6250	61
4		0.650	66.7	71.9	0.41	97	6010	65

prolonged HRT. With low effluent concentrations, high decolorization and degradation percentages are achieved in relatively shorter HRT.

From Table 3.3, it is inferred that at high HRT, decolorization and COD is found to be at its maximum irrespective of initial substrate concentration. When HRT is reduced, the percentage decolorization and COD also reduces, but the significant drop in reduction occurs when the HRT is reduced from 6 days to 4 days. For lower and higher initial concentrations, the optimal HRT for better percentage COD removal efficiency is found to be 6 days.

The decolorization and COD removal efficiency in ASBR is better than the ASBR operated without the addition of sorbent (Venkatesh Babu, 2016). The reason for this is the addition of sorbent, which leads to both sorption and degradation occuring simultaneously in ASBR. This enhancement in color reduction is due to the sorption of dye molecules on the surface of sorbent. The results confirm that in this ASBR, both biodegradation and adsorption take place. The addition of ground nut shell powder is an effective way to allow continuous operation of the biological process in the presence of dye molecules and other inhibitory chemicals present in textile dyeing effluent, as indicated by increased COD as well as color removal efficiencies. This is because the inhibitory chemical compounds in the effluent will affect the performance of microorganisms during the biodegradation process.

In the powdered activated carbon treatment (PACT) system, the powdered activated carbon (PAC) particles are predominantly associated physically with the floc. The inhibitory species are concentrated in the floc, while the concentration in the bulk solution is reduced. Thus, PAC can stimulate biological activity by preventing the inhibitory substrates from exerting their toxic effect (Sublette et al., 1982). On the other hand, the biomass is in intimate contact with the wastewater to be treated, thus optimizing the conditions for enhanced biodegradation of pollutants. As the adsorbed material is biodegraded, it releases active sites on the carbon surface, which allows further adsorption of substrate (Iwami et al., 1992). The renewal of adsorptive capacity of activated carbon through the action of microorganisms is defined as bioregeneration (Sublette et al., 1982).

3.3.4 SVI AND MIXED LIQUOR VOLATILE SUSPENDED SOLID CONCENTRATION IN ASBR

In the ASBR, the SVI is found to be rather low, in the range of 80–120 mL/g (Figure 3.6), which shows good settling property of the sludge in the ASBR. Janczukowicz et al. (2001) obtained the SVI of 30–60 mL/g, in a SBR at the wastewater treatment plant in University of Olsztyn et al. (2011) and observed sludge of the SVI value of 65–105 during the dairy wastewater treatment in an SBR. The MLVSS concentrations in the ASBR at various HRTs and initial substrate concentrations are observed and depicted in Figure 3.6. From the figure it is observed that the MLVSS concentration is relatively stable in the range of 6,200–6,450 mg/L in the ASBR. When the HRT is reduced, there is a drop in MLVSS during the initial period, and then an increase and later it became relatively stable within a few days. The reaction time is reduced at lower HRT and hence leads to high MLVSS concentrations. At longer HRT (8 d), the reaction time is longer, and hence giving more time for the complex organic molecules to be converted to soluble monomer molecules by the anaerobic microorganisms. Wong Shih Wei (2007) observed that more suspended solids could be converted into soluble compounds during the longer HRT, thus reducing the MLVSS concentration.

3.3.5 EFFECT OF HRT AND SUBSTRATE CONCENTRATION ON VOLATILE FATTY ACID

From Table 3.3, it is observed that at short HRT the value of VFA is high when compared to long HRT. This may be due to the short residence time of substrates in short HRT systems, which leads to accumulation of VFAs. From the figure it is also observed that the increase in substrate concentration increases the VFA. This is because of the high substrate concentrations (high organic loadings)

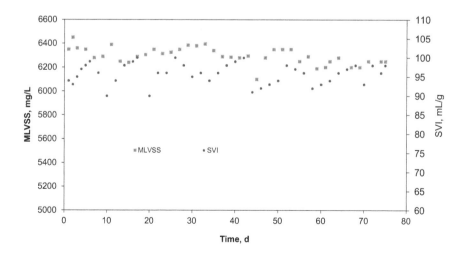

FIGURE 3.6 SVI and MLVSS during the operation of ASBR at various OLR.

resulting in a high level of production of VFAs, especially in short HRT systems, in comparison to long HRT systems where organic loadings are low and the residence time for substrate is high. These are well supported by Udeme James Ndon (1995) and Sen and Demirer (2003). The high dye concentration caused a temporary inhibition of microbial metabolism, resulting in the VFAs being present in the effluent. However, recovery is almost immediate, with the effluent VFA concentration returning to <100 mg/L. It is concluded that the reactor operation is stable since the effluent VFA is <200 mg/L throughout the test period (Willetts, 2000).

3.3.6 Gas Production and F/M Ratio in ASBR

The biogas production at various HRTs in the ASBR is monitored, and it is depicted in Figure 3.7. A decrease in HRT increases daily production of biogas. This increase is due to an increase in the loading rate and hence greater supply of degradable substrate. Based on Monod kinetics, the rise is also likely attributable to the more improved contact between the substrate and the microbial population. The specific biogas yield, which is the amount of gas produced per unit of the degraded substrate, decreases below a HRT of 6 days. In practice, if HRT is progressively reduced, with the HRT using higher daily volumetric loading and unloading, a point would be reached where there might be an imbalance of microbial populations, resulting in failure of the digester. For the textile dye wastewater, the maximum biogas yields are approximately 0.135 mL biogas/mg COD. The recommended F/M ratio for efficient treatment is 0.1 to 1 g COD/g MLVSS.d (Ndon and Dague, 1997). From the table, the F/M ratio from 0.14 mg COD/mg MLVSS.d to 0.45 mg COD/mg MLVSS.d indicates that the system's MLVSS are generally high, which resulted in high SRTs and the high performance of the ASBR systems observed in this research.

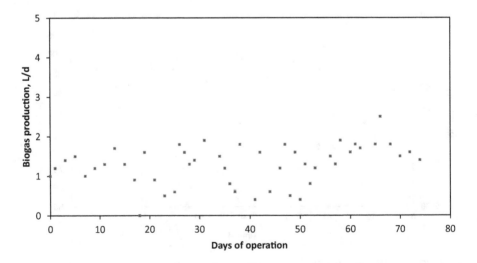

FIGURE 3.7 Biogas production during the operation of ASBR at various OLR.

3.4 CONCLUSIONS

In an anaerobic sequential batch reactor, ground nut shell powder is added as a sorbent along with microorganisms. This leads to simultaneous degradation and sorption processes. The addition of sorbent in the ASBR system is found to be effective because of its high removal efficiency of color and COD, low excess sludge production and low SVI conditions. From the results it is found that the addition of sorbent increases the performance of ASBR by 14–16% in terms of color and COD removal. A maximum decolorization of 87.2% is achieved at an OLR of 0.110 kg COD/m^{3}d. A low SVI (92–104 mL/g) value indicates good settling of sludge in the ASBR. Another advantage of using sorbent in the ASBR system is that the sorbent acts as a medium for the biofilm to attach to but will not adsorb the dyes on the surface. From the results, it is observed that the ASBR with the addition of sorbent could be effectively utilized for the treatment of textile dyeing industry wastewater.

REFERENCES

APHA. 1992. *Standard Methods for the Examination of Water and Wastewater*, 16th ed. American Public Health Association, New York.

Chiavola, A., Baciocchib, R., Gavasci, R. 2010. Biological treatment of PAH contaminated sediments in a sequencing batch reactor. *Journal of Hazardous Materials* 184(1–3): 97–104.

Durai, G., Rajasimman, M., Rajamohan, N. 2011. Aerobic digestion of tannery wastewater in a sequential batch reactor by salt-tolerant bacterial strains. *Applied Water Science* 1(1–2): 35–40.

El-Gohary, F., Tawfik, A. 2009. Decolorization and COD reduction of disperse and reactive dyes wastewater using chemical-coagulation followed by sequential batch reactor (SBR) process. *Desalination* 249(3): 1159–64.

Hudson, N., Doyle, J., Lant, P., Roach, N., Bruyn, B., Staib, C. 2001. Sequencing batch reactor technology: The key to a BP refinery [Bulwer Island] upgraded environmental protection system – A low cost lagoon based retro-fit. *Water Science and Technology* 43: 339–46.

Iwami, N., Imai, A., Inamori, Y., Sudo, R. 1992. Treatment of a landfill leachate containing refractory organics and ammonium nitrogen by the microorganism-attached activated carbon fluidized bed process. *Water Science and Technology* 26(9–11): 1999–2002.

James Ndon, Udeme. 1995. *Anaerobic Sequencing Batch Reactor Treatment of Low Strength Wastewater. Retrospective Theses and Dissertations.* Iowa State University.

Janczukowicz, W., Szewczyk, M., Krzemieniewski, M., Pesta, J. 2001. Settling properties of activated sludge from a sequencing batch reactor (SBR). *Polish Journal of Environmental Studies* 10: 15–20.

Metcalf, Eddy. 2003. *Wastewater Engineering Treatment and Reuse*, 4th ed. McGraw-Hill Companies, Inc.

Mohan, S.V., Chandrashekara, N.R., Krishna, K.P., Madhavi, B.T.V., Sharma, P.N. 2005. Treatment of complex chemical wastewater in a sequencing batch reactor (SBR) with an aerobic suspended growth configuration. *Process Biochemistry* 40(5): 1501–8.

Nandi, B.K., Goswami, A., Das, A.K., Mondal, B., Purkait, M.K. 2008. Kinetic and equilibrium studies on the adsorption of crystal violet dye using kaolin as an adsorbent. *Separation Science and Technology* 43(6): 1382–403.

Ndon, U.J., Dague, R.R. 1997. Effects of temperature and hydraulic retention time on anaerobic sequencing batch reactor treatment of low strength waster. *Water Resource* 31: 2455–66.

Neczaj, E., Kacprzak, M., Kamizela, T., Lach, J., Okoniewska, E. 2008. Sequencing batch reactor system for the co-treatment of landfill leachate and dairy wastewater. *Desalination* 222(1–3): 404–9.

Noordin, M.Y., Venkatesh, V.C., Sharif, S., Elting, S., Abdullah, A. 2004. Application of response surface methodology in describing the performance of coated carbide tools when turning AISI 1045 steel. *Journal of Materials Processing Technology* 145(1): 46–58.

Oliveira, R.P., Ghilardi, J.A., Ratusznei, S.M., Rodrigues, J.A.D., Zaiat, M., Foresti, E. 2008. Anaerobic sequencing batch biofilm reactor applied to automobile industry wastewater treatment: Volumetric loading rate and feed strategy effects. *Chemical Engineering and Processing: Process Intensification* 47(8): 1374–83.

Palm, J., Jenkins, D. 1980. Relationship between organic loading, dissolved oxygen concentration and sludge settleability in the completely mixed activated sludge process. *Journal of Water Pollution Control Federation* 10: 2484–506.

Papadimitriou, C.A., Samaras, P., Sakellaropoulos, G.P. 2009. Comparative study of phenol and cyanide containing wastewater in CSTR and SBR activated sludge reactors. *Bioresource Technology* 100(1): 31–7.

Pavlostathis, S.G., Giraldo-Gomez, M. 1991. Toxic inhibition of anaerobic biodegradation. *Journal of Water Pollution Control Federation* 52: 472–82.

Rajasimman, M., Venkatesh Babu, S., Rajamohan, N. 2017. Biodegradation of textile dyeing industry wastewater using modified anaerobic sequential batch reactor – Start-up, parameter optimization and performance analysis. *Journal of the Taiwan Institute of Chemical Engineers* 72: 171–81.

Rezaee, A., Khavanin, M.A., Ansari, M. 2008. Treatment of work camp wastewater using a sequencing batch reactor followed by a sand filter. *American Journal of Environmental Sciences* 4(4): 342–6.

Santhi, T., Manonmani, S., Smitha, T., Mahalakshmi, K. 2009. Adsorption of malachite green from aqueous solution onto a waste aqua cultural shell powders (prawn waste): Kinetic study. *Rasayan Journal of Chemistry* 2: 813–24.

Sathian, S., Rajasimman, M., Rathnasabapathy, C.S., Karthikeyan, C. 2014. Performance evaluation of SBR for the treatment of dyeing wastewater by simultaneous biological and adsorption processes. *Journal of Water Process Engineering* 4: 82–90.

Sawyer, C.N., McCarty, P.L., Parkin, G.F. 1994. *Chemistry for Environmental Engineering*, 4th ed. McGraw-Hill, p. 658.

Sen, S., Demirer, G.N. 2003. Anaerobic treatment of synthetic textile waste water containing a reactive azo dye. *Journal of Environmental Engineering* 129: 595–601.

Sublette, K.L., Sinder, E.H., Sylvester, N.D. 1982. A review of the mechanism of powdered activated carbon enhancement of activated sludge treatment. *Water Research* 16(7): 1075–82.

Tsang, Y.F., Hua, F.L., Chua, H., Sin, S.N., Wang, Y.J. 2007. Optimization of biological treatment of paper mill effluent in a sequencing batch reactor. *Journal of Biochemical Engineering* 34(3): 193–9.

Tyagi, R.D. 1990. *Wastewater Treatment by Immobilized Cells*. CRC Press, Boca Raton, FL.

Venkatesh Babu, S. 2016. Treatment of Dye Wastewater in a Biological Reactor. Ph.D. thesis, Annamalai University, Tamilnadu, India.

Wei, W.S. 2007. *Anaerobic Sequencing Batch Reactor for the Treatment of Municipal Wastewater*. M.E. (Thesis), National University of Singapore, Singapore.

Willetts, J.R.M., Ashbolt, N.J., Moosbrugger, R.E., Aslam, M.R. 2000. The use of a thermophilic anaerobic system for pretreatment of textile dye wastewater. *Water Science and Technology* 42(5–6): 309–16.

Wu, W.M., Jain, M.K., Thiele, J.H., Zeikus, J.G. 1995. Effect of storage on the performance of methanogenic granules. *Water Research* 29(6): 1445–52.

Zinatizadeh, A.A.L., Mansouri, Y., Akhbari, A., Pashaei, S. 2011. Biological treatment of a synthetic dairy wastewater in a sequencing batch biofilm reactor: Statistical modeling using optimization using response surface methodology. *Chemical Industry and Chemical Engineering Quarterly* 17(4): 485–95.

4 Waste Sea Shells for Biodiesel Production – Current Status and Future Perspective

Subramaniapillai Niju and M. Balajii

CONTENTS

4.1 Introduction 53
4.2 Waste Shells..54
    4.2.1 Environmental Impacts of Waste Sea Shells.......................................54
    4.2.2 Beneficial Uses of Waste Sea Shells..55
        4.2.2.1 Biopolymer Synthesis 55
        4.2.2.2 Astaxanthin Extraction 55
        4.2.2.3 In Constructions...55
        4.2.2.4 Wastewater Treatment..56
        4.2.2.5 Adsorbent...56
        4.2.2.6 Biodiesel Synthesis 56
    4.2.3 Characterization of Waste Sea Shells...56
4.3 Catalyst Preparation and Modification Technique 65
4.4 Production of Biodiesel Using Waste Shells..71
4.5 Future Perspective of Waste Shells...78
4.6 Conclusion 82
References..82

4.1 INTRODUCTION

Until now, industry and transportation sectors of the world have been highly reliant on non-renewable energy sources such as crude oil, petrol, and other fossil-derived fuels. Increasing and continuous consumption of fossil fuels has triggered detrimental impacts on the environment and resulted in global warming. Limited reserves of non-renewable energy and increasing energy demand have stimulated the world to look for an alternative fuel (Karmee 2018; Shalmashi and Khodadadi 2018). In India, biodiesel is gaining huge attention as a supreme alternative due to its advantages over other fossil-based resources. Biodiesel is renewable, biodegradable and non-toxic (Catarino et al. 2017; Qin et al. 2010; Ozturk et al. 2010). In addition, it is considered environmentally friendly because it causes less emission of greenhouse gases and particulate matter. Also, it was popularly known as cleaner fuel, as it does not contain sulfur (Atapour and Kariminia 2013; Wang, Xie, and Zhong 2017; Ghanei et al. 2016). Generally, biodiesel can be produced by pyrolysis, micro-emulsion, blending with diesel fuel, and transesterification. Among these methods, transesterification is the most common, cost-efficient, and easy method for biodiesel production (Wang, Xie, and Zhong 2017).

Transesterification is a process in which triglycerides present in any kind of lipid feedstock, primarily vegetable oil (edible and non-edible) or animal fat, reacts with alcohol in the presence of a catalyst to produce fatty acid methyl esters (FAME) and glycerol as a byproduct (Atapour and Kariminia 2013). The use of edible vegetable oil as feedstock for biodiesel production has caused deforestation, increased food prices, and also amplified the production cost of biodiesel. Recently, non-edible oils and used cooking oil have gained the attention of researchers as feedstock for biodiesel (Niju, Anushya, and Balajii 2018a; Wang, Xie, and Zhong 2017; Alhassan and Kumar 2016; Demirbas et al. 2016). The catalysts used for transesterification mostly depend on the free fatty acid content of the lipid feedstock. Homogeneous catalysts are used conventionally for transesterification reactions under mild temperatures and produce high conversion rates in less time. But, these catalysts are difficult to recover, generate wastewater, and cause saponification (Buasri et al. 2014). Hence, heterogeneous catalyst-based transesterification is gaining increasing attention as it exhibits numerous advantages, such as being non-corrosive, environmentally friendly, easier to separate from liquid byproducts, and reusable. Also, recent research was mainly focused on the development of cost-effective heterogeneous catalysts from different waste materials (Thushari and Babel 2018; Taufiq-Yap, Lee, and Lau 2012).

Among several wastes, the waste sea shells accumulated along the seashore can be mitigated by utilizing them as catalysts for biodiesel production, since they cause environmental pollution. Mostly, these materials contain calcium carbonate, which can be transformed and used effectively as a heterogeneous catalyst for transesterification. The present study aims to provide an informative review on sea shell-derived heterogeneous catalyst and its application for biodiesel production.

4.2 WASTE SHELLS

4.2.1 Environmental Impacts of Waste Sea Shells

Waste materials generated from different sources have been recycled and utilized in a beneficial way to avoid problems caused by their disposal. One of the potential and prospective wastes is waste sea shell. In India, 35–45% by weight of seafood raw material goes to waste. About 8.5 million tons of waste was generated by the shellfish processing industry every year, with shrimp processing accounting for more than one hundred thousand tons of industrial waste. The disposal of such wastes in landfills has become a serious environmental problem. Waste shells include oyster, clam, scallop, and mussel shells. Most of the sea shells are disposed into public water bodies or reclaimed lands. If the waste dumped is left untreated for a long time, it can cause severe environmental pollution and produce noxious odors due to the decay of organic materials or due to gases such as H_2S, NH_3, and a few amines caused by the decomposition of salts. Furthermore, it can negatively affect the lives and health of local people. The treatment of waste shells also requires huge expense, and thus research has been conducted to explore the potential utilization of these waste sea shells.

4.2.2 BENEFICIAL USES OF WASTE SEA SHELLS

4.2.2.1 Biopolymer Synthesis

Sea shell waste is a potential raw material which can be used in the production of value-added products. Alabaraoye, Achilonu, and Hester (2018) utilized sea shells as a source for the production of the biopolymer (chitin). Shell wastes utilized in this process include mollusks (mussel and oyster shell) and crustacean (prawn and crab). They are made up of about 30–40% protein, about 30–50% calcium carbonate and calcium phosphate, and 20–30% chitin. Isolation and purification of chitin from these waste shells includes two steps: (1) removal of proteins (deproteinization) by treatment with sodium hydroxide and (2) removal of minerals (demineralization) by treatment with hydrochloric acid. The yield of chitin produced ranged from 31.11 to 69.65%. Chitin produced was of high quality and was used in medicine, food, biotechnology, agricultural, and cosmetic industries.

4.2.2.2 Astaxanthin Extraction

Dalei and Sahoo (2015) employed shell waste of shrimp and crustaceans in the recovery of astaxanthin, which is a carotenoid that belongs to a class of phytochemicals known as tetraterpenoids. The extraction was done by using six different organic solvents, and the phytochemical constituents in the extract were screened. Thin layer chromatography was used to identify the carotenoid fractions in the extract. Hexane and acetone extract contained the highest number of carotenoid fractions, which include astaxanthin, astaxanthin mono, and diester. Due to its potent antimicrobial activity, the crustacean waste extract can be used in pharmaceutical industries.

4.2.2.3 In Constructions

The most prevalent and important use of the waste shell is in the construction of buildings, and it is considered a cheaper method to meet the growing demand for construction because of its abundant quantity. Richardson and Fuller (2013) used waste shells as a partial aggregate replacement in concrete, which decreased the requirement of quarried aggregate. Two aspects of performance of the concrete when it was made with a proportion of sea shells, namely compressive strength and porosity, were examined. The shells were characterized and confirmed to contain no components that might adversely react with cement. The shells were broken and added to the concrete mix. The porosity and compressive strength of the concrete produced by sand replacement and coarse aggregate replacement were compared with that of the plain concrete. It was concluded that using sea shells as a replacement in cement reduced the porosity of the aggregate replacement at a small percentage, with no change in compressive strength. Mo et al. (2018) investigated the use of sea shells in concrete. The durability, setting time, and air content were studied in detail, and they concluded that the workability and strength of concrete made with seashell waste, particularly as a partial fine aggregate replacement, could be attained as long as the maximum replacement levels are limited to below 20%.

4.2.2.4 Wastewater Treatment

Shell wastes also play an important role in wastewater treatment to remove unwanted substances like hormones, pharmaceuticals, or fertilizers. The tertiary water treatment involves the use of titanium dioxide, which is expensive. Hydroxyapatite from calcium derived from waste shells can be used as an alternative to titanium dioxide. Jones et al. (2011) investigated the use of waste mussel shell as a replacement of limestone for removal of phosphate. The phosphate-removing capacity was studied by adding a known amount of calcined shell to a 10 mg orthophosphate solution in deionized water in a 2 L stirred batch reactor. The standard vanadomolybdophosphoric acid colorimetric method was used to measure the concentration of phosphate in the initial solution and reaction supernatant. It was observed that around 90% removal of phosphate was achieved by calcined shells.

4.2.2.5 Adsorbent

Jung et al. (2012) recycled waste shells as a sorbent for acidic gases and nitrogen oxide from industry. It is a significant method to preserve the marine ecosystem. Shells were pretreated with calcination and hydration to increase the surface area and pore volume. Pretreatment also converts calcium carbonate present in the shells to calcium oxide and calcium hydroxide, which react with acidic gases. Temperature and time for hydration reaction of absorbent were optimized as about 80–90°C and 24 h. The experiments were carried out in a fixed-bed quartz reactor placed in a hot water bath at 1°C. The sample was added and the temperature was stabilized in N_2 flow followed by injection of SO_2, O_2, and NO_x by the mass flow controller. Gas concentration was measured by hygrometer and SO_2/NO_x analyzer. The SO_2 removal activity and reaction rate of calcined/hydrated waste oyster shells was higher than that of calcined/hydrated limestone, and they can be used as a substitute for commercial limestone.

4.2.2.6 Biodiesel Synthesis

In recent decades, the use of waste sea shell as heterogeneous catalyst for biodiesel production has been significantly studied, owing to their numerous advantages over homogeneous catalyst which include less pollution, easier separation, and non-toxicity. Other significant uses of waste sea shells include soil conditioner, sludge conditioner, and as a flavorant.

4.2.3 CHARACTERIZATION OF WASTE SEA SHELLS

Scanning electron microscope analysis was employed to obtain information on the external morphology and molecular size of the waste shell from detailed high-resolution images of the material. The morphology of various sea shells was found to be spherical (Hangun-Balkir 2016), cubic (Nur Syazwani, Rashid, and Taufiq Yap 2015), angular (Raut, Jadhav, and Nimbalkar 2016), vesicular (Li, Jiang, and Gao 2015; Nurdin et al. 2015), rod-like particles (Viriya-empikul et al. 2010; Niju, Anushya, and Balajii 2018a), and hexagonal (Niju, Niyas, Sheriffa Begum, and Anantharaman 2015). A few shells were irregular and uneven in shape (Hadiyanto, Lestari, and Widayat 2016a; Shankar and Jambulingam 2017; Madhu et al. 2016; Hu, Wang, and Han 2011; Hadiyanto et al. 2017). The porous structure was observed, and the particles were bonded together as aggregates. The molecular size of particles ranges from a few micrometers (Niju, Sheriffa Begum, and Anantharaman 2016a) to less than 100 micrometers (Hadiyanto et al. 2016a). Porosity and molecular size increase after calcination and modification (Niju, Sheriffa Begum, and Anantharaman 2016a).

X-ray diffraction is a technique used in characterizing the molecular structure of a crystalline material. The diffraction pattern produced is the fingerprint of the material being analyzed. The diffraction pattern can be used to identify the compound present in the material. Powder diffractogram of known compounds is compiled into a database using Joint Committee on Powder Diffraction Standard (JCPDS). The XRD analysis of waste sea shells was performed, and the characteristic peaks obtained were compared with JCPDS. The peaks appeared from diffractogram of calcined waste shells at 2θ of 32°, 37°, 53°, 64°, 67°, 79°, 88°, 92°, 104°, and 111° (Buasri et al. 2014) and were characteristic peaks of calcium oxide, while the uncalcined sea shells appeared to have peaks which were characteristic of calcium carbonate (Li, Jiang, and Gao 2015). Thus, the calcination process converts calcium carbonate to calcium oxide.

Fourier transform infrared (FTIR) analysis produces an infrared absorption spectrum which is used to identify the organic material. The chemical bonds from the spectrum are characteristic of a particular compound and are used to identify the material. The absorption band of CaO appeared around 600–800 cm^{-1} and 1400 cm^{-1} (Madhu et al. 2016; Suryaputra et al. 2013). The stretching and bending vibration of calcium hydroxide ($Ca(OH)_2$) formed by the reaction of calcium oxide with atmospheric air were observed around 3600–3700 cm^{-1} and 1460 cm^{-1} (Nurdin et al. 2015; Margaretha et al. 2012). However, the characteristic peak corresponding calcium carbonate ($CaCO_3$) decreases after the calcination process (Margaretha et al. 2012).

Brunauer-Emmett-Teller (BET) analysis evaluates the surface area and pore volume of the material using adsorption and desorption techniques. The increase in the surface area of the waste sea shell increases the active site for the reaction. The modification techniques employed in catalyst preparation are adopted to enhance the rate of the reaction by increasing the surface area. The unmodified shell had the surface area around 1–7 m^2/g (Chen et al. 2016; Mazaheri et al. 2018; Viriya-empikul et al. 2010), while the modified sea shell had higher surface area (Chen et al. 2016; Niju, Sheriffa Begum, and Anantharaman 2016a), which produced a high yield of biodiesel.

XRD and FTIR peaks of the waste shells so far reported in the literature confirmed the presence of CaO formed by calcination of $CaCO_3$. It also reported the presence of peaks corresponding to $Ca(OH)_2$ and $CaCO_3$ due to the sensitivity of calcium oxide to ambient air. BET analysis affirmed that the modification by impregnation increased the surface area and pore volume of the CaO catalyst. Porosity and molecular size increase after calcination and modification (Niju, Sheriffa Begum, and Anantharaman 2016a) were also assured by SEM analysis. The summary of characterization results reported on various sea shells is presented in Table 4.1.

TABLE 4.1
Summary of Characterization Results Reported on Various Sea Shells

Name of the Sea Shell Used as a Catalyst	Surface Observations Using SEM	Characteristic Peaks of CaO Identified Using XRD	BET Analysis Surface Area (m²/g)	BET Analysis Pore Volume (cm³/g)	Functional Group Identification Using FTIR	References
Abalone	Unmodified CaO (U-CaO) – Layered stack structure Modified CaO (M-CaO) – Porous and fluffy structure	32.1°, 37.2°, 53.9°, 64.2° and 67.4°	U-CaO-3.8 M-CaO-14.7	U-CaO-0.009 M-CaO-0.0094	–	(Chen et al. 2016)
Amusium cristatum	–	34°, 38°, and 54°	–	–	Two infrared bands around 868 and 1420 cm⁻¹	(Suryaputra et al. 2013)
Anadara granosa and *Paphia undulata*	Molecular size of 1. *Anadara granosa* – less than 85 μm 2. *Paphia undulata* – less than 75 μm	*Anadara granosa* – 32.27°, 37.43°, 53.92°, 64.20°, and 67.44° *Paphia undulata* – 32.26°, 37.40°, 53.90°, 64.17°, and 67.42°	–	–	Band at 3644 cm⁻¹	(Hadiyanto et al. 2016b)
Anadara granosa	Irregular particles Molecular size less than 100 μm	32°, 37°, 53°, 64°, and 67°	–	–	–	(Hadiyanto, Lestari, and Widayat 2016a)

(*Continued*)

TABLE 4.1 (CONTINUED)
Summary of Characterization Results Reported on Various Sea Shells

Name of the Sea Shell Used as a Catalyst	Surface Observations Using SEM	Characteristic Peaks of CaO Identified Using XRD	BET Analysis Surface Area (m²/g)	BET Analysis Pore Volume (cm³/g)	Functional Group Identification Using FTIR	References
Scallop shell	–	32°, 37°, and 52°	7.33	–	–	(Sirisomboonchai et al. 2015)
Lobster shells	Spherical particles with the size range from 0.20 to 5.0 μm of width	32.40°, 37.50°, and 54.10°	–	–	Band at 1450 cm^{-1} (O-Ca-O)	(Hangun-Balkir 2016)
Chicoreus brunneus	–	32.30°, 37.38°, 53.8°, and 54.10°	1.56	0.0044	Absorption band at wave numbers <600 cm^{-1} (CaO stretching) Absorption peak at 3,641 cm^{-1} (OH stretching)	(Mazaheri et al. 2018)
Combusted mollusks	–	32.2° and 37.4°	–	–	–	(Abang Chi et al. 2016)
Crab	Constructive surface area, non-uniform	32.45° and 37.54°	365	23.6	–	(Shankar and Jambulingam 2017)
Crab	Irregular, maximum Particles were in the range 1–1.8 μm	32°, 37.3°, 53.8°, 64.4°, and 67.3°.	16.4775	0.032475	Band at 700 cm^{-1} (CaO)	(Madhu et al. 2016)
Cyrtopleura costata (angel wing shell)	Cluster of well-developed cubic crystal with obvious edges	32.49°, 37.64°, 54.13°, 64.41°, and 67.63°	1.83	–	3359.39 cm^{-1} (Ca(OH)$_2$)	(Nur Syazwani, Rashid, and Taufiq Yap 2015)

(Continued)

TABLE 4.1 (CONTINUED)
Summary of Characterization Results Reported on Various Sea Shells

Name of the Sea Shell Used as a Catalyst	Surface Observations Using SEM	Characteristic Peaks of CaO Identified Using XRD	BET Analysis Surface Area (m²/g)	BET Analysis Pore Volume (cm³/g)	Functional Group Identification Using FTIR	References
Enamel venus	Irregular in shape and some of them bonded together as aggregates	32°, 36°, 54°, 65°, and 67°	—	—	—	(Buasri et al. 2015)
Mussel	Irregular	34°, 38°, 54°, 65°, and 67°	23.2	—	—	(Hu, Wang, and Han 2011)
Green mussel shell	Uneven structure and shape. The particle sizes in the range of 5 to 10 μm	31.64°, 36.6°, 52.9°, 66.36°, 78.38°, and 88.62°	—	—	C=O and C-O stretching are clearly observed at region of 1700–1800 cm^{-1} and 1170–1200 cm^{-1}, respectively	(Hadiyanto et al. 2017)
Mud crab shell	Angular in shape, size ranging from 1 to 10 μm	29.7147°	—	—	—	(Raut, Jadhav, and Nimbalkar 2016)
Mussel shell	Calcined shells were rod-like molecules, and some of them were bonded together as aggregates, vesicular structure	32.318°, 37.455°, and 53.930°	—	—	1470, 1040, and 820 cm^{-1} (CO_3^{2-})	(Li, Jiang, and Gao 2015)
Meretrix venus shell Golden apple snail shell	Rod-like particles, the largest size of aggregated particles was observed	—	Meretrix venus – 0.9 Golden apple snail shell – 0.5	Meretrix venus – 0.004 Golden apple snail shell – 0.002	—	(Viriya-empikul et al. 2010)

(Continued)

TABLE 4.1 (CONTINUED)
Summary of Characterization Results Reported on Various Sea Shells

Name of the Sea Shell Used as a Catalyst	Surface Observations Using SEM	Characteristic Peaks of CaO Identified Using XRD	BET Analysis Surface Area (m²/g)	BET Analysis Pore Volume (cm³/g)	Functional Group Identification Using FTIR	References
Mud clam shell	The porous structure, form of aggregates	—	68.57	0.100	1.1650 cm^{-1} and 2350 cm^{-1} correspond to C=O 2.950 cm^{-1} and 875 cm^{-1} correspond to C-O	(Ismail et al. 2016)
Mussel shell	—	32°, 37°, 54°, 63° and 67°	—	—	—	(Rezaei, Mohadesi, and Moradi 2013)
Mussel shell base catalyst	Vesicular structure, irregular in shape	34°, 38°, 55°, 64° and 67°	0.68	0.02	Stretching and bending vibration of Ca(OH)$_2$ –3753.72 cm^{-1}, 1462.9 cm^{-1}	(Nurdin et al. 2015)
Mussel, cockle, and scallop	Porous structure, irregular in shape, bonded together as aggregates, less porous	33°, 37°, 54°, 64° and 67°	Mussel – 89.91 Cockle – 59.87 Scallop – 74.96	Mussel – 0.130 Cockle – 0.087 Scallop – 0.097	—	(Buasri et al. 2013)
White bivalve clam shell	Modified CaO (M-CaO) – rod-like particles with sizes ranging from 53.9 to 62.6 nm of width. Unmodified CaO (U-CaO) – sizes ranging from 1.71 to 2.42 μm of width	—	U-CaO –1.3477(m²/g) M-CaO –10.5642 (m²/g)	—	—	(Niju, Sheriffa Begum, and Anantharaman 2016a)

(Continued)

TABLE 4.1 (CONTINUED)
Summary of Characterization Results Reported on Various Sea Shells

Name of the Sea Shell Used as a Catalyst	Surface Observations Using SEM	Characteristic Peaks of CaO Identified Using XRD	BET Analysis Surface Area (m²/g)	BET Analysis Pore Volume (cm³/g)	Functional Group Identification Using FTIR	References
KF-impregnated clam shells	Hexagonal and irregular shaped cluster of particles	28.74°, 51.15° (KCaF₃) 18.16°, 34.33°, 47.27° (Ca(OH)₂)	7.95	—	Sharp stretching band at 3646 cm⁻¹ (presence of hydroxyl group attached to calcium oxide)	(Niju, Niyas, Sheriffa Begum, and Anantharaman 2015)
Clam shell	1.71–2.42 μm of width	32.46°, 37.64°, 54.12°, 64.36°, and 67.24°	1.34	—	—	(Niju, Sheriffa Begum, and Anantharaman 2016b)
Pomacea sp.	—	32.2°, 37.3°, and 53.8°	17	0.04	Peaks of C-O stretching And bending modes of CaCO₃ are observed at 2,513, 1,420, 867, and 3,117 cm⁻¹	(Margaretha et al. 2012)
River snail	Large particles with high porosity	32.2°, 37.3°, and 53.9°	2.664	6.611 *10⁻³	1050–1100 cm⁻¹ (CaO stretching) 3000–3600 cm⁻¹ (OH stretching)	(Roschat et al. 2016)
River snail	Rough and crack with high porosity	32.23°, 37.41°, 53.93°, 64.26°, 67.48°, 79.1°, and 88.56°	—	—	—	(Kaewdaeng, Sintuya, and Nirunsin 2017)
Scallop shell	Coarse particles of micron size	32° and 37°	—	—	—	(Kouzu, Kajita, and Fujimori 2016)

(*Continued*)

TABLE 4.1 (CONTINUED)
Summary of Characterization Results Reported on Various Sea Shells

Name of the Sea Shell Used as a Catalyst	Surface Observations Using SEM	Characteristic Peaks of CaO Identified Using XRD	BET Analysis Surface Area (m²/g)	BET Analysis Pore Volume (cm³/g)	Functional Group Identification Using FTIR	References
Scallop	Irregular in shape and some of them bonded together as aggregates	32.26°, 37.34°, 53.80°, 64.40°, 67.65°, 78.56°, 90.11°, 92.43°, 104.58°, and 111.82°	74.96	0.097	–	(Buasri et al. 2014)
Scylla serrata	–	32°, 37°, 55°, 63°, 67°, 80°, 87°, and 92°	–	–	–	(Boey, Maniam, and Hamid 2009)
Shrimp shell	–	–	–	–	–	(Yang, Zhang, and Zheng 2009)
Turbo jourdani	Before calcination – granular particle size of a smooth crystalline structure. After calcination – rugged surface	32.32°, 37.47°, 54.05°, 64.39°, and 67.62°	–	–	–	(Boonyuen et al. 2018)
Waste sea shells	–	32.30°, 37.38°, and 54.10°	–	–	Absorption band at 1742 cm⁻¹ (ester –C=O bond)	(Perea, Kelly, and Hangun-Balkir 2016)
Waste cockle shell	–	–	–	–	Broadband in the region of 1700 to 1421 cm⁻¹	(Hadi et al. 2017)

(*Continued*)

TABLE 4.1 (CONTINUED)
Summary of Characterization Results Reported on Various Sea Shells

Name of the Sea Shell Used as a Catalyst	Surface Observations Using SEM	Characteristic Peaks of CaO Identified Using XRD	BET Analysis		Functional Group Identification Using FTIR	References
			Surface Area (m^2/g)	Pore Volume (cm^3/g)		
Zebra mussel shells	Unorganized and crystalline	–	–	–		(Johnson 2017)
Conch shells	Porous structure exhibiting regular rod-like particles	32.22°, 37.36°, 53.92°, 64.19°, and 67.41°	1.19	–	At 874 and 1474 cm^{-1} represents CO_3^{2-}, while the peaks between 3600 and 3700 cm^{-1} indicates –OH groups in $Ca(OH)_2$	(Niju, Anushya, and Balajii 2018a)

4.3 CATALYST PREPARATION AND MODIFICATION TECHNIQUE

The general schematic workflow of catalyst preparation and modification techniques employed on various sea shells is shown in Figure 4.1. Initially, the waste shells were washed repeatedly with deionized water to remove dust and impurities. Subsequently, they were dried in a hot air oven at 100–110°C for 24 h (Shankar and Jambulingam 2017) and crushed, followed by calcination in the furnace. During calcination, $CaCO_3$ present in the shell is converted into CaO. The shells were calcined at different temperature (Hadi, Idrus, Ghafar, and Salleh 2017) to optimize the calcination temperature and time at which maximum conversion of biodiesel was obtained. The optimum temperature at which a sample was completely calcined was at 800°C for waste cockle shell, river snail shell, and angel wing shell (Hadi et al. 2017; Nur Syazwani, Rashid, and Taufiq Yap 2015; Roschat et al. 2016), and 900°C for combusted mollusk shell (Abang Chi et al. 2016) and conch shell (Niju, Anushya, and Balajii 2018a). The calcination time for the waste shells was between 2–6 h and the calcined sample was further modified by different techniques to enhance the biodiesel yield. Table 4.2 represents the summary of catalyst preparation and modification techniques reported on various sea shells.

Hadiyanto et al. (2016b) used fly ash as a support material with the shell of *Anadara granosa*. Fly ash containing inorganic oxides, such as SiO_2 (41–55%) and Al_2O_3 (20–25%), is derived from the waste of coal combustion. Wet impregnation was used to prepare fly ash-supported CaO catalyst. Fly ash-supported catalyst increased the biodiesel yield up to 94%. Reddy et al. 2017 prepared CaO-based catalyst from *Anadara granosa* by wet impregnation method. The calcined shells were refluxed with distilled water for 6 h at 60°C to hydrate the CaO. The sample was filtered and dried in a hot air oven at 120°C. The ball mill was employed to finely

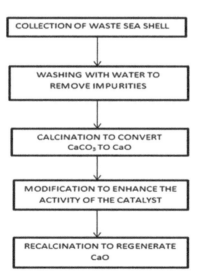

FIGURE 4.1 Schematic workflow of catalyst preparation and modification technique employed.

TABLE 4.2
Summary of Catalyst Preparation and Modification Techniques Employed on Various Waste Sea Shells

Catalyst Source	Calcination Temperature (°C)	Calcination Time (h)	Modification Type	Condition Impregnation	Condition Recalcination	References
Anadara granosa	800	3	Fly ash-supported wet impregnation	4 h 70°C	800°C for 2 h	(Hadiyanto, Lestari, and Widayat 2016a)
Anadara granosa	900	2	Wet impregnation	6 h 60°C	3 h at 600°C	(Reddy et al. 2017)
Scallop shell	1000	2	Heating the mixture of CSS, glycerol, and methanol	6 h 60°C	1000°C at 2 h	(Sirisomboonchai et al. 2015)
Chicoreus brunneus	900	2.5	Wet impregnation	60°C 24 h	800, 900, 1000, and 1100°C at 3 h	(Mazaheri et al. 2018)
Crab	900		CaO coated ZSM-5	80°C 40 min	550°C for 10 h	(Shankar and Jambulingam 2017)
Mussel	900	4	Wet impregnation		3 h at 600°C	(Hu, Wang, and Han 2011)
Green mussel shell	900	3	Activated carbon as carrier and immersion in NaOH		500°C for 5 h	(Hadiyanto et al. 2017)
Mud clam shell	800	3	Wet impregnation	24 h	900°C for 2 h	(Ismail et al. 2016)
Mussel shell base catalyst (MS-BC)	1000	1	Impregnation process using potassium hydroxide	80°C 2 h	400°C for 12 h	(Nurdin et al. 2015)
White bivalve clam shell	900	4	Wet impregnation	60°C 6 h	600°C for 3 h	(Niju, Meera Sheriffa Begum, and Anantharaman 2016)
Clam shells	900	4	Treated with aqueous KF solution		600°C for 4 h	(Niju, Niyas, Sheriffa Begum, and Anantharaman 2015)
Malleus malleus shells	900	4	Hydration-dehydration method		600°C for 3 h	(Niju et al. 2018b)
Pistachio shell	800	3	Wet impregnation	50°C 5 h	600°C for 4 h	(Sinha and Murugavelh 2016)
Scallop shell	900		Blending with basic-$MgCO_3$			(Kouzu, Kajita, and Fujimori 2016)

(Continued)

TABLE 4.2 (CONTINUED)
Summary of Catalyst Preparation and Modification Techniques Employed on Various Waste Sea Shells

	Calcination		Modification		Condition	
Catalyst Source	Temperature (°C)	Time (h)	Type	Impregnation	Recalcination	References
Shrimp shell	450 (carbonization)		Modification with KF	—	250°C (activation)	(Yang, Zhang, and Zheng 2009)
Abalone	800	4	Treated in ethanol solution and Na_2SiO_3 as precursor	—	—	(Chen et al. 2016)
Anadara granosa and *Paphia undulate*	800	3	Wet impregnation	—	—	(Hadiyanto et al. 2016b)
Sea Shells Subjected to Calcination Treatment Alone						
Amusium cristatum	900	2				(Suryaputra et al. 2013)
Snail shell	850	4				(Mohan 2015)
Snail shell	800	4				(Sani et al. 2017)
Turbo jourdani	900	5				(Boonyuen et al. 2018)
Waste sea shells	900	2				(Perea, Kelly, and Hangun-Balkir 2016)
Waste cockle shell	800	2				(Hadi 2017)
Zebra mussel shells	800	6				(Johnson 2017)
Cockles shell	800	2				(Hadi 2017)
Combusted mollusks	900					(Abang Chi et al. 2016)
Lobster shells	900	3				(Hangun-Balkir 2016)
Crab	900					(Madhu et al. 2016)
Cyrtopleura costata (angel wing shell)	700 to 900 (range)	2				(Nur Syazwani, Rashid, and Taufiq Yap 2015)
Enamel venus	900	2				(Buasri et al. 2015)

(Continued)

TABLE 4.2 (CONTINUED)
Summary of Catalyst Preparation and Modification Techniques Employed on Various Waste Sea Shells

	Calcination		Modification			
				Condition		
Catalyst Source	Temperature (°C)	Time (h)	Type	Impregnation	Recalcination	References
Mud crab shell	900	3				(Raut, Jadhav, and Nimbalkar 2016)
Mussel shell	900					(Li, Jiang, and Gao 2015)
Meretrix venus shell and Golden apple snail shell	800	4				(Viriya-empikul et al. 2010)
Mussel shell	950, 1000, and 1050	2				(Rezaei, Mohadesi, and Moradi 2013)
Mussel, cockle, and scallop	700–1000	4				(Buasri et al. 2013)
Clam shell	900	4				(Niju, Sheriffa Begum, and Anantharaman 2016b)
Pomacea sp.	900	2				(Margaretha et al. 2012)
River snail	600 to 1000	3				(Roschat et al. 2016)
River snail	800	4				(Kaewdaeng, Sintuya, and Nirunsin 2017)
Scallop	1000	4				(Buasri et al. 2014)
Scylla serrate	900	2				(Boey, Maniam, and Hamid 2009)
Mud crab shell	900	3				(Raut, Jadhav, and Nimbalkar 2016)
Conch shell	900	3				(Niju, Anushya, and Balajii 2018a)

ground the calcium particles at a speed of 200 rpm. The solid calcium particles were then recalcined at 600°C for 3 h to convert the calcium hydroxide to calcium oxide. This calcination-hydration-dehydration method increases the surface area and basic sites of the CaO catalyst. This increased the rate of transesterification reaction and produced a high biodiesel yield of 96.2%.

Sirisomboonchai et al. (2015) synthesized calcium glyceride-based catalyst from scallop shell. The mixture of calcined scallop shell, methanol, and glycerol were heated under mechanical agitation at 60°C for 3 h. The author compared the activity of the calcium glyceroxide with calcium oxide obtained from the calcination of scallop shell at 1000°C for 2 h. The calcium glyceride had higher catalytic activity with a yield of 82%, but its activity was comparatively lower than CaO. However, glyceroxide in the surface of CaO prevented its reaction with carbon dioxide and water in the air. Mazaheri et al. (2018) produced *C. brunneus*-derived catalyst by the calcination-hydration-dehydration method. The calcined shell was refluxed in water at 60°C for 24 h. The solid particles were filtered and dried in a hot air oven at 70° C for 24 h. The planetary ball mill was used to ground the calcium particles at 200 rpm for 3 h and recalcined at different temperatures (800, 900, 1000, and 1100°C) for 3 h. This enhanced the activity of the catalyst and produced biodiesel with a yield of 93% at recalcination temperature of 1100°C.

Hu, Wang, and Han (2011) used the calcination-hydration-dehydration method to synthesize CaO catalyst from freshwater mussel shell. The calcined solid particles were impregnated in deionized water and recalcined at 600°C for 3 h. The catalyst produced has a porous surface due to gaseous water molecules released from the decomposition of $Ca(OH)_2$. It also exhibits high surface area, which resulted in high catalytic activity. Hadiyanto et al. (2017) prepared CaO catalyst from green mussel shell supplemented by activated carbon as support material. The weight ratio of activated carbon to catalyst was optimized to 2:3. The catalyst and activated carbon were mixed and sodium hydroxide (NaOH) solution was added and stored overnight. The volume of NaOH solution added was also optimized, and the highest yield was obtained at 30 wt%. The catalyst was then dried in a hot air oven at 100°C for 5 h and recalcined at 500°C for 5 h.

Ismail et al. (2016) developed CaO catalyst from mud clam shell by impregnation in water. The calcined shell was refluxed in distilled water for 24 h. The calcium particles were filtered, dried in an oven at 120°C, and recalcined at 900°C for 2 h. The fine CaO produced has a smaller particle size and higher surface area. The higher surface area promotes greater chance of collisions between the reactants and the catalyst, which increases the rate of transesterification reaction, thereby increasing the yield of biodiesel. Nurdin et al. (2015) synthesized mussel shell-based catalyst by impregnation in potassium hydroxide (KOH). 50 ml of aqueous KOH was added to CaO suspended in 5 wt% ammonia solutions at 80°C for 2 h. The catalyst was filtered, washed, and calcined at 400°C for 12 h.

Niju, Sheriffa Begum, and Anantharaman (2016a) enhanced the activity of the catalyst derived from the bivalve clam shell using the calcination-hydration-dehydration method. The calcined shell was refluxed in water at 60°C for 6 h. The filtered solid particles were calcined at 600°C for 3 h. The surface area of the catalyst was 7.83 times higher than calcined catalyst. High surface area and strong basic strength increased the biodiesel yield up to 94.25%. Niju, Niyas, Sheriffa Begum, and Anantharaman (2015) prepared potassium fluoride (KF) impregnated catalyst derived from clam shell. The calcined shell was treated with aqueous KF in a mass ratio of 0.25. The solid particles were recalcined at 600°C for 4 h under static air conditions. The structure of CaO, modified due to the formation of $KCaF_3$ upon impregnation in KF, thereby increased the surface area and basic strength of the catalyst. The modified catalyst produced 95.77% yield.

Sinha and Murugavelh (2016) developed CaO catalyst from pistachio shell by the calcination-hydration-dehydration method. The calcined pistachio shell was refluxed in water at 50°C for 5 h and filtered. The fine solid particles were dried in an oven at 130°C for 8 h and recalcined at 600°C for 4 h. The modified catalyst with high catalytic activity produced methyl ester conversion of 92%. Kouzu, Kajita, and Fujimori (2016) modified the catalyst derived from scallop shell by blending with basic-$MgCO_3$. The catalyst was blended with $MgCO_3$ reagent using mortar in a ratio ranging from 3:7 to 9:1. The bending ratio of 5:5 produced a highly active catalyst. The blending decreased the crystalline size of lime phase and the size of the particle consisting of CaO matrix, thus enhancing the activity of the catalyst.

Yang, Zhang, and Zheng (2009) developed CaO catalyst modified with KF from shrimp shell. The incompletely carbonized shell was loaded with KF and impregnated, followed by drying at 120°C for 12 h and activation at 250°C for 2 h. The modification with KF increased the basicity of the catalyst and enhanced the activity of the catalyst. Chen et al. (2016) used ethanol as a modification agent for CaO catalyst derived from abalone shells. The solid particles obtained by calcination were treated with ethanol solution at three different temperatures. The modified catalyst exhibited high surface area, strong basicity, and decreased crystalline size. The catalyst treated with ethanol at 100°C produced maximum methyl ester conversion of 96.2%. Niju, Anushya, and Balajii (2018a) utilized conch shells as heterogeneous catalyst for the production of biodiesel from *Moringa oleifera* oil. Authors calcined the conch shell in a muffle furnace at 900°C for 3 h to convert the shell into CaO-based catalyst.

From the overall reports, it was observed that the waste shell was calcined to convert calcium carbonate to calcium oxide. The calcination temperature at which maximum conversion of biodiesel was obtained is optimized. The optimized temperature is about 800–900°C. The calcined shell had low catalytic activity and thus modified by different methods to increase the rate of the transesterification reaction. The calcination-hydration-dehydration method was the most commonly employed method to modify the catalyst. The calcined solid particles were impregnated in distilled water at 60°C from 6 h to 24 h in the literature reported and increased the yield. Some catalysts are modified to protect them from reaction with water and carbon dioxide and not for increasing the yield (Sirisomboonchai et al. 2015). Support materials like fly ash and activated carbon were also used to increase surface area and thus the activity of the catalyst. CaO leaching can also be reduced by using support material which readily increases the reusability capacity of the catalyst.

4.4 PRODUCTION OF BIODIESEL USING WASTE SHELLS

The schematic workflow of biodiesel production using waste sea shells is shown in Figure 4.2. Chen et al. (2016) employed abalone shell-derived CaO catalysts for the production of biodiesel from palm oil by transesterification reaction. The reaction was carried out in a three-neck round bottom flask equipped with a stirrer, condenser, and a thermocouple thermometer. The catalyst concentration, methanol to oil ratio, and the reaction time were optimized. The catalyst was recovered by centrifugation and the filtrate was separated by funnel separator. The optimum methanol to oil ratio was 9:1 and 12:1 for catalyst modified by ethanol and unmodified CaO catalyst, respectively. The catalyst concentration was optimized to 7 wt% for both modified and unmodified CaO catalyst, and reaction time was optimized to 2.5 h and 3 h for modified and unmodified CaO catalyst, respectively. The maximum biodiesel yield obtained was 96.2% with modified CaO catalyst.

Suryaputra et al. (2013) derived CaO catalyst from waste capiz shell and utilized it as a catalyst for biodiesel production from palm oil. The biodiesel was produced by transesterification reaction in a round bottom flask equipped with a reflux condenser, heating mantle controller, and mechanical stirrer. The methanol to oil ratio employed was 8:1, while the catalyst was added at the different ratio and the mixture was heated at 60°C with continuous stirring at 700 rpm for the different time period. After completion of the reaction, the catalyst was removed from the mixture by centrifugation and the filtrate was separated into two layers of biodiesel and glycerol using a funnel separator. The optimum catalyst concentration and reaction time to obtain maximum methyl ester conversion of 93% was 3 wt% and 6 h respectively.

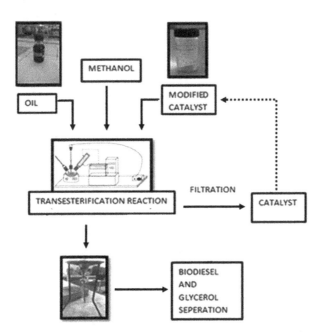

FIGURE 4.2 Production of biodiesel using waste shells.

Hadiyanto et al. (2016b) developed fly ash-supported catalyst derived from mollusk shells of *Anadara granosa* and *Paphia undulate* and utilized it for biodiesel synthesis. The biodiesel was produced by methanolysis of oil in a round bottom flask equipped with a magnetic stirrer and reflux condenser. 35 ml of palm oil was added to constant methanol to oil ratio (12:1), with the different catalyst concentration (2–10%). The mixture was stirred uniformly at 70°C for 2 h. The catalyst was removed from the reaction mixture using a vacuum pump. The biodiesel and methanol were separated by using a funnel separator. Gas chromatography was used to analyze the biodiesel produced. Optimal transesterification conditions of 6 wt% catalyst loading, 12:1 methanol to oil mole ratio, 60°C reaction temperature, and 2 h reaction time resulted in a biodiesel yield of 92 and 94% for CaO derived from *A. granosa* and *P. undulata* shells, respectively.

Hadiyanto et al. (2016a) prepared a catalyst from *Anadara granosa* shells supported by fly ash and employed it to synthesize biodiesel from palm oil. Transesterification reaction was carried out in a three-neck round bottom flask equipped with a magnetic stirrer, thermometer, and reflux condenser. Different catalyst concentrations (2, 6, and 10%) and constant methanol:oil (12:1) was added to 31 g of palm oil in a round bottom flask with constant stirring at 70°C for 2 h. The vacuum pump was used to filter the catalyst, and centrifugation at 3000 rpm for 25 min was used to separate the fatty acid methyl ester phase and glycerol phase. The maximum biodiesel yield of 92% was obtained at a catalyst concentration of 6 wt%. Sirisomboonchai et al. (2015) produced biodiesel from waste cooking oil using calcium glyceroxide-modified catalyst derived from scallop shells. The closed reactor was used to carry out transesterification reaction to prevent evaporation of methanol. The reaction was carried out with different methanol/oil molar ratio (3:1, 6:1, and 12:1), catalyst loading amount (1, 2, 5, and 10 wt% based on the oil weight), and reaction time (30, 60, 120, and 180 min), and the product was separated by centrifugation at 6000 rpm for 15 min. High-performance liquid chromatography equipped with a silica gel column and a refractive index detector was used to analyze the biodiesel. Transesterification conditions were optimized as catalyst concentration (5 wt%), methanol/oil molar ratio (6:1), reaction temperature (65°C), and reaction time (120 min) with a maximum yield of 86%.

Hangun-Balkir (2016) employed CaO catalyst derived from lobster shells to synthesize biodiesel from *Camelina sativa* oil. Biodiesel was produced by transesterification reaction, which was performed using a round bottom flask with a magnetic stirrer, thermometer, and reflux condenser. Methanol to oil ratio (12:1) and 1% (w/w) catalyst was mixed with oil and vigorously stirred at 65°C for 3 h. The catalyst was recovered through filtration, and biodiesel was separated using a separating funnel. The biodiesel yield was estimated using NMR and FTIR spectroscopy techniques. The author also studied the effect of different parameters on biodiesel synthesis. Optimum biodiesel yield was attained at 12:1 (alcohol:oil) molar ratio with 1 wt% of the prepared heterogeneous catalyst in 3 h at 65°C. Mazaheri et al. (2018) produced rice bran oil-based biodiesel using calcium oxide catalyst derived from *Chicoreus brunneus* shell. Transesterification was carried out in a three-neck, round bottom flask equipped with a thermometer, condenser, and magnetic heating mantle. The mixture of oil, methanol, and catalyst was heated at 65°C for 120 min with continuous stirring at 150 rpm. Centrifugation at 2000 rpm for 7 min was employed to recover the catalyst. Methanol was removed by vacuum evaporator. Finally, the biodiesel was washed with distilled water. Transesterification reaction was optimized by carrying out the reaction using catalyst produced at a range of calcination temperatures (800, 900, 1000, and 1100°C), methanol to oil molar ratio (20:1, 25:1, 30:1, and 35:1), catalyst loading (0.2, 0.4, 0.6, and 0.8 wt% of rice bran oil), and reaction times (48, 72, 96, and 120 min). Artificial neural network and ant colony optimization were used to model and optimize the reaction conditions. Maximum biodiesel yield was obtained using the optimal conditions of 35:1 methanol to rice bran oil molar ratio and 0.5 wt% catalyst loading.

Abang Chi et al. (2016) derived a solid base catalyst from combusted mollusk shell and utilized it for biodiesel synthesis. A two-neck glass reactor equipped with condenser partially submerged in a water bath was used to carry out the reaction. The mixture of palm oil, methanol (12:1), and catalyst (8 wt%) was refluxed under an ultrasonic reactor at 65°C for 15 min. The non-ultrasonic reaction was carried out using a magnetic stirrer at 65°C for 3 h. Centrifugation was used to separate the catalyst, and methanol was evaporated. Final products were separated into three phases by centrifugation: the upper layer was biodiesel, the middle layer was glycerol, and the lower layer was a mixture of solid CaO and a small amount of glycerol. The methyl ester conversion obtained was 96.77% using ultrasonic irradiation. Shankar and Jambulingam (2017) used CaO-impregnated Na-ZSM-5 as a solid base catalyst for the transesterification of neem oil into biodiesel. Neem oil (1 wt%), methanol (3–15 mole), and CaO-ZSM-5 catalyst (5 wt%) were mixed in a round bottomed flask fitted with a reflux condenser and heated at a constant temperature (75, 100, and 120°C) with concurrent stirring. The progress of transesterification is monitored by centrifuging the aliquots withdrawn every 2 h. Biodiesel yield of 95% was achieved and optimum reaction was identified at a reaction temperature of 75°C, reaction time of 6 h, methanol-to-neem oil molar ratio of 12:1, and catalyst dosage of 0.2 g.

Madhu et al. (2016) prepared crab shell-derived catalyst and utilized it to synthesize biodiesel from crude *Millettia pinnata* oil. Esterification reactions were conducted to reduce the acid value of the oil using a three-necked round bottom flask placed in a serological water bath attached with a magnetic stirrer at 650 rpm. Optimum reaction conditions were identified at reaction temperature 65°C, reaction time 2 h, methanol-to-oil molar ratio 8:1, and catalyst concentration 2.5%. Nur Syazwani, Rashid, and Taufiq Yap (2015) developed a low-cost solid catalyst derived from waste *Cyrtopleura costata* for biodiesel production using microalgae oil. Transesterification reaction was carried out by mixing 0.1 g of *N. oculata* crude lipid with methanol to lipid weight ratio range from 1:90 to 1:210 g/g weight ratios and different catalyst loading (3, 5, 7, 9, and 11 wt%) at 65°C with stirring speed of 300 rpm and reaction time (0.5, 1, 2, 3, 4, and 5 h). Centrifugation was used to separate catalyst and liquid mixtures. The optimum reaction was identified at a reaction temperature 65°C, reaction time 1 h, methanol-to-oil weight ratio 150:1, and catalyst concentration 9 wt%. The methyl ester conversion obtained at the above-mentioned condition was around 65%.

Buasri et al. (2015) utilized waste enamel venus shell as an environmentally friendly catalyst for synthesis of biodiesel. A 500 ml glass reactor equipped with a condenser and mechanical stirrer at atmospheric pressure, placed inside a household microwave oven, was used to carry out the reaction. The calculated amount of *Jatropha* curcas oil, different concentration of prepared catalyst (2–6 wt%), and methanol of different molar ratio (9:1–21:1) were mixed by stirring in a reactor. The reaction was operated at 150–800 W with a varied reaction time of 2–6 min under microwave irradiation, and it was instantly stopped by rapid cooling in an ice bath. Biodiesel produced was analyzed with gas chromatography-mass spectroscopy (GC-MS) equipped with a flame ionization detector. Maximum biodiesel yield of 93% was obtained at the optimal conditions of reaction time 5 h, microwave power 800 W, methanol/oil molar ratio 18:1, and catalyst loading 4 wt%. Hu, Wang, and Han (2011) developed a waste freshwater mussel shell as an economic catalyst for biodiesel production from Chinese tallow oil. The transesterification reaction was performed using a 100 mL glass reactor equipped with a condenser and a mechanical stirrer at 70°C with different catalyst concentration and methanol to oil ratio. Final products were separated using filtration, and a maximum biodiesel yield of above 90% was achieved under the conditions of reaction temperature of 70°C with a methanol/oil molar ratio of 12:1, catalyst concentration of 5%, and a reaction time of 1.5 h.

Hadiyanto et al. (2017) produced biodiesel using a heterogeneous catalyst derived from green mussel shell. The esterification process was carried out by mixing shell catalyst (1.5%, 2.5%, 5%, and 7.5%), methanol (1:2 (w/w)), and pretreated cooking oil in a batch reactor stirred at 600 rpm for 30 min. The mixture was warmed at 65°C for 3 h. The optimum yield of biodiesel of 95.12% was achieved at a catalyst load of 7.5%. Raut, Jadhav, and Nimbalkar (2016) derived a heterogeneous catalyst crab shell and utilized it for the transesterification of biofuel. The transesterification reaction was conducted in a 500 ml three-necked round bottom flask with a stirring speed of 600 rpm. The 3 wt% of crab shell catalyst for different methanol to oil (soybean oil) ratio was introduced to a round bottom flask at 65°C for 2 h. The catalyst was removed from the mixture by filtration. Furthermore, the resultant was separated into biodiesel and glycerol by using a separating funnel. Optimal conditions such as agitation speed 600 rpm, reaction time 2 h, methanol to oil ratio 15:1 (methanol/oil), and catalyst loading 3 wt% resulted in a maximum biodiesel yield of 78.84%.

Li, Jiang, and Gao (2015) developed a heterogeneous catalyst utilizing mollusk shells to produce biodiesel. Esterification was carried out to reduce the acid value of the oil, followed by transesterification. The reaction was conducted in a flask equipped with a magnetic stirrer using different catalyst concentration, temperature, and methanol to oil (*Jatropha curcas* oil) ratio. The catalyst was separated by centrifugation. At optimum reaction conditions of a temperature of 65°C, methanol to oil molar ratio 9:1, reaction time 3 h, and catalyst loading 3 wt% of oil, a high biodiesel yield of more than 99.0% was obtained. Viriya-empikul et al. (2010) utilized a heterogeneous catalyst derived from waste shells of egg, golden apple snail, and *Meretrix venus* for transesterification of palm olein oil. Transesterification reaction was carried out in a three-necked round bottom flask with constant temperature at 60°C and catalyst concentration of 10 wt%. After completion of the reaction, the catalyst was separated by centrifugation and excessive methanol was evaporated. All catalysts produced a high biodiesel yield of 90% in 2 h.

Ismail et al. (2016) produced biodiesel from castor oil by using calcium oxide derived from mud clam shell. A three-neck round bottom flask was used to carry out transesterification. The parameters involved in the reaction methanol to oil ratio (1:10 to 1:18), catalyst loading (0.5% to 7% w/w), temperature (50°C to 70°C), and time (0.5 h to 3 h) were optimized. The optimum conditions such as 1:14 oil to methanol molar ratio, 3% w/w catalyst concentration, 60°C reaction temperature, and 2 h reaction time produced maximum biodiesel yield of 96.7%. Rezaei, Mohadesi, and Moradi (2013) optimized biodiesel produced utilizing waste mussel shell. A two-neck flask with a thermometer and condenser was used to carry out the reaction using different catalyst concentrations, molar ratios of methanol to oil, and calcination temperatures. The catalyst was recovered by centrifugation, and the final products were separated by using a separating funnel. The Box-Behnken design was used to optimize the reaction parameters. A calcination temperature of 1050°C, catalyst concentration of 12 wt%, and methanol to oil ratio of 24:1 produced biodiesel with a purity and yield of 100% and 94.1%, respectively.

Nurdin et al. (2015) utilized mussel shell base catalyst for the production of biodiesel from castor oil. The effect of reaction parameters catalyst amount (1–5% (wt/wt)), catalyst recyclability (1–5 cycles), time (1–5 h), and temperature (40°–80°C) on the yield of biodiesel was studied. The reaction was conducted under batch conditions with methanol to oil ratio of 6:1. The maximum biodiesel yield (91.17%) was achieved by the catalyst loading of 2% (wt/wt), a time of 3 h, a temperature of 60°C, and a methanol oil ratio of 6:1. Buasri et al. (2013) developed CaO-based catalyst utilizing waste shells of mussel, cockle, and scallop to synthesize biodiesel from palm oil. The transesterification reaction was conducted using a glass reactor, and the parameters involved in the reaction time (2–6 h), reaction temperature (50°–70°C), methanol to oil molar ratio (6–18), catalyst loading (5–25 wt%), and reusability of catalyst (1–4 times) were optimized. The optimum conditions produced a maximum yield of biodiesel (95%) for all waste shell-derived catalysts under reaction time 3 h, reaction temperature 65°C, methanol to oil molar ratio 9, and catalyst loading 10 wt% with 1 atm pressure in a glass reactor.

Niju, Sheriffa Begum, and Anantharaman (2016a) derived highly active CaO catalyst from natural white bivalve clam shell. A three-neck round bottom flask was used to carry out transesterification reaction with the different amount of catalyst, methanol to oil ratio, reaction temperature, and reaction time to obtain the maximum yield of biodiesel. The biodiesel and glycerol were separated by a separating funnel and the catalyst was recovered by filtration. The maximum yield of biodiesel obtained was 94.25% for the CaO obtained from calcination–hydration–dehydration treatment at 7 wt% catalysts, methanol to oil ratio of 12:1, reaction temperature of 65°C, and reaction time of 1 h. Niju, Niyas, Sheriffa Begum, and Anantharaman (2015) utilized KF-impregnated clam shells for biodiesel production from waste frying oil by transesterification. The reaction was undertaken in a 1 L reactor equipped with condenser, mechanical stirrer, and condenser. The reaction was repeated by varying the operating variables such as the methanol to oil molar ratio, amount of catalyst, reaction temperature, and time to obtain the maximum yield of biodiesel. Maximum biodiesel yield of 95.77% was achieved at methanol to oil molar ratio of 9:1, catalyst amount of 4 wt%, reaction temperature of 65°C, and reaction time of 2 h.

Niju, Sheriffa Begum, and Anantharaman (2016b) developed a clam shell-derived catalyst for the continuous production of biodiesel from waste frying oil. The reaction was carried out in a reactive distillation (RD) system packed with clam shell-based CaO as heterogeneous catalyst. It consists of a packed bed distillation column equipped with a hot water jacket at 65°C. The glass beads of 1–2 mm diameter were used to pack the column up to a certain height, and the rest of the column was loaded with CaO of 2–4 mm particle size. The lower end of the column was connected to the reboiler at 65°C, and the top of the column was connected with a condenser to recover methanol. Methanol and oil were mixed using a magnetic stirrer and then fed to the catalytic section of the column continuously through a peristaltic pump. The methanol was completely vaporized by the reboiler, and the vapors were recirculated in the RD column and utilized in the reactive zone. All the catalysts are retained in the column itself, and hence the products collected in the reboiler. The product mixture was withdrawn using a peristaltic pump and separated using glycerol-ester separator. Maximum methyl ester conversion of 94.41% was obtained at a reactant flow rate of 0.2 ml/min, methanol to oil ratio of 6:1, and catalyst bed height of 180 mm. Sinha and Murugavelh (2016) compared the production of biodiesel using catalyst developed from pistachio shells with alkaline catalyst and egg shell-derived catalyst. Transesterification reaction was performed in a round bottom flask furnished with a temperature indicator and magnetic stirrer. To obtain maximum yield of biodiesel, the reaction was repeated with different amounts of catalyst weight percentage (2, 3, 4, and 5 wt% based on the oil weight), methanol to oil molar ratios (6:1, 9:1, 12:1, and 15:1), time (40, 50, 60, and 70 min), and at various temperatures (40, 50, 60, and 70°C). The final products were separated into layers of methyl ester and glycerol using a funnel separator. The temperature, reaction time, catalyst loading, and methanol to oil ratio was optimized to be 60°C, 60 min, 3%, and 12:1, respectively.

Margaretha et al. (2012) developed calcium oxide-based catalyst derived from *Pomacea sp.* shell. The transesterification reaction was performed in a round bottom flask at a temperature of 60°C and a stirring speed of 700 rpm with the amount of catalyst varied from 1% to 5% w/w, and oil to methanol ratios were 1:5 to 1:11. Methanol and glycerol were separated from biodiesel produced by a rotary vacuum evaporator and funnel separator, respectively. The highest yield of 95.61% was obtained at a reaction temperature of 60°C, a reaction period of 4 h, the ratio of methanol-oil at 7:1, and the amount of catalyst of 4% w/w. Roschat et al. (2016) synthesized biodiesel using river snail shell-derived heterogeneous catalyst. The transesterification reaction conditions were varied as follows: methanol to oil molar ratio of 6:1–18:1, amount of catalyst to palm oil of 1–7 wt%, and constant magnetic stirring speed of 300 rpm. The co-solvent method was used to accelerate the catalyzed reaction. The amount and type of the co-solvent were also varied. A maximum yield of $98.5 \pm 1.5\%$ was achieved under the optimal conditions of catalyst to oil ratio of 5 wt%, methanol to oil molar ratio of 12:1, a reaction temperature of 65°C, 10% v/v of Tetrahydrofuran in methanol, and a reaction time of 90 min.

Kaewdaeng, Sintuya, and Nirunsin (2017) utilized calcium oxide as a catalyst derived from river snail shell for biodiesel production. Transesterification was performed with different reaction conditions to obtain a maximum yield of biodiesel. Methanol to oil ratio (6:1, 9:1, and 12:1), catalyst weight ratio (1, 2, and 3 wt%), and reaction time (1, 2, and 3 h) were employed in the reaction. The optimum conditions which produced a maximum yield of biodiesel (92.5%) were methanol to oil ratio 9:1, catalyst concentration of 3 wt%, and reaction time of 1 h. Kouzu, Kajita, and Fujimori (2016) synthesized calcined scallop shell for rapeseed oil transesterification to produce biodiesel. A glass batch reactor system was used to conduct transesterification. The reaction conditions employed were a catalyst concentration of 9 wt%, a molar ratio of 12, and a reaction temperature of 60°C. Biodiesel yield of 96% was achieved by modifying the catalyst with $MgCO_3$.

Buasri et al. (2014) utilized a scallop waste shell for biodiesel production from palm oil. Biodiesel was synthesized by transesterification reaction carried out using a three-necked round bottom flask. The effect of reaction parameters was also studied by repeating it with different reaction conditions. The Taguchi approach was used for the process optimization of transesterification. The maximum conversion of 95.44% was achieved at the reaction time of 3 h, reaction temperature of 65°C, CaO catalyst amount of 10 wt%, and methanol to oil molar ratio of 9. Boey, Maniam, and Hamid (2009) developed a catalyst from waste crab shell (*Scylla serrata*) for palm olein transesterification. Several parameters involved in the transesterification reaction performed in a two-necked glass reactor were studied and optimized. The maximum conversion of methyl ester (95.44%) was achieved at the reaction time of 3 h, reaction temperature of 65°C, catalyst amount of 10 wt%, and methanol to oil molar ratio 9.

Yang, Zhang, and Zheng (2009) prepared shrimp shell catalyst for biodiesel production. Transesterification reaction was performed in a two-neck round bottom flask with varying methanol to oil ratio, shrimp shell catalyst concentration, and reaction time at 65°C for 3 h. Decompressed distillation and filtration was used to separate excessive methanol and catalyst, respectively. The resultant mixture was separated into biodiesel and glycerol in a separating funnel. When the reaction was carried out at 65°C with a catalyst amount of 2.5 wt%, methanol to rapeseed oil molar ratio of 9:1, and a reaction time of 3 h, the highest conversion of 89.1% was achieved. Mohan (2015) optimized biodiesel produced utilizing catalyst derived from snail shell. The methanol to oil ratio (7:1, 8:1, 9:1.10:1, and 11:1), calcium oxide catalyst concentration (6, 8, 10, 12, and 14 wt%), temperature (55, 60, 65, 70, and 75°C), and time (1, 2, 3, 4, and 5 h) were used to carry out the transesterification reaction in a three-neck round bottom flask. The optimization of process parameters was performed using the central composite design of response surface methodology. The optimum condition for maximum biodiesel yield of 96% was observed at a temperature at 65°C, catalyst amount of 10 wt%, a methanol to oil molar ratio of 9:1, and a reaction time of 3 h.

Sani et al. (2017) synthesized a snail shell catalyst for biodiesel production. A four-neck round bottom flask with a water-cooled condenser, magnetic stirrer, and a thermometer was used to perform the reaction. The reaction was carried out with 9 wt% catalyst at 60°C for 3 h. After the reaction was completed, the catalyst was screened by using a filter paper, and the final product was separated into biodiesel and glycerol by gravity. A maximum biodiesel conversion of 84.14% was observed. Perea, Kelly, and Hangun-Balkir (2016) utilized waste shells for biodiesel production. Transesterification reaction was carried out in a three-neck round bottom flask with 12:1 (alcohol:oil) molar ratio with 1 wt% waste seashell catalysts in 2 h at 65°C.

NMR and FTIR spectroscopy was used to analyze the biodiesel produced. Mussel shell, clam shell, and oyster shell produced high biodiesel yields of 95%, 93%, and 91% respectively.

From the reported literature, it was observed that the feedstock is the major reason for the increase in the cost of biodiesel production. The use of *Camelina sativa, Jatropha curcas, Millettia pinnata, Moringa oleifera*, castor, soybean, rapeseed, palm oil, and waste cooking oil as feedstock has been reported in the literature. Palm oil is the most commonly used feedstock because of its low cost, and the feedstock requires land for production and increases the demand and cost of food. Thus, the use of edible oil for biodiesel production is considered uneconomical. On the other hand, using waste cooking oil as a cost-effective feedstock for biodiesel production is gaining attention in recent years, which reduces the cost of biodiesel production and also prevents the environmental impacts caused by the disposal of used cooking oil. Transesterification was the most effective way to produce biodiesel. A batch reactor was utilized to carry out the reaction, which is a three-neck round bottom flask equipped with magnetic stirrer, thermometer, and condenser. Furthermore, continuous production of biodiesel was also reported in the literature (Niju, Sheriffa Begum, and Anantharaman 2016b). The transesterification reaction conditions were optimized to maximize the biodiesel yield. Central composite design and Box-Behnken design of response surface methodology, artificial neural networks, and Taguchi methods were used to statistically optimize the reaction conditions. The optimum transesterification temperature is 60–65°C for most of the waste shells. The optimum time (1–6 h), catalyst concentration (1–12 wt%), and methanol to oil (8:1–24:1) molar ratio vary for the different shells. The biodiesel produced was analyzed by gas chromatography-mass spectrometry, high-performance liquid chromatography, proton NMR, and FTIR spectroscopy. The physical properties of biodiesel produced, like flash point, relative density, kinematic viscosity, total acid number, and iodine value, were determined and compared with the properties of ASTM petroleum standards. The summary of optimal biodiesel process conditions reported using various sea shells as catalyst sources is presented in Table 4.3.

4.5 FUTURE PERSPECTIVE OF WASTE SHELLS

Waste shells utilized for biodiesel production are considered the best choice to eliminate environmental impacts caused by disposal of these shells in landfills and reclaimed land. They also extremely reduced the high cost involved in the development of a heterogeneous catalyst. The synthesis of biodiesel is also cost-effective and simple. The most important issue to be considered while employing waste shell as a catalyst is CaO leaching. The solubility of CaO in methanol is less than in a glycerol-methanol mixture due to the formation of calcium diglyceroxide, which is more soluble in methanol. The leaching blocks the active sites in CaO, which reduces the conversion of biodiesel when the catalyst is recycled. The presence of calcium ions in the ester phase also increases the amount of water used in rinsing the products. Anchoring catalyst with a support material or using mixed oxides reduces leaching to a certain extent and increases the stability of CaO catalyst.

TABLE 4.3
Summary of Biodiesel Production Using Various Sea Shells as a Source of Catalyst

		Transesterification Process Parameters					
Catalyst Source	Lipid Feedstock	Methanol to Oil Ratio	Amount of Catalyst (wt%)	Reaction Temperature (°C)	Reaction Time (h)	Yield/Conversion (%)	References
Abalone	Palm oil	9:1	7	–	2.5	96.2	(Chen et al. 2016)
Capiz	Palm oil	8:1	3	60	6	93	(Suryaputra et al. 2013)
Anadara granosa and *Paphia undulate*	Palm oil	12:1	8	60	2	A. granosa – 92 P. undulate – 94	(Hadiyanto et al. 2016b)
Anadara granosa	Palm oil	12:1	6	70	2	92	(Hadiyanto, Lestari, and Widayat 2016a)
Scallop	Waste cooking oil	6:1	5	65	2	86	(Sirisomboonchai et al. 2015)
Lobster shells	*Camelina sativa* oil	12:1	1	65	3	–	(Hangun-Balkir 2016)
Chicoreus brunneus shell	Rice bran oil	35:1	0.5	65	2	–	(Mazaheri et al. 2018)
Combusted mollusk shell	Palm oil	12:1	8	65	3	96.77	(Abang Chi et al. 2016)
Crab shell	Neem oil	12:1	5	75	2	95	(Shankar and Jambulingam 2017)
Crab shell	Crude *Millettia pinnata* oil	8:1	2.5	65	2	–	(Madhu et al. 2016)
Cyrtopleura costata (Angel wing shell)	Microalgae oil	150:1 (weight ratio)	9	65	1	above 65	(Nur Syazwani, Rashid, and Taufiq Yap 2015)

(*Continued*)

TABLE 4.3 (CONTINUED)
Summary of Biodiesel Production Using Various Sea Shells as a Source of Catalyst

		Transesterification Process Parameters					
Catalyst Source	Lipid Feedstock	Methanol to Oil Ratio	Amount of Catalyst (wt%)	Reaction Temperature (°C)	Reaction Time (h)	Yield/Conversion (%)	References
Enamel venus	*Jatropha Curcas* oil	18:1	4	–	5	93	(Buasri et al. 2015)
Green mussel	Waste cooking oil	12:1	5	70	1.5	90	(Hu, Wang, and Han 2011)
Crab shell	Soybean oil	15:1	3	65	2	78.84	(Raut, Jadhav, and Nimbalkar 2016)
Mollusk	*Jatropha curcas* oil	9:1	3	65	3	99.0	(Li, Jiang, and Gao 2015)
Golden apple	Palm olein oil	10:1	10	60	2	90	(Viriya-empikul et al. 2010)
Mud Clam	Castor oil	14:1	3	60	2	96.7	(Ismail et al. 2016)
Mussel	Soya bean oil	24:1	12	60	8	94.1	(Rezaei, Mohadesi, and Moradi 2013)
Mussel	Castor oil	6:1	2	60	3	91.17	(Nurdin et al. 2015)
Mussel, cockle, and scallop	Palm oil	9:1	10	65	3	95	(Buasri et al. 2013)
Bivalve clam	Waste frying oil	12:1	7	65	1	94.25	(Niju, Sheriffa Begum, and Anantharaman 2016a)
Clam	Waste frying oil	9:1	4	65	2	95.77	(Niju, Niyas, Sheriffa Begum, and Anantharaman 2015)
Clam	Waste frying oil	6:1	–	65	–	94.41	(Niju, Sheriffa Begum, and Anantharaman 2016b)

(Continued)

TABLE 4.3 (CONTINUED)
Summary of Biodiesel Production Using Various Sea Shells as a Source of Catalyst

			Transesterification Process Parameters				
Catalyst Source	Lipid Feedstock	Methanol to Oil Ratio	Amount of Catalyst (wt%)	Reaction Temperature (°C)	Reaction Time (h)	Yield/Conversion (%)	References
Pistachio	Waste cotton cooking oil	12:1	3	60	1	91	(Sinha and Murugavelh 2016)
River snail	Palm oil	12:1	5	65	1.30	98.5	(Roschat et al. 2016)
Pomacea sp.	Palm oil	7:1	4	60	4	95.61	(Margaretha et al. 2012)
River snail	Waste cooking oil	9:1	3	65	1	92.5	(Kaewdaeng, Sintuya, and Nirunsin 2017)
Scallop	Rapeseed oil	12:1	9	60	4	96	(Kouzu, Kajita, and Fujimori 2016)
Scallop waste	Palm oil	9:1	10	65	3	95.44	(Buasri et al. 2014)
Crab shell	Palm olein oil	9:1	10	65	3	95.44	(Boey, Maniam, and Hamid 2009)
Shrimp shell	Refined rapeseed oil	9:1	2.5	65	3	89.1	(Yang, Zhang, and Zheng 2009)
Snail shell	Neem oil	9:1	10	65	4	96	(Mohan 2015)
Snail shell	Castor oil	9:1	9	60	3	84.14	(Sani et al. 2017)
Waste mussel, clam, and oyster shells	*Camelina sativa* oil	12:1	1	65	2	Mussel – 95 Clam – 93 Oyster – 91	(Perea, Kelly, and Hangun-Balkir 2016)
Conch shell	*Moringa oleifera* oil	8.66:1	8.02	65	2 h 10 min	97.06	(Niju et al. 2018a)
Malleus malleus shell	Waste cooking oil	11.85:1	7.5	65	1 h 26 min	93.81	(Niju et al. 2018b)

Cation exchange resins can be employed to remove leached Ca species, but life expectancy prevents its reuse. It also increases purification cost and generates wastewater. The reaction of CaO with H_2O and CO_2 also reduces its catalytic activity. The mass transfer resistance caused by the presence of three phases (alcohol, glycerol, and catalyst) reduces the rate of the reaction and increases the cost of production. Use of support materials like alumina, silica, and oxide and intensifying technologies like microwave, ultrasonic, or co-solvent methods could overcome mass transfer limitation. Reusability of the calcium oxide catalyst can be increased by modification to calcium methoxide or calcium diglyceroxide. The sensitivity of the CaO catalyst to ambient conditions could be overcome by the use of calcium diglyceroxide and glycerolate. Another major issue is the high cost involved in the synthesis of biodiesel. The cost is effectively reduced by using waste cooking oil and non-edible oil. However, CaO catalyst is not suitable for oil with high free fatty acid content. The oil has to be treated to reduce the fatty acid content to an acceptable level. Many modification technologies should be employed to increase the life expectancy of the CaO catalyst. The current research is on pelleting CaO or using carriers as support to increase the number of reuse cycles of the catalyst and stability of the catalyst.

4.6 CONCLUSION

Thus, waste sea shells are a potential raw material which can be used as an aggregate replacement in concrete, wastewater treatment, sorbent of acidic gases, carotenoid synthesis, biopolymer production, and biodiesel synthesis. This chapter discusses in detail the utilization of waste shells as heterogeneous catalyst for biodiesel production. Additionally, this chapter reviews various modification techniques employed in increasing catalytic activity of the CaO catalyst derived from waste shells by increasing surface area and regenerating active sites involved in the transesterification reaction. Moreover, the several issues involved in using CaO catalyst derived from waste shells and major developments to overcome these issues are discussed. It was concluded that waste shells have huge potential to be utilized as a heterogeneous catalyst for biodiesel production.

REFERENCES

Alabaraoye, Ernestine, Mathew Achilonu, and Robert Hester. 2018. "Biopolymer (Chitin) from Various Marine Seashell Wastes: Isolation and Characterization." *Journal of Polymers and the Environment* 26(6): 2207–18.

Alhassan, Yahaya, and Naveen Kumar. 2016. "Single Step Biodiesel Production from *Pongamia pinnata* (Karanja) Seed Oil Using Deep Eutectic Solvent (DESs) Catalysts." *Waste and Biomass Valorization* 7(5): 1055–65.

Atapour, Mehdi, and Hamid Reza Kariminia. 2013. "Optimization of Biodiesel Production from Iranian Bitter Almond Oil Using Statistical Approach." *Waste and Biomass Valorization* 4(3): 467–74.

Boey, Peng-Lim, Gaanty Pragas Maniam, and Shafida Abd Hamid. 2009 "Biodiesel Production via Transesterification of Palm Olein Using Waste Mud Crab (Scylla Serrata) Shell as a Heterogeneous Catalyst." *Bioresource Technology* 100(24): 6362–68.

Boonyuen, Supakorn, Siwaporn Meejoo Smith, Monta Malaithong, Apisit Prokaew, Benya Cherdhirunkorn, and Apanee Luengnaruemitchai. 2018. "Biodiesel Production by a Renewable Catalyst from Calcined *Turbo Jourdani* (Gastropoda: Turbinidae) Shells." *Journal of Cleaner Production* 177: 925–29.

Buasri, Achanai, Nattawut Chaiyut, Vorrada Loryuenyong, Phatsakon Worawanitchaphong, and Sarinthip Trongyong. 2013. "Calcium Oxide Derived from Waste Shells of Mussel, Cockle, and Scallop as the Heterogeneous Catalyst for Biodiesel Production." *The Scientific World Journal* 2013, 7.

Buasri, Achanai, Teera Sriboonraung, Kittika Ruangnam, Pattarapon Imsombati, and Vorrada Loryuenyong. 2015. "Utilization of Waste Enamel Venus Shell as Friendly Environmental Catalyst for Synthesis of Biodiesel." *Key Engineering Materials* 659: 237–41.

Buasri, Achanai, Phatsakon Worawanitchaphong, Sarinthip Trongyong, and Vorrada Loryuenyong. 2014. "Utilization of Scallop Waste Shell for Biodiesel Production from Palm Oil – Optimization Using Taguchi Method." *APCBEE Procedia* 8: 216–21.

Catarino, M., M. Ramos, A. P. Soares Dias, M. T. Santos, J. F. Puna, and J. F. Gomes. 2017. "Calcium Rich Food Wastes Based Catalysts for Biodiesel Production." *Waste and Biomass Valorization* 8(5): 1699–707.

Chen, Guan Yi, Rui Shan, Bei Bei Yan, Jia Fu Shi, Shang Yao Li, and Chang Ye Liu. 2016. "Remarkably Enhancing the Biodiesel Yield from Palm Oil upon Abalone Shell-Derived CaO Catalysts Treated by Ethanol." *Fuel Processing Technology* 143: 110–17.

Chi, Abang, Dayang Aisah, Norhasnan Sahari, Rabuyah Ni, and Zaini Assim Zani. 2016. "Combusted Molluscs Shell as Solid Base Catalyst for Transesterification to Produce Biodiesel." *Transactions on Science and Technology* 3(1): 19–24.

Dalei, Jikasmita, and Debasish Sahoo. 2015. "Extraction and Characterization of Astaxanthin from the Crustacean Shell Waste from Shrimp Processing Industries." *International Journal of Pharmaceutical Sciences and Research* 6(6): 2532–37.

Demirbas, Ayhan, Abdullah Bafail, Waqar Ahmad, and Manzoor Sheikh. 2016. "Biodiesel Production from Non-Edible Plant Oils." *Energy Exploration and Exploitation* 34(2): 290–318.

Ghanei, R., R. Khalili Dermani, Y. Salehi, and M. Mohammadi. 2016. "Waste Animal Bone as Support for CaO Impregnation in Catalytic Biodiesel Production from Vegetable Oil." *Waste and Biomass Valorization* 7(3): 527–32.

Hadi, Norulakmal Nor, Nur Afini Idrus, Faridah Ghafar, and Marmy Roshaidah Mohd Salleh. 2017. "Waste Cockle Shell as Natural Catalyst for Biodiesel Production from Jatropha Oil." *AIP Conference Proceedings* 1901: 1–6.

Hadiyanto, Hadiyanto, Asha Herda Afianti, Ulul Ilma Navi'A, Nais Adetya, Widayat Widayat, and Heri Sutanto. 2017. "The Development of Heterogeneous Catalyst C/CaO/NaOH from Waste of Green Mussel Shell (*Perna Varidis*) for Biodiesel Synthesis." *Journal of Environmental Chemical Engineering* 5(5): 4559–63.

Hadiyanto, Hadiyanto, Sri Puji Lestari, and Widayat Widayat. 2016a. "Preparation and Characterization of *Anadara Granosa* Shells and $CaCO_3$ as Heterogeneous Catalyst for Biodiesel Production." *Bulletin of Chemical Reaction Engineering and Catalysis* 11(1): 21–6.

Hadiyanto, Hadiyanto, Sri Puji Lestari, Abdullah Abdullah, Widayat Widayat, and Heri Sutanto. 2016b. "The Development of Fly Ash-Supported CaO Derived from Mollusk Shell of *Anadara Granosa* and *Paphia Undulata* as Heterogeneous CaO Catalyst in Biodiesel Synthesis." *International Journal of Energy and Environmental Engineering* 7(3): 297–305.

Hangun-Balkir, Yelda. 2016. "Green Biodiesel Synthesis Using Waste Shells as Sustainable Catalysts with *Camelina sativa* Oil." *Journal of Chemistry* 2016.

Hu, Shengyang, Yun Wang, and Heyou Han. 2011. "Utilization of Waste Freshwater Mussel Shell as an Economic Catalyst for Biodiesel Production." *Biomass and Bioenergy* 35(8): 3627–35.

Ismail, Syahriza, Abukar Sheikh Ahmed, Reddy Anr, and Salehhuddin Hamdan. 2016. "Biodiesel Production from Castor Oil by Using Calcium Oxide Derived from Mud Clam Shell." *Journal of Renewable Energy* 2016: 1–8.

Johnson, Ian. 2017. "A 'Green' Catalyst Derived from Zebra Mussel Shells for the Production of Biodiesel from Waste Vegetable Oil." Allegheny College. https://dspace.allegheny.edu/bitstream/handle/10456/42905/CompFinal.pdf?sequence=1.

Jones, Mark Ian, L. Y. Wang, Arjan Abeynaike, and Darrell A. Patterson. 2011. "Utilisation of Waste Material for Environmental Applications: Calcination of Mussel Shells for Waste Water Treatment." *Advances in Applied Ceramics* 110(5): 280–86.

Jung, Jae-jeong, Jong-hyeon Lee, Gang-woo Lee, Yoo Kyung-seun, and Byung-hyun Shon. 2012. "Reuse of Waste Shells as a SO2/NOx Removal Sorbent." In: *Materials Recycling – Trends and Perspectives*, 301–22. doi:10.5772/33887.

Kaewdaeng, Sasiprapha, Panlop Sintuya, and Rotjapun Nirunsin. 2017. "Biodiesel Production Using Calcium Oxide from River Snail Shell Ash as Catalyst." *Energy Procedia* 138: 937–42.

Karmee, Sanjib Kumar. 2018. "Enzymatic Biodiesel Production from *Manilkara zapota* (L.) Seed Oil." *Waste and Biomass Valorization* 9(5): 725–30.

Kouzu, Masato, Akio Kajita, and Akitoshi Fujimori. 2016. "Catalytic Activity of Calcined Scallop Shell for Rapeseed Oil Transesterification to Produce Biodiesel." *Fuel* 182: 220–26.

Li, Y., Y. Jiang, and J. Gao. 2015. "Heterogeneous Catalyst Derived from Waste Shells for Biodiesel Production." *Energy Sources, Part A: Recovery, Utilization and Environmental Effects* 37(6): 598–605.

Madhu, Devarapaga, Supriya B. Chavan, Veena Singh, Bhaskar Singh, and Yogesh C. Sharma. 2016. "An Economically Viable Synthesis of Biodiesel from a Crude *Millettia pinnata* Oil of Jharkhand, India as Feedstock and Crab Shell Derived Catalyst." *Bioresource Technology* 214: 210–17.

Margaretha, Yosephine Yulia, Henry Sanaga Prastyo, Aning Ayucitra, and Suryadi Ismadji. 2012. "Calcium Oxide from *Pomacea sp.* Shell as a Catalyst for Biodiesel Production." *International Journal of Energy and Environmental Engineering* 3(1): 1–9.

Mazaheri, Hoora, Hwai Chyuan Ong, Haji Hassan Masjuki, Zeynab Amini, Mark D. Harrison, Chin Tsan Wang, Fitranto Kusumo, and Azham Alwi. 2018. "Rice Bran Oil Based Biodiesel Production Using Calcium Oxide Catalyst Derived from *Chicoreus Brunneus* Shell." *Energy* 144: 10–9.

Mo, Kim Hung, U. Johnson Alengaram, Mohd Zamin Jumaat, Siew Cheng Lee, Wan Inn Goh, and Choon Wah Yuen. 2018. "Recycling of Seashell Waste in Concrete: A Review." *Construction and Building Materials* 162: 751–64.

Mohan, S. K. 2015. "Studies on Optimization of Biodiesel Production – Snail Shell as Eco-Friendly Catalyst by Transesterification of Neem Oil." *International Journal of Innovative Research in Technology, Science & Engineering* 1(10): 5–10.

Niju, Subramaniapillai, Chelladurai Anushya, and Muthusamy Balajii. 2018a. "Process Optimization for Biodiesel Production from *Moringa oleifera* Oil Using Conch Shells as Heterogeneous Catalyst." *Environmental Progress and Sustainable Energy*. doi:10.1002/ep.13015.

Niju, Subramaniapillai, Rathinavel Rabia, K. Sumithra Devi, M. Naveen Kumar, and Muthusamy Balajii. 2018b. "Modified *Malleus Malleus* Shells for Biodiesel Production from Waste Cooking Oil: An Optimization Study Using Box–Behnken Design." *Waste and Biomass Valorization*. doi:10.1007/s12649-018-0520-6.

Niju, Subramaniapillai, Kader Mohamed Meera Sheriffa Begum, and Narayanan Anantharaman. 2016a. "Enhancement of Biodiesel Synthesis over Highly Active CaO Derived from Natural White Bivalve Clam Shell." *Arabian Journal of Chemistry* 9(5): 633–39.

Niju, Subramaniapillai, Kader Mohamed Meera Sheriffa Begum, and Narayanan Anantharaman. 2016b. "Clam Shell Catalyst for Continuous Production of Biodiesel." *International Journal of Green Energy* 13(13): 1314–19.

Niju, Subramaniapillai, Muhammed Niyas, Kader Mohamed Meera Sheriffa Begum, and Narayanan Anantharaman. 2015. "KF-Impregnated Clam Shells for Biodiesel Production and Its Effect on a Diesel Engine Performance and Emission Characteristics." *Environmental Progress & Sustainable Energy* 34(4): 1166–73.

Nur Syazwani, Osman, Umer Rashid, and Yun Hin Taufiq-Yap. 2015. "Low-Cost Solid Catalyst Derived from Waste *Cyrtopleura Costata* (Angel Wing Shell) for Biodiesel Production Using Microalgae Oil." *Energy Conversion and Management* 101: 749–56.

Nurdin, Said, Nurul A. Rosnan, Nur S. Ghazali, Jolius Gimbun, Abdurahman H. Nour, and Siti F. Haron. 2015. "Economical Biodiesel Fuel Synthesis from Castor Oil Using Mussel Shell-Base Catalyst (MS-BC)." *Energy Procedia* 79: 576–583.

Ozturk, Gulsen, Aylin Kafadar, M. Duz, Abdurrahman Saydut, and Candan Hamamci. 2010. "Microwave Assisted Transesterification of Maize (*Zea mays* L.) Oil as a Biodiesel Fuel." *Energy, Exploration and Exploitation* 28(1): 47–58.

Perea, Adrienne, Therese Kelly, and Yelda Hangun-Balkir. 2016. "Utilization of Waste Seashells and *Camelina sativa* Oil for Biodiesel Synthesis." *Green Chemistry Letters and Reviews* 9(1): 27–32.

Qin, Shenjun, Yuzhuang Sun, Xiaocai Meng, and Shouxin Zhang. 2010. "Production and Analysis of Biodiesel from Non-Edible Seed Oil of *Pistacia chinensis*." *Energy Exploration and Exploitation* 28(1): 37–46.

Raut, Shubham N., Aniket P. Jadhav, and Rashmi P. Nimbalkar. 2016. "Preparation of Heterogeneous Catalyst Derived from Natural Resources for the Transesterification of Biofuel." *International Research Journal of Engineering and Technology* 3(9): 196–204. www.irjet.net.

Reddy, A. N. R., A. A. Saleh, M. S. Islam, and S. Hamdan. 2017. "Active Heterogeneous CaO Catalyst Synthesis from *Anadara* Granosa (Kerang) Seashells for Jatropha Biodiesel Production." *MATEC Web of Conferences* 87(January): 02008.

Rezaei, R., M. Mohadesi, and G. R. Moradi. 2013. "Optimization of Biodiesel Production Using Waste Mussel Shell Catalyst." *Fuel* 109: 534–41.

Richardson, Alan Elliott, and Thomas Fuller. 2013. "Sea Shells Used as Partial Aggregate Replacement in Concrete." *Structural Survey* 31(5): 347–54.

Roschat, Wuttichai, Theeranun Siritanon, Teadkait Kaewpuang, Boonyawan Yoosuk, and Vinich Promarak. 2016. "Economical and Green Biodiesel Production Process Using River Snail Shells-Derived Heterogeneous Catalyst and Co-Solvent Method." *Bioresource Technology* 209: 343–50.

Sani, J., Saini Samir, Rikoto II, A. D. Tambuwal, A. Sanda, S. M. Maishanu, and M. M. Ladan. 2017. "Production and Characterization of Heterogeneous Catalyst (CaO) from Snail Shell for Biodiesel Production Using Waste Cooking Oil." *Innovative Energy and Research* 06(02): 2–5.

Shalmashi, Anvar, and Fahimeh Khodadadi. 2018. "Ultrasound-Assisted Synthesis of Biodiesel from Peanut Oil Using Response Surface Methodology." *Energy and Environment*. doi:JST.JSTAGE/jos/59.235 [pii].

Shankar, Vijayalakshmi, and Ranjitha Jambulingam. 2017. "Waste Crab Shell Derived CaO Impregnated Na-ZSM-5 as a Solid Base Catalyst for the Transesterification of Neem Oil into Biodiesel." *Sustainable Environment Research* 27(6): 273–78.

Sinha, Duple, and Somasundaram Murugavelh. 2016. "Comparative Studies on Biodiesel Production from Waste Cotton Cooking Oil Using Alkaline, Calcined Eggshell and Pistachio Shell Catalyst." *International Conference on Energy Efficient Technologies for Sustainability*, 130–33. doi:10.1109/ICEETS.2016.7582912.

Sirisomboonchai, Suchada, Maidinamu Abuduwayiti, Guoqing Guan, Chanatip Samart, Shawket Abliz, Xiaogang Hao, Katsuki Kusakabe, and Abuliti Abudula. 2015. "Biodiesel Production from Waste Cooking Oil Using Calcined Scallop Shell as Catalyst." *Energy Conversion and Management* 95: 242–47.

Suryaputra, Wijaya, Indra Winata, Nani Indraswati, and Suryadi Ismadji. 2013. "Waste Capiz (*Amusium Cristatum*) Shell as a New Heterogeneous Catalyst for Biodiesel Production." *Renewable Energy* 50: 795–99.

Taufiq-Yap, Yun Hin, Hwei Voon Lee, and Poh Lin Lau. 2012. "Transesterification of Jatropha Curcas Oil to Biodiesel by Using Short Necked Clam (*Orbicularia orbiculata*) Shell Derived Catalyst." *Energy Exploration and Exploitation* 30(5): 853–66.

Thushari, Indika, and Sandhya Babel. 2018. "Preparation of Solid Acid Catalysts from Waste Biomass and Their Application for Microwave-Assisted Biodiesel Production from Waste Palm Oil." *Waste Management and Research: The Journal of the International Solid Wastes and Public Cleansing Association, ISWA* 36(8): 719–28.

Viriya-empikul, Nawin, Pawnprapa Krasae, Buppa Puttasawat, Boonyawan Yoosuk, Nuwong Chollacoop, and Kajornsak Faungnawakij. 2010. "Waste Shells of Mollusk and Egg as Biodiesel Production Catalysts." *Bioresource Technology* 101(10): 3765–67.

Wang, Chunming, Senyang Xie, and Meixin Zhong. 2017. "Effect of Hydrothermal Pretreatment on Kitchen Waste for Biodiesel Production Using Alkaline Catalyst." *Waste and Biomass Valorization* 8(2): 369–77.

Yang, Linguo, Aiqing Zhang, and Xinsheng Zheng. 2009. "Shrimp Shell Catalyst for Biodiesel Production." *Energy and Fuels* 23(8): 3859–65.

5 An Intensified and Integrated Biorefinery Approach for Biofuel Production

Devadasu Sushmitha and Srinath Suranani

CONTENTS

Abbreviations	88
5.1 Introduction	88
5.2 A View on Biorefineries in India	90
5.2.1 The Godavari Sugar Mills Ltd.	91
5.3 Necessity for Biorefinery Approach	92
5.4 Challenges in Biorefinery	92
5.5 Conceptualization	93
5.6 Biorefinery Phases	93
5.7 Prevailing Technologies for Biorefinery	94
5.8 Intensified Biorefinery Processes	95
5.8.1 Intensified Biorefinery Processes Based on Sophisticated and Prominent Technology	95
5.8.1.1 Microwave Irradiation	95
5.8.1.2 Pyrolysis	95
5.8.1.3 Acoustic Cavitation	95
5.8.1.4 Hydrodynamic Cavitation	96
5.8.1.5 Gamma Ray	96
5.8.1.6 Electron Beam Irradiation	97
5.8.1.7 High-Pressure Autoclave Reactor	97
5.8.1.8 Steam Explosion	97
5.8.1.9 Photochemical Oxidation	98
5.8.2 Intensification of Processes with Novel Synthetic Routes	98
5.8.3 Intensification of Biorefinery Processes in Terms of Modifying the Equipment	98
5.8.3.1 Second-Generation Biofuel from a Fifth-Generation Bioreactor	99
5.9 Benefits	99
5.10 Applicability	100

5.11 Recent Advancements and Future Scope of Biorefinery 100
5.12 Conclusions and Perspective 101
References 101

ABBREVIATIONS

keV	Kiloelectron volt
MeV	Megaelectron volt
Rad	Radian
Mrad	Milliradian
Gy	Gray
kGy	Kilogray
MMT	million metric tons
kHz	Kilohertz
ε''	Quotient of dielectric loss
ε'	Dielectric constant
tan δ	Loss tangent
OMCs	Oil marketing companies
UV	Ultraviolet
EBI	Electron beam irradiation
SHF	Separate hydrolysis and fermentation
SSF	Simultaneous saccharification and fermentation
SSCF	Simultaneous saccharification and co-fermentation
DMC	Direct microbial conversion
CVC	Capture value creation
HMF	5-(hydroxymethyl)furfural
DDGS	Distiller's dried grains and solubles
Bio CNG	Purified form of biogas
N_2O	Nitrous oxide
Nox	Oxides of nitrogen
NH_3	Ammonia
CO	Carbon monoxide
H_2	Hydrogen
C	Total carbon content
°C	Centigrade
%	Percentage
w/w	Weight by weight
w/v	Weight by volume
g/l	Gram per liter
g/kg	Gram per kilogram
M	Molarity

5.1 INTRODUCTION

Depletion of fossil fuels has made the search for alternative fuels necessary. Biofuel is one of the alternative fuels, which could easily be produced from renewable and

A Biorefinery Approach for Biofuel Production 89

sustainable raw materials. Investigation has been ongoing for several years, and gradual development in the technologies has made the process very interesting and easy by integrated and intensified approaches along with multiple value-added byproducts. This chapter discusses the state of the art of biorefineries and their challenges.

First-generation biofuels are obtained from food crops, which limits the requirement of food; to compensate for that, second-generation biofuels have come into the picture, which utilize only non-food crops. Third-generation biofuels use non-arable lands. Algae have a typical characteristic that converts algae oil to diesel. Fourth-generation biofuel does not require a crop; a microorganism is incorporated into a solar converter that absorbs CO_2 and sunlight in a water-based medium and produces fuel. Fifth-generation biofuels are those that utilize newly developed bioreactors for biofuel production.

Lignocellulosic biomass consists of structured carbohydrates like cellulose, hemicelluloses, and unstructured carbohydrates, lignin. Cellulose was covered by hemicellulose and lignin. Cellulose was the main source for the production of bioethanol.

The main steps involved in any bioethanol production process are: (1) pretreatment of lignocellulosic biomass; (2) saccharification of pretreated biomass (enzyme production); and (3) fermentation (isolation of microorganisms).

Lignocellulosic biomass could be any plant material which is not a food crop. Selection of lignocellulosic biomass is mostly dependent upon the high cellulose and hemicellulose content.

Pretreatment processes could be physical, chemical, thermochemical, or enzymatic processes. The most utilized physical process could be the dry milling or wet milling process; among these two, wet milling is the most popular process because of the high costs of byproducts obtained during biofuel production. In this process, lignocellulosic biomass is soaked for 10–20 days, and in the dry milling process, biomass is soaked for 1 or 2 days and further milling takes place. This process makes cellulose available from lignocellulosic biomass by removing lignin. This soluble lignin could be used as a byproduct for the production of phenol formaldehyde resins. However, soluble lignin can also be utilized for applications of anticorrosion/anti-oxidant, UV-shielding, etc.

The second step in bioethanol production is enzymatic hydrolysis of lignocellulose for the production of reducing sugars, which in turn fermented with microorganisms to produce ethanol. High costs are emphasized in large-scale production for two separate reactors (i.e., one for saccharification and another for fermentation). Efforts can be made for the utilization of fifth-generation bioreactor concepts for simultaneous saccharification and ethanol fermentation. The feed stock required for bioethanol production are represented in Figure 5.1.

Biofuels could be liquid fuels, solid fuels, and gaseous fuels, as represented in Figure 5.2. Liquid fuels are generally flammable and are used for energy production and with proper utilization can be converted to mechanical energy, such as bioethanol, bio-oil, biocrude, green gasoline, biodiesel, etc. Gaseous fuels are hydrogen, methane, and synthesis gas (CO and H_2). Solid fuels like biochar can be obtained by direct burning or by co-firing.

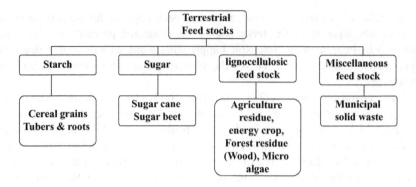

FIGURE 5.1 Feed stocks utilized for bioethanol production.

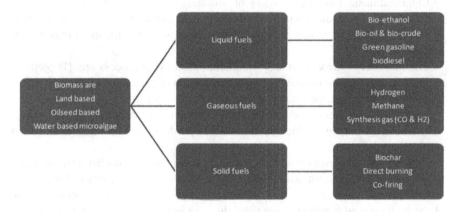

FIGURE 5.2 Classification of biofuels based on liquid, gaseous, and solid fuels.

5.2 A VIEW ON BIOREFINERIES IN INDIA

The present plan of all biorefineries is to increase the blending percentage of ethanol in transportation fuels. Since ethanol blend increases the octane number from 2 to 3 higher than gasoline, this in turn reduces carbon monoxide emissions by 20–30%.

India has a huge potential for biomass, which could without much of a stretch be depended upon to satisfy a large portion of energy needs. There is a need of 50 MMT (million metric tons) of fuel yearly in India; however, with full usage of the available biomass, India could generate twofold the existing fuel utilization costs per annum. After the first generation of biofuels, the utilization of non-foody crops was made compulsory for biofuel production.

In 2015 the Government of India made a decision to blend ethanol in transportation fuels in as many states as possible and targeted oil marketing companies (OMCs) to blend 10% of bioethanol. In November 2012 a fuel doping program started with 5% blending of ethanol. It was known as the Motor Spirit Act and enacted on January 2, 2013. The Government has likewise enabled OMCs to get ethanol created from other non-nourishment feedstocks, similar to cellulosic and lignocellulosic materials.

Ethanol mixing is the act of mixing petroleum with ethanol. Internationally, many nations, including India, have turned to ethanol mixing to diminish vehicle outflows. The introduction of the refinery makes venturing up to a 20% ethanol mixing program possible and can lead to decreases in the import costs of transportation fuel [6].

5.2.1 The Godavari Sugar Mills Ltd.

The Godavari Sugar Mills Ltd. were incorporated by the Late Shri. Karamshi Jethabhai Somaiya (Padma Bhushan) in 1939 and turned into the Godavari Biorefineries Ltd. in 2009. This is one of the highest ranked sugar industries and the biggest producers of alcohol and chemicals among other industries in India for over seven decades. The value chain of a biorefinery is represented in Figure 5.3. This figure gives a better idea of how a variety of value-added chemicals can be obtained from waste or renewable resources. Starting with biofertilizers, bagasse crop production to bioethanol production and value-added chemicals is done in one place. This variety of byproducts reduces disadvantages in techno-economic considerations for bioethanol production.

The development of biorefinery concepts should consider the green chemistry principles formulated by Paul T. Anastas and John C. Warner. Their 12 rules of green chemistry are:

1. Prevention: It is smarter to forecast danger and to arrange to treat it beforehand.
2. Atom economy: Plan engineered techniques to amplify use of all material utilized into conclusive items/products.
3. Less danger: Manufacturing strategies should be practicable, and products should be of low human lethality and ecological effect.
4. Safer synthetic substances: Chemical substance configuration should safeguard adequacy while diminishing harmfulness.
5. Safer solvents: Maintain a strategic distance from assistant materials by solvents or extractants, generally making them harmless.
6. Energy effectiveness: Energy necessities ought to be limited by direct amalgamation at surrounding temperature and pressures.
7. Renewable feedstocks: Crude materials should be selected, where easily available, and be inexhaustible.
8. Reduce derivatives/subordinates: Pointless derivatization (unnecessary changes in physical or chemical process) ought to be reduced/maintained a strategic distance from where conceivable.
9. Smart catalysis: Specifically catalyzed procedures are better than stoichiometric procedures.
10. Degradable structure: Chemicals are intended to be harmless, degradable items when discarded and not to create pollution from industries.
11. Real-time examination for contamination or pollution prevention: Frequent monitoring of the process to ensure no contamination or any unwanted products that may further turn to unsafe materials.
12. Hazard and mishap prevention: Materials/substances utilized in processing ought to be chosen to limit danger and hazardous mishaps; for example, discharges, blasts, and flames.

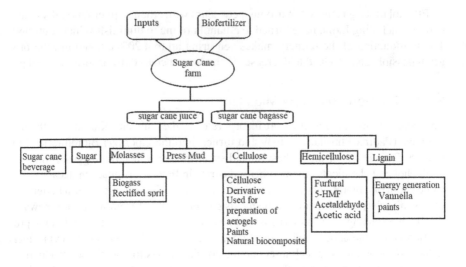

FIGURE 5.3 Typical flowsheet for value chain in biorefinery.

5.3 NECESSITY FOR BIOREFINERY APPROACH

The assessed energy required for India is 40% in household and transport sectors and 20% in modern industries. The present utilization of crude petroleum and gas is practically 90% in essential and transportation sectors, with the remaining 10% in modern synthetic chemicals generation. Higher prices of crude oil in the global market and subsequent concerns about energy security have driven developing nations to investigate alternative and modest energy sources to take care of the higher demand for energy. Biorefinery is one of the most promising answers for agrarian economies [2].

Generally in India, kitchen wastes and food wastes are left unutilized. Therefore, waste management has to come from every individual. Even the government has to implement some rules to segregate and utilize wastes. However, old practices like incineration of wastes produce plenty of greenhouse gases and carbon dioxide, as well as N_2O (nitrous oxide), NOx (oxides of nitrogen), NH_3 (ammonia), and organic C measured as total carbon.

Kamisato in Japan and San Francisco, California, are considered zero-waste cities. They incinerate wastes to produce heat energy and also provide valuable compost. Farmers are utilizing these composts to grow power crops with high carbon content. The whole carbon present in the atmosphere is returning back to the soil as a compost. This carbon cycle process reduces climate change by balancing the carbon content.

5.4 CHALLENGES IN BIOREFINERY

- Discovery, integration, and evaluation of various routes for exploitation of diverse raw materials like biomass and urban waste applications.
- Genetic modification of crops and addition of catalysts for improvement in crop generation of clean fuel.

- Converting the existing technologies of pulp and paper production to bioethanol and various chemicals.
- Incorporation of interdisciplinary optimized technical procedures in a sustainable biorefinery approach.
- Intensified pretreatment/fractionation processes for better delignification, or to its individual compounds of lignocellulosic biomass.
- Enzymatic saccharification of pretreated or fractionated biomass and recovery/recycle/reuse of enzymes.
- Production of in-house enzymes and enhancement of enzyme activity after its use.
- Reuse of utilized water.
- Simultaneous saccharification and fermentation.
- Immobilization of enzymes and their reuse.

5.5 CONCEPTUALIZATION

Biorefinery is undifferentiated from the customary oil refineries utilizing fractional refining processes for getting distinctive portions or segments from a similar crude/raw material, for example, unrefined petroleum. Biorefinery includes the integration of various biomass treatments and processing techniques into one framework, which results in the creation of various byproducts from a single biomass. This makes the whole chain progressively practical financially and furthermore diminishes the waste produced [2].

The results include high-volume, low-energy liquid fuels that could serve the transport industry's needs with high-efficacy chemicals that could contribute to the practicability of good projects. During the process, heat generated from lignin combustion could be used to meet process heat requirements. The byproducts, such as valuable chemicals, composts and fertilizers, cellulose nanofibers, alpha cellulose in application of paints, etc., provide additional revenue streams that can be obtained (Figure 5.4) [2].

5.6 BIOREFINERY PHASES

The phases of biorefinery are classified into three depending upon the raw material and product. The first phase of biorefinery basically involves one raw material

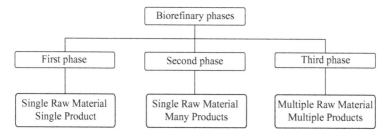

FIGURE 5.4 Classification of biorefinery phases based on quantity of raw materials and products.

and one product, i.e. bagasse as a raw material and ethanol as a main product. The second phase of biorefinery involves one raw material and several byproducts, i.e. fractionation of biomass to cellulose hemicellulose and lignin in its purest form and further processing it to value-added chemicals, including bio-ethanol. Hemicellulose can be converted to acetaldehyde, acetic acid, HMF 5-(hydroxymethyl)furfural, and furfural.

The third phase of biorefinery involves several raw materials and several products, i.e. two or more kinds of energy crops as feed materials, and fractionation and utilizing raw materials to get valued chemical precursors.

5.7 PREVAILING TECHNOLOGIES FOR BIOREFINERY

Right now, there are different advanced and accessible processes to produce valuable chemicals from non-food lignocellulosic biomass/crops. However, all such processes spin around two primary sorts of procedures, either biochemical or thermal [3].

The biochemical procedure includes utilization of microorganisms with novo enzymes in aerobic or anaerobic processes for the production of biogas. These processes may include a very old procedure called fermentation, which results in the production of ethanol. In other methods, the product bioethanol is treated chemically, followed by transesterification reaction to generate biodiesel.

On the other hand, the thermochemical process includes more direct burning or incineration in the presence of oxygen or air, known as combustion. Another process is heating at high temperatures in the presence of inert gas atmosphere, called gasification or pyrolysis. These thermochemical processes generate heat, vitality rich gas, and fluids. These products can be utilized as they are or can be additionally handled to produce amazing biofuels or synthetic compounds.

Corn processing is done with three sorts of mills. Dry mills are proposed to convey fuel ethanol and distiller's dried grains and solubles (DDGS); wet mills make a slate of starch auxiliaries and mammal feed. A third kind of mill, generally known as a dry plant, produces corn dinners, corn meal, and corn flours. The volume of corn arranged through this third kind of dry mill is little conversely with the other two sorts of mills, so our discussion will focus on the first two [1].

This section emphasizes the most efficient pretreatment methodologies in a cost-effective way for the fractionation of lignocellulosic biomass processes with maximum ethanol production. The present industries focus on integrated biorefinery approaches accounting for all effective possible techno-economic ways of ethanol production. This study mostly focuses on the efficient pretreatment methodologies such as dilute acid pretreatment (single stage and double stage), ammonia fiber explosion, hot water pretreatment, fractionation, saccharification, and fermentation for ethanol production. Considering the above targets in lignocellulosic processing, with added low capital and operating costs, is challenging. The best way has been found to be fractionation of lignocellulosic biomass and simultaneous saccharification and fermentation of fractionated hemicelluloses and cellulose to C5 and C6 sugars, with further processing to ethanol. Pervaporation was given the best results both in terms of ethanol yield and operating costs.

5.8 INTENSIFIED BIOREFINERY PROCESSES

Biorefinery process intensification can be classified according to the utilization of emerging technologies in terms of equipment and novel synthesis routes. Third, intensification of the process could be achieved by modifying the existing equipment or combining two processes.

5.8.1 Intensified Biorefinery Processes Based on Sophisticated and Prominent Technology

This section outlines emerging technologies for intensified processes such as pretreatment with microwave irradiation, pyrolysis, wet oxidation, dry oxidation pretreatment, saccharification and fermentation using ultrasound, hydrodynamic cavitation, gamma ray electron beam irradiation, pulsed electric field, high hydrostatic pressure, high pressure homogenization, high pressure autoclave reactor, and steam explosion.

5.8.1.1 Microwave Irradiation

The heating characteristics of a microwave-irradiated reactor are directly related to the dielectric properties of a material. The most important dielectric properties of a material are tan δ, ε', and ε''. There is a relation among these where tan δ is loss tangent and can be obtained as a quotient of dielectric loss (ε'') and the dielectric constant (ε'). Loss tangent can be defined as the ability of a specific substance to convert electromagnetic energy into heat at a given temperature and frequency. High absorption of microwaves consequently gives fast and efficient heating with a high tan δ reaction media. It is important to note that microwave heating also depends on the penetration depth, relaxation time, etc.

There is a difference between microwave heating and conventional heating. Conventional methods of heating take place from the outer surface and then reach the core of the processing material, whereas microwave heating is a volumetric heating that takes place from the core of the material. In general, microwave reactors use borosilicate glass (tan $\delta = 0.0010$), which is transparent to microwave irradiation [8]. The scientific microwave has a flexible probe for temperature measurement with an IR sensor [9].

5.8.1.2 Pyrolysis

Traditional pyrolysis of lignocellulosic biomass involves heating in the absence of oxygen or in the presence of inert gas. This cleaves hydrogen and alkali ether bonds that are combined lignin and carbohydrate [10]. Especially in lignin, pyrolysis normally includes the utilization of catalyst, for example, zeolite minerals and metallic mixes, to improve the yields and selectivity of required products from lignin pyrolysis. The metallic mixtures have been accounted for the increment of lignin products and decreases the inhibitory products [11]. Different catalysts, for example, zeolite, are viable in advancing the breaking responses of oxygen-containing mixes which lessen the scorch development [12, 13].

5.8.1.3 Acoustic Cavitation

Acoustic cavitation is most likely to occur under low frequency and high signal amplitude ultrasonic waves. Generally, sonication can be transmitted by piezoelectric transducers that can convert alternative current to ultrasonic energy waves.

The enhancement in the extraction of bioactive compounds achieved using sonication is attributed to cavitation in the solvent, a process that involves nucleation, growth, and collapse of bubbles in a liquid, driven by the passage of ultrasonic waves [14].

Different types of sonicators are classified according to the way sonication waves are liberated in processing media. They are the longitudinal horn, double frequency horn, triple frequency horn, bath sonicator, probe type ultrasonic horn (horns of different sizes, for various intensities), continuous ultrasound flow cell, etc. [15]. Application of these different types of acoustic devices are used according to their particular purposes.

Ultrasonic waves deliver energy that could be utilized for efficient pretreatment that increases the delignification to 3–4 folds. Optimum utilization of an ultrasound intensity can increase the enzyme activity. Simultaneous saccharification and fermentation (SSF) of bioethanol yields can be enhanced with sonication [16].

The hydrolysate obtained after dilute acid hydrolysis of eight different feed stocks was subjected to enzymatic saccharification followed by fermentation with *S. cerevisiae* and *C. shehata* using ultrasound at 35 kHz at 10% duty cycle, 220 g/kg of bioethanol per kg of raw biomass with 86.8, and 133 g/kg of pentose and hexose fermentation [17].

5.8.1.4 Hydrodynamic Cavitation

Hydrodynamics play a crucial role in the efficient pretreatment of lignocellulosic biomass, which can be obtained by cavitation reactors (like stator and rotor). But special designs of reactors are necessary for intense mixing and to perfect cleave of lignin and carbohydrate bonds.

Kim et al. (2015) performed experiments for pretreatment as well as batch SSF, the results of which gave a maximum glucose yield of 326.7 g/kg biomass at optimized conditions of 3.0% NaOH at a solid-to-liquid (S/L) ratio of 11.8% for 41.1 min, and an ethanol concentration of 25.9 g/l and ethanol yield of 90% [18]. Terán Hilares et al. (2016) tried pretreatment as well as saccharification with a 1 mm orifice and reported optimized conditions of 0.48 M of NaOH, 4.27% of S/L ratio, and 44.48 min; 52.1% of glucan content, 60.4% of lignin removal, and 97.2% of enzymatic digestibility were achieved [19]. Nakashima worked with narrow throat type hydrodynamic cavitation reactor along with sodium percarbonate for pretreatment. The efficacy of pretreatment and high-throughput was observed in terms of glucose and xylose production, with minimum inhibitor concentration, that is, furfural [20].

5.8.1.5 Gamma Ray

Gamma beam radiation is acquired from radioisotopes (Cobalt-60 or Cesium-137) and has also been tried as a lignocellulosic pretreatment. Ionizing radiation can reach inside the lignocellulosic structure, causing cleavage of the lignin and structured carbohydrate bonds. This causes further reaction by the development of free radicals which decay rapidly at amorphous regions at the end of radiation. Exposure of radiation for a particular period can degrade the crystalline regions.

1200 kilogray (kGy) gamma radiation on rapeseed straw was treated with gamma radiation and obtained better cleavage of lignin and cellulose bonds, with good thermal stability and better enzymatic treatment [21].

A Biorefinery Approach for Biofuel Production 97

The impact of 891 kilogray (kGy) gamma irradiation was used for better bioconversion of lignocellulosic biomass to microcrystalline cellulose. Gamma irradiation pretreatment was compared with acidic aqueous ionic solvents such as 1% HCL and 1% H_2SO_4 [22].

5.8.1.6 Electron Beam Irradiation

Electron beam processing or electron beam irradiation (EBI) is a procedure that utilizes beta radiation with high energy to treat lignocellulosic biomass and also for a variety of purposes. This may occur under maximum temperatures and also in an inert environment like nitrogen. Conceivable utilizations for electron beam radiation include cleavage of lignin and carbohydrate bonds in lignocellulosic biomass and in cross-linking of polymers.

Electron energies typically vary from the keV to MeV range, depending on the depth of penetration required. The irradiation dose is usually measured in grays but also in Mrads (1 Gy is equivalent to 100 rad).

An electron beam accelerator with different dosages of 100–1000 kGy was used for cellulose degradation [23]. Water-soaked rice straw is treated with electron beam irradiation at a dose of 1 MeV at 80 kGy for 120 hours of hydrolysis to obtain 70.4%, and on simultaneous saccharification and fermentation for 48 hours gave an ethanol yield and productivity of 9.3 g/l, 57%, respectively [24].

5.8.1.7 High-Pressure Autoclave Reactor

Required selective fractionation of biomass can be obtained at higher temperatures and pressures, for example water extractives and hemicellulose can be removed at temperatures from 120 to 180°C and at a pressure of 1–5 bars.

Freeze-dried oligosaccharides are rich in xylose contents, which are derived from xylan-rich raw materials like lignocellulosic biomass, for example sugarcane bagasse. Xylans can be converted to xylose, xylitol, ethanol, additives in paper making, furfural, succinic acid, films, bio-based polymers, and hydrogels [25].

Eucalyptus is used for hydrothermal treatment at 181°C for 37.5 minutes with a liquid to solid ratio of 6:1 w/w, with a xylan yield of 84.6% for a xylan removal of 58.4% [26].

Another paper reported xylan yield and xylan removal at 79% and 54.4 respectively obtained with hydrothermally treated eucalyptus sawdust for a set of operating parameters such as 150°C, 113 minutes, with a liquid to solid ratio 5:1 w/w [27].

5.8.1.8 Steam Explosion

Steam explosion of empty fruit bunches using autohydrolysis for a biorefinery approach provides information about the solubilization of hemicellulose in the form of xylose, an arabinose obtained as a hydrolysate, which are rich in fermentable sugars that are readily available for fermentation. There is an increase in porosity and crystallinity after steam explosion of empty fruit bunches [28].

Bioethanol, along with co-product bio-oil, is produced with steam-exploded wheat straw through a biorefinery approach. Cellulose and hemicellulose were utilized for ethanol fermentation for a better yield by increasing the inoculum size from 1 g/l to 3 g/l. Pyrolytic conversion was done to the remaining lignin to produce bio-oil [29].

5.8.1.9 Photochemical Oxidation

Photochemical oxidation involves UV light, hydrogen peroxide, and ozone for pretreatment of biomass. However, this process, in combination with steam-exploded biomass, reported higher fermentable sugars and better crystallinity of treated biomass.

A combination of ultrasound and ozonation can improve the lignin yield but is not as effective as microwave and seam-exploded treated biomass. In a comparison of techno-economic analysis, steam-exploded or high-pressure autoclave reactors gave the best results with minimum operating and maintenance costs [30].

5.8.2 Intensification of Processes with Novel Synthetic Routes

Some novel synthetic routes are available for the production of biofuels. One of the ways is through genetic modification of crops or by the addition of catalysts to improve the crop production, especially those that are rich in carbohydrates.

Traditional methods of ethanol production involve saccharification and fermentation, but another strategy for producing ethanol is from dimethyl ether (DME) and syngas, which is through consolidated carbonylation of dimethyl ether and hydrogenation of methyl acetic acid derivation. The transformation of DME is up to 100%, and the selectivity of ethanol is as high as 48.5%. Methanol is also a main product, with 47.6%, which can be reused as a reactant of DME synthesis, prompting bringing down the expense of this novel ethanol production technique [31].

Simultaneous saccharification and co-fermentation (SSCF) is a consolidated bioprocessing or direct microbial conversion (DMC) which involves all the reactions required for conversion of biomass to ethanol in a single step. This step involves a single microorganism for enzyme synthesis for saccharification as well as fermentation. This process has many advantages in terms of compatibility for enzyme production, as it requires the same working environment for microorganisms, thus reducing the capital or operating production costs.

Traditional methods for fermentation use microorganism which are capable of assimilating the fermentable sugar glucose faster than xylose, resulting in inhibitor formation. This can be overcome by engineering a strain which is capable of assimilating both the glucose and xylose at the same time. Similarly, in case enzymes survive and work better at 50°C, yeast works better at 27–30°C. Present SSCF processes engineered a novel thermo-tolerant yeast strain, *K. marxianus*, for co-display of endoglucanase and beta-glucosidase on its cell surface. This novel strain not only works at higher temperatures, i.e. 48°C, but also could generate 0.47 g/l of ethanol for one gram of carbohydrate [32].

5.8.3 Intensification of Biorefinery Processes in Terms of Modifying the Equipment

Intensification of biorefinery processes in terms of modifying the equipment or combining any two efficient technologies, such as ultrasound and microwave irradiation for efficient pretreatment, pretreatment of biomass involving ozonolysis, photocatalysis, oxidative catalysis, advanced oxidation processes, electrochemical process, and Fenton reactions.

A Biorefinery Approach for Biofuel Production

5.8.3.1 Second-Generation Biofuel from a Fifth-Generation Bioreactor

The second-generation production process is more complex than the first, because the carbon that is available is not in the form that is required for easy access to microorganisms such as starch or sugar.

Initially, fractionation of lignocellulosic biomass to cellulose, hemicellulose, and lignin is followed by saccharification of cellulose to sugars and anaerobic fermentation of sugars to bioethanol [5]. Generally, this process requires two types of equipment for enzymatic hydrolysis and fermentation [7]. But this could be performed on a single type of equipment. To quicken the procedure and save money on the expense of buying two unique plants, endeavors are being made to join these two procedures. This procedure with concurrent hydrolysis and aging is called simultaneous saccharification and fermentation (SSF).

The specialty of joining these two processing steps is to plan a bioreactor that meets the stringent prerequisites with respect to the careful blending of solids amid the hydrolysis step, similar to in a perfect bioreactor, as it gives impeccable development conditions and bioprocess control amid the anaerobic maturation (Figure 5.5).

Finally, a good biorefinery process should consider the following points:

- Efficient pretreatment or fractionation of lignocellulosic biomass to its purest individual compounds, inhibiting degradation of structural carbohydrates.
- Application of each byproduct to a useful product.
- Reducing the inhibitory compounds during simultaneous saccharification and fermentation.
- Reducing water and enzyme requirements by reutilizing, especially with the enhancement of enzyme activity.
- The process should be economically feasible for production of pure biofuels.

5.9 BENEFITS

Biorefineries can contribute to the optimal energy potential of organic waste and can also solve waste management and GHG emissions problems. Wastes can be

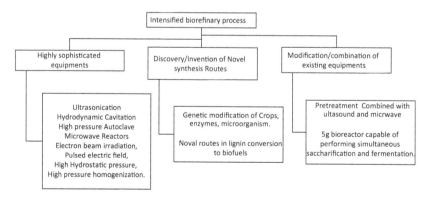

FIGURE 5.5 Classification of intensified biorefinery processes based on technology and processes.

converted into either gaseous or liquid fuels through appropriate enzymatic/chemical treatments. Biorefining pretreatment processes produce products such as paper pulp, cellulose nanofibers, alpha cellulose derivatives, additives in paints, adhesion promoters, perfumes, fuel enhancers, value-added chemical solvents such as acetic acid, acetaldehyde, furfural, HMF, etc., unlike traditional processes. The appropriateness of this procedure is additionally improved by the likelihood of utilizing an assortment of biomass assets, regardless of whether they come from plants or creatures.

5.10 APPLICABILITY

Biorefinery is still at an early stage in most parts of the world. Problems such as the availability of raw materials, the feasibility of the product requirement areas, and development of the pilot model to real plant model hinder development of biorefinery at a commercial scale. The United States National Renewable Energy Laboratory (NREL) leads biorefinery investigation with forward-moving discoveries and innovations. In spite of the fact that the innovation is still in its earliest stages, it is the way to the ideal utilization of waste and the normal assets that individuals have constantly attempted to accomplish. It is presently up to governments and organizations to energize or finance innovative work around this.

5.11 RECENT ADVANCEMENTS AND FUTURE SCOPE OF BIOREFINERY

Fractionation or pretreatment of lignocellulosic biomass are available methods; among these two, the best way is fractionation of biomass followed by synthesis of individual compounds to required products. For example, lignocellulosic biomass can be fractionated to cellulose hemicellulose, lignin, and traces of silica and ashes. Hemicellulose can be fractionated at high temperatures and pressures by solvolysis. Xylan-rich hemicelluloses are fractionated and further processed to obtain furfural and HMF 5-hydroxy methyl furfural. Other byproducts such as succinic acid, acetic acid, and acetaldehyde are obtained as inhibitory products, which can be manipulated for any one such inhibitory product as a byproduct for value-added chemical. Alpha cellulose can be utilized in paints for better corrosion activity as well as lignin and its derivatives, which are used for rinse-resistant materials. Synthesis of lignin emulsion polymers, which have a high zeta potential, can be used as better UV shielding coatings.

Consolidated biorefinery is one of the advanced technologies where the integration of reducing sugar production and the bioethanol fermentation process can be achieved in a single pot. Although several promising processing techniques are available for saccharification of fermentable sugars as well as ethanol production, like separate saccharification and fermentation [SSF], simultaneous saccharification and fermentation [SSF], and simultaneous saccharification and co-fermentation [SSCF]. Several problems involved in SSF are that it requires separate reactors and processing conditions for saccharification and fermentation. In order to circumvent

A Biorefinery Approach for Biofuel Production

the problems in SSF, a technique was developed where both simultaneous saccharification and fermentation SSF can be done on a single piece of equipment. Due to different enzyme operating temperatures of 500°C, and yeast at 30°C, the glucose is easily reacted and further converted to inhibitory compounds, which are difficult to remove. SSCF has many advantages, as mentioned in Section 5.8.2.

5.12 CONCLUSIONS AND PERSPECTIVE

The production of biofuel became mandatory as per the instructions laid by the government of increasing bioethanol blend from E5 to E10 to reduce greenhouse gas emissions. The recent developments in biorefinery concepts have decreased the cost of bioethanol production. Intensified approaches for fractionation and pretreatment of biomass have been incorporated for a sustainable biorefinery approach. Consolidated microbial conversion could perform the experiments in a single pot utilizing the same microorganism for in-house enzyme production as well as fermentation; this could be achieved by tailoring the microorganism to survive at higher temperatures.

In general, this chapter has discussed trending and emerging technologies for pretreatment of biomass and overall challenges of biorefineries in India. The utilization of fifth-generation bioreactors in second-generation bioethanol production saves energy as well as processing time.

REFERENCES

1. Eggeman, T., and D. Verser. 2006. "The importance of utility systems in today's biorefineries and a vision for tomorrow." In: *Twenty-Seventh Symposium on Biotechnology for Fuels and Chemicals*. Humana Press. doi:10.1007/978-1-59745-268-7_30.
2. Goyal, Setu. 2016. Bioenergy consultant powering clean energy future. https://www.bioenergyconsult.com/tag/chemicals.
3. Edgar, Thomas F., and Kody M. Powell. "Energy intensification using thermal storage." *Current Opinion in Chemical Engineering* 9 (2015): 83–88.
4. Somaiya, Shantilal K. 2011. *Sharing Wisdom in Search of Inner and Outer Peace*. Somaiya Publications Pvt Ltd. ISBN: 978-81-7039-280-4.
5. Skevis, George. "Liquid biofuels: Bioalcohols, biodiesel and biogasoline and algal biofuels." *Handbook of Combustion Online* (2010): 1–43.
6. Basantha Kumara Behera, Ajith Varma. *Bioenergy for Sustainability and Security*. Capital Publishing Company, New Delhi (2019).
7. Hodge, David B., M. Nazmul Karim, Daniel J. Schell, and James D. McMillan. "Soluble and insoluble solids contributions to high-solids enzymatic hydrolysis of lignocellulose." *Bioresource Technology* 99(18) (2008): 8940–8948.
8. De Souza, R. O. M. A. "Theoretical aspects of microwave irradiation practices." In: *Production of Biofuels and Chemicals with Microwave*. Springer, Dordrecht (2015): 3–16.
9. Sushmitha, Devadasu, and Srinath Suranai. "Microwave-assisted alkali-peroxide treated sawdust for delignification and its characterization." *Waste Valorization and Recycling* 2 (2018).
10. Brebu, Mihai, and Cornelia Vasile. "Thermal degradation of lignin—A review." *Cellulose Chemistry and Technology* 44(9) (2010): 353.
11. Maldhure, Atul V., and J. D. Ekhe. "Pyrolysis of purified kraft lignin in the presence of $AlCl_3$ and $ZnCl_2$." *Journal of Environmental Chemical Engineering* 1(4) (2013): 844–849.

12. Li, Bosong, Wei Lv, Qi Zhang, Tiejun Wang, and Longlong Ma. "Pyrolysis and catalytic pyrolysis of industrial lignins by TG-FTIR: Kinetics and products." *Journal of Analytical and Applied Pyrolysis* 108 (2014): 295–300.
13. Kim, Jae-Young, Jae Lee, Jeesu Park, Jeong Kwon Kim, Donghwan An, In Kyu Song, and Joon Weon Choi. "Catalytic pyrolysis of lignin over HZSM-5 catalysts: Effect of various parameters on the production of aromatic hydrocarbon." *Journal of Analytical and Applied Pyrolysis* 114 (2015): 273–280.
14. Savin, Igor I., Sergey N. Tsyganok, Andrey N. Lebedev, Dmitry V. Genne, and Elena S. Smerdina 2007. "Ultrasonic chemical reactors." In: *Electron Devices and Materials, 2007. EDM'07. 8th Siberian Russian Workshop and Tutorial on*. IEEE: 289–292. doi:10.1109/SIBEDM.2007.4292989.
15. Bussemaker, Madeleine J., and Dongke Zhang. "Effect of ultrasound on lignocellulosic biomass as a pretreatment for biorefinery and biofuel applications." *Industrial and Engineering Chemistry Research* 52(10) (2013): 3563–3580.
16. Volynets, Bohdan, Farhad Ein-Mozaffari, and Yaser Dahman. "Biomass processing into ethanol: Pretreatment, enzymatic hydrolysis, fermentation, rheology, and mixing." *Green Processing and Synthesis* 6(1) (2017): 1–22.
17. Borah, Arup Jyoti, Mayank Agarwal, Arun Goyal, and Vijayanand S. Moholkar. "Physical insights of ultrasound-assisted ethanol production from composite feedstock of invasive weeds." *Ultrasonics Sonochemistry* 51 (2019): 378–385.
18. Kim, Ilgook, Ilgyu Lee, Seok Hwan Jeon, Taewoon Hwang, and Jong-In Han. "Hydrodynamic cavitation as a novel pretreatment approach for bioethanol production from reed." *Bioresource Technology* 192 (2015): 335–339.
19. Terán Hilares, Ruly, Júlio César dos Santos, Muhammad Ajaz Ahmed, Seok Hwan Jeon, Silvio Silvério da Silva, and Jong-In Han. "Hydrodynamic cavitation-assisted alkaline pretreatment as a new approach for sugarcane bagasse biorefineries." *Bioresource Technology* 214 (2016): 609–614.
20. Nakashima, Kazunori, Yuuki Ebi, Naomi Shibasaki-Kitakawa, Hitoshi Soyama, and Toshikuni Yonemoto. "Hydrodynamic cavitation reactor for efficient pretreatment of lignocellulosic biomass." *Industrial and Engineering Chemistry Research* 55(7) (2016): 1866–1871.
21. Zhou, Hua, Renli Zhang, Wang Zhan, Liuyang Wang, Lijun Guo, and Yun Liu. "High biomass loadings of 40 wt% for efficient fractionation in biorefineries with an aqueous solvent system without adding adscititious catalyst." *Green Chemistry* 18(22) (2016): 6108–6114.
22. Liu, Yun, Hua Zhou, Shihui Wang, Keqin Wang, and Xiaojun Su. "Comparison of γ-irradiation with other pretreatments followed with simultaneous saccharification and fermentation on bioconversion of microcrystalline cellulose for bioethanol production." *Bioresource Technology* 182 (2015): 289–295.
23. Jusri, N. A. A., Amizon Azizan, N. Ibrahim, R. Mohd Salleh, and M. F. Abd Rahman. 2018. "Pretreatment of cellulose by electron beam irradiation method." In: *IOP Conference Series: Materials Science and Engineering*. IOP Publishing, Vol. 358: 1–12.
24. Bak, Jin Seop. "Electron beam irradiation enhances the digestibility and fermentation yield of water-soaked lignocellulosic biomass." *Biotechnology Reports* 4 (2014): 30–33.
25. Huang, Caoxing, Ben Jeuck, Jing Du, Qiang Yong, Hou-min Chang, Hasan Jameel, and Richard Phillips. "Novel process for the coproduction of xylo-oligosaccharides, fermentable sugars, and lignosulfonates from hardwood." *Bioresource Technology* 219 (2016): 600–607.
26. Aditiya, H. B., T. M. I. Mahlia, W. T. Chong, Hadi Nur, and A. H. Sebayang. "Second generation bioethanol production: A critical review." *Renewable and Sustainable Energy Reviews* 66 (2016): 631–653.

27. Alfaro, A., A. Rivera, A. Pérez, R. Yáñez, J. C. García, and F. López. "Integral valorization of two legumes by autohydrolysis and organosolv delignification." *Bioresource Technology* 100(1) (2009): 440–445.
28. Medina, Jesus David Coral, Adenise Woiciechowski, Arion Zandona Filho, Poonam Singh Nigam, Luiz Pereira Ramos, Carlos Ricardo Soccol, and C. R. Soccol. "Steam explosion pretreatment of oil palm empty fruit bunches (EFB) using autocatalytic hydrolysis: A biorefinery approach." *Bioresource Technology* 199 (2016): 173–180.
29. Tomas-Pejo, E., J. Fermoso, E. Herrador, H. Hernando, S. Jiménez-Sánchez, M. Ballesteros, C. González-Fernández, and D. P. Serrano. "Valorization of steam-exploded wheat straw through a biorefinery approach: Bioethanol and bio-oil co-production." *Fuel* 199 (2017): 403–412.
30. Mahamud, M. R., and D. J. Gomes. "Enzymatic saccharification of sugar cane bagasse by the crude enzyme from indigenous fungi." *Journal Of Scientific Research* 4(1) (2012): 227–238.
31. Zhang, Yi, Xiaoguang San, Noritatsu Tsubaki, Yisheng Tan, and Jianfeng Chen. "Novel ethanol synthesis method via C1 chemicals without any agriculture feedstocks." *Industrial and Engineering Chemistry Research* 49(11) (2010): 5485–5488.
32. Vohra, Mustafa, Jagdish Manwar, Rahul Manmode, Satish Padgilwar, and Sanjay Patil. "Bioethanol production: Feedstock and current technologies." *Journal of Environmental Chemical Engineering* 2(1) (2014): 573–584.

6 Hydrothermal Carbonization for Valorization of Rice Husk

B. Sai Rohith, Naga Prapurna, Kuldeep B. Kamble, Rajmohan K. S. and S. Srinath

CONTENTS

6.1	Introduction	106
6.2	Technologies Involved in Conversion of Biomass	106
	6.2.1 Gasification	108
	6.2.2 Pyrolysis	108
	6.2.3 Dry Torrefaction	109
	6.2.4 Hydrothermal Carbonization	109
6.3	Decomposition Reactions and Mechanisms Involved in HTC	111
6.4	Influence of Reaction Parameters on the HTC Process	111
	6.4.1 Reaction Temperature	112
	6.4.2 Operating Pressure	112
	6.4.3 Reaction Time	112
	6.4.3 pH	112
	6.4.4 Solid Load	113
6.5	Hydrochar Applications	113
	6.5.1 Soil Amendment	113
	6.5.2 Renewable Energy Resource	113
	6.5.3 Activated Carbon Adsorbent	114
	6.5.4 Carbon Sequestration	114
	6.5.5 Additional Applications	114
6.6	Hydrothermal Carbonization of Rice Husk	114
6.7	Thermogravimetric Analysis of Hydrochar	116
	6.7.1 Kissinger-Akahira-Sunose Method (KAS)	117
	6.7.2 Friedman Method	117
	6.7.3 Flynn-Wall-Ozawa Method	117
	6.7.4 Coats-Redfern Method	117
6.8	Modeling of HTC Process	117
	6.8.1 Two-Step Kinetic Model for the HTC Process	118
6.9	Challenges and Future Scope	118
References		119

6.1 INTRODUCTION

Global energy consumption is estimated to increase 56% by 2040, with China and India consuming the highest among all countries. India's total energy consumption was 775 Mtoe in 2015 and has been expected to increase by about 65% by 2040 [1]. Worldwide fossil fuels are the largest sources for supplying energy, and its high consumption is creating several problems for economically developing nations. There are various environmental effects caused due to the consumption of fossil fuels. To avoid the difficulties mentioned earlier, it is suggested that we should look for clean and effective energy resources. Biomass can be taken as an alternative source, as it produces biofuel through numerous procedures (Figure 6.1) [2].

The biomass availability in India is around 500 million metric tons per year. Biomass is the organic matter derived from wood, animals, forest residuals and municipal solid wastes (MSW), which can be viewed as a potential energy source. Its huge availability and capability to grow anywhere frame it as a starting power point for supportable energy production (Figures 6.2 and 6.3) [3].

Biomass derived from living matter consists of three main compounds, as shown in Figure 6.4. Although it is considered a natural power resource, it is not represented as an optimal source of energy due to its fibrous nature, wetness, presence of volatile rich matter and its low heating value. To overcome these problems, biomass undergoes preliminary treatments like pyrolysis, anaerobic digestion, fermentation, dry torrefaction etc. to improve its combustion characteristics

6.2 TECHNOLOGIES INVOLVED IN CONVERSION OF BIOMASS

The mechanisms which allow the specific biomass to be converted into the required fuel are as follows:

1. Thermal conversion: It uses heat in the presence or absence of oxygen to transform biomass to a required source of energy. It includes processes like combustion, pyrolysis and torrefaction.
2. Thermochemical conversion: Thermochemical conversion combines both heat and chemical processes to produce energy products from biomass material. Gasification and hydrothermal carbonization come under this conversion technology.
3. Biological conversion: Biological conversion involves the action of bacterial organisms and other microbes, which helps in breaking down the complex biomass molecules and produces the desired fuels. It comprises anaerobic digestion and fermentation.
4. Chemical conversion: Chemical conversion is associated with the usage of chemical agents to convert biomass into fuels.

Thermochemical processes have more advantages as compared to other conversion processes because they require less reaction time and offer high conversion efficiency. Some of the currently used conversion processes are shown in Figure 6.5.

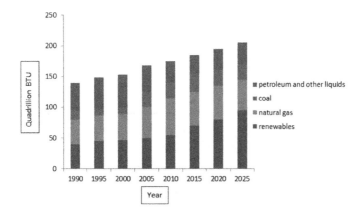

FIGURE 6.1 World energy consumption by energy source.

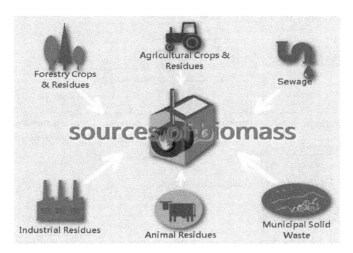

FIGURE 6.2 Various sources of biomass.

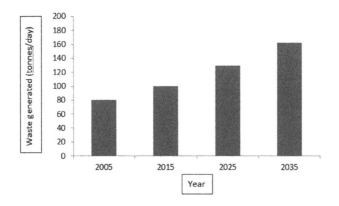

FIGURE 6.3 Estimation of waste generated.

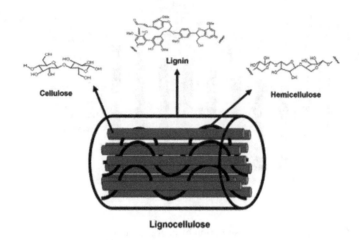

FIGURE 6.4 Structural representation of biomass.

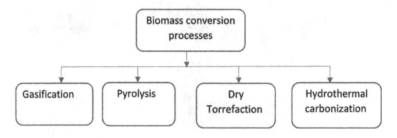

FIGURE 6.5 Biomass conversion technologies.

6.2.1 GASIFICATION

Gasification includes biomass burnt within a controlled air supply to produce gaseous products like CO, CO_2, N_2, CH_4, H_2 and H_2O, accompanied by minor quantities of tar, char and ash. This process takes place at a very high temperature range and the product gas is termed as synthetic gas (syn gas). Syn gas is utilized in gas turbines and boilers. Other products can be used as precursors for synthesis of various chemicals [4]. This process produces very small sums of biochar due to the conversion of organic material to gases and char. The biochar produced from the gasification process consists of alkali and alkaline metals and toxic substances which include polyaromatic hydrocarbons [5].

6.2.2 PYROLYSIS

This process occurs in a lack of oxygen which decomposes the biomass feedstock by maintaining the temperature between 300–650°C. The biomass is transformed into biochar, which has enhanced carbon characteristics, and a small sum of bio-oil, along with non-condensable gases, is released. The gases comprise methane, carbon dioxide, carbon monoxide and hydrogen. Slow pyrolysis, conventional pyrolysis, fast pyrolysis and flash pyrolysis are the available processes. The slow pyrolysis process has a greater yield of solid products in comparison with other processes [6].

6.2.3 Dry Torrefaction

In dry torrefaction, the biomass is treated thermally in a nonreactive environment at a temperature span of 180–320°C. It is termed a mild pyrolysis process. During this process, moisture gets evaporated and the low calorific material (volatiles) is transformed into gaseous form. The torrefaction process results in an energy loss of 10–15%. This process improves the physiochemical properties of the feedstock, and it is a pre-processing method before pyrolysis process. The characteristics of the torrefied product lie between that of biomass and biochar [7].

6.2.4 Hydrothermal Carbonization

This is a biomass conversion technology where wet organic matter gets directly transformed into highly enriched carbonized material known as hydrochar. HTC is achieved by means of thermal and chemical changes that take place in the process. This process involves immersion of the feedstock in water and operates at 180–260°C.

The pressure is kept high (10–40) to keep up the water in the aqueous phase. In these operating conditions, the pressurized water produces more ions than at ambient conditions, which allows water to act as a solvent, reactant and catalyst [8]. Water alters its characteristics when it crosses the subcritical conditions (374°C, 220bar) [9]. In subcritical conditions, HTC converts the wet organic material into hydrochar, enhanced in its carbon content. A temperature-pressure phase diagram of water is displayed in Figure 6.6.

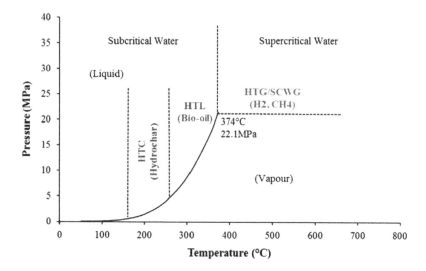

FIGURE 6.6 Temperature-pressure phase diagram of water [2].

As HTC is performed in the presence of water, biomass feedstock can have high moisture content. Thus, the HTC process can ignore pre-heating, which consumes more energy and is an expensive process, especially when the thermal pretreatment process is carried through pyrolysis and dry torrefaction [10]. In the pyrolysis process the biochar produced from biomass has high carbon content, but HTC has an advantage of having high hydrochar yield compared to other conversion processes. Another benefit of HTC over other processes is that it functions at lower temperatures and produces few gaseous pollutants [11]. In recent times, the HTC process aims to produce more solid products, which have valuable industrial and environmental applications [12].

The hydrochar has higher heating values, with a chemical structure identical to natural coal [12]. This similarity is shown appropriately by the Van Krevelen diagram (Figure 6.7). Depending upon the desired product, operating time can be from minutes to hours. At maximum operating conditions, the carbon content of the char is high but results in less hydrochar yield.

HTC has an advantage in that the process is complete in hours and can produce small size carbon particles, which have higher heating values, exceptional thermal and chemical stability and excellent adsorption capacity. The aqueous products have further applications, and the properties of the products rely upon the operating conditions [10].

The basic aim of thermochemical pretreatment is to reduce large biomass molecules to small molecules. During hydrolysis, the existence of subcritical water reduces the activation energy of cellulose and hemicellulose so that it accelerates the degradation of biomass [13]. As water is cheap, present in the wet organic matter and it does not contain any toxic chemicals, it is considered a reacting medium. The corrosive chemicals and the toxic solvents produced during the HTC process are soluble in water. The inorganic elements (Ca, Mg, K, Fe etc.) present in the lignocellulosic biomass are driven out in the form of ash during combustion [14].

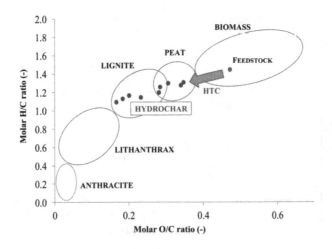

FIGURE 6.7 Van Krevelen diagram.

6.3 DECOMPOSITION REACTIONS AND MECHANISMS INVOLVED IN HTC

In the HTC process various chemical reactions take place at subcritical conditions. Hydrothermal carbonization is associated with three main stages: hydrolysis, polymerization and aromatization. During hydrolysis, hemicellulose (at T greater than 180°C), cellulose (at T greater than 200°C) and lignin (at T greater than 220°C) converts into small products due to the addition of water [13].

When hemicellulose undergoes hydrolysis, it produces various byproducts which are later converted to 5-hydroxy-methyl-furfural, and after a certain period of time it transforms into methanoic acid and levulinic acid. The cellulose gets hydrolyzed into D-glucose, producing 5-HMF, and then into methanoic acid and levulinic acid. Lignin usually forms phenolic components, which can be viewed in Figure 6.8.

The decrease in the hydrogen-to-carbon and oxygen-to-carbon ratios during the HTC process can be understood by dehydration and decarboxylation mechanisms. In this process, dehydration is an influential mechanism compared to decarboxylation. In the HTC process, ease of formation of aromatic compounds can occur at high operating conditions and is responsible for low carbon content of hydrochars [15, 16]. The occurrence of aromatic bonds during the HTC process could lower the carbon content of the hydrochars. The other mechanisms which occur during the hydrothermal carbonization process are methylation, pyrolytic, transformation and Fischer–Tropsch reactions.

6.4 INFLUENCE OF REACTION PARAMETERS ON THE HTC PROCESS

The goal of HTC is to maximize the yield of hydrochar. The product composition and characteristics depend upon the reaction temperature, solid load, operating pressure, reaction time, pH and composition of biomass feedstock.

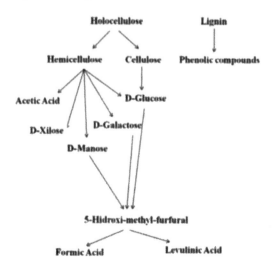

FIGURE 6.8 Products formed from HTC process [13].

6.4.1 Reaction Temperature

Reaction temperature is treated as a dominant parameter in this process, as it enhances dehydration and decarboxylation reactions and also influences the reaction kinetics. The augmentation of the temperature enhances the hydrochar's carbon content, which in turn lowers its yield [17–19]. However, the increase in temperature decreases the oxygen and hydrogen content of the biomass and thereupon decreases the oxygen to carbon ratio. In the HTC process, a reaction usually starts at 160°C, and the rate of the reaction increases with a rise in temperature [18]. Moreover, at higher temperatures the process allows production of biogas and more liquid products, hence it lowers the production of hydrochar. At low temperatures, condensation and depolymerization take place, and at high temperatures polymerization and aromatization prevail [20].

6.4.2 Operating Pressure

Pressure is generally treated as the ancillary parameter of this process, since it is controlled by the temperature. For further continuing the hydrothermal carbonization process, the pressure has to be maintained appropriately, which does not allow water to change its phase. Le Chatelier's principle gives a detailed explanation about the shifting of equilibrium when pressure is altered. During the HTC process, the dehydration and decarboxylation reactions appear to be weakened by higher pressures [21, 22].

6.4.3 Reaction Time

The reaction time of the HTC process is useful for designing and operating the process. When reaction rates are not known, it is very difficult to estimate how much reaction time is required to complete the process. Generally, this process takes place starting from a few minutes to numerous hours. Moreover, the experiments that were conducted have shown that increasing the residence time inside the reactor usually enhances the carbon content and in turn increases the high heating value [23, 24]. By analyzing the properties and characteristics of the products, the residence time enhances the influential reactions, thus eliminating the usage of catalysts for higher conversion rates of biomass feedstock. Reaction time usually reduces the total hydrochar yield, enhancing formation of higher amounts of water-soluble compounds [13].

6.4.3 pH

During the HTC process, the pH usually falls due to the occurrence of acetic acid, formic acid and lactic acid [25, 26]. The effect of several acids and other products alters the characteristic properties of the products [27]. Generally, an acidic environment enhances the carbon content, while the effect on other mechanisms in the process is not known. The increase of pH in this process allows product formation with high H/C ratios (Table 6.1) [28, 29].

TABLE 6.1
Saturated Steam Table

Saturation Temperature (°C)	Pressure (bar gauge)	Density of Water (kg/m^3)
180	9.01	887.1
190	11.53	876.1
200	14.52	864.7
210	18.05	852.8
220	22.17	840.3
230	26.94	827.3
240	32.43	813.5
250	38.72	799.1
260	46.92	784.2

6.4.4 SOLID LOAD

The solid load can be referred as the ratio of biomass feedstock to water (B/W). This will help in lowering the reaction time by enhancing the rate, which raises the concentration of the monomers in turn, allowing the polymerization reaction to start earlier. As a matter of fact, to increase the production of coal it is necessary that the feedstock should be completely submerged in water such that it increases the ease of stirring and pumping [13].

The HTC process is widely used compared to other conversion processes (pyrolysis, biological treatments etc.). HTC allows us to acquire homogenous products despite any biomass taken.

6.5 HYDROCHAR APPLICATIONS

The main important use of hyrdrochar is as a source of soil amendment. However, in recent times hydrochar has a broad range of applications in the field of power production, agriculture usage, carbon capturing, wastewater treatment etc.

6.5.1 SOIL AMENDMENT

Hydrochar can be used as an ion-bonding component which in turn improves the quality of the soil. It is used as a soil conditioner to enhance soil fertility. Hydrochar has a high surface area and its porous structure allows it to improve water retention when applied to the soil. Moreover, it enhances the adequate supply of nutrients to plants and lowers nutrient misplacement by leaching [30].

6.5.2 RENEWABLE ENERGY RESOURCE

Hydrochar's major application is that it can be used as a combustible in combustion plants, thermal power plants, cement and steel factories and gasification. In economically growing nations, cooking can be done by using hydrochar rather than wood or coal, which will be helpful in replacing deforestation [31].

6.5.3 Activated Carbon Adsorbent

The adsorption properties of activated char are very high due to its high surface-to-volume ratio, and it can adsorb various organic pollutants and metals from water. Sometimes hydrochar is further energized to increase its adsorbent properties. The oxygen functional groups present on hydrochar's surface allow it to increase its adsorption capacity [32].

6.5.4 Carbon Sequestration

Hydrochar involves carbon capturing and has the ability to store CO_2 for a long period of time. Hydrochar can be stored in a reservoir to bind atmospheric CO_2 through photosynthesis. Enhancing the amount of carbon improves the quality and fertility of soil, reduces nutrient misplacement, eliminates soil erosion and increases water conservation and crop yield. It is investigated that the chemical stability of the charcoal allows the carbon to persist in the soil for a long time [33].

6.5.5 Additional Applications

In modern times, hydrothermal carbonization of specific biomass by adding certain required compounds allows the production of nano-structured carbonized material. The properties of these nanoparticles have several applications which involve catalyst and adsorbent production and carbon fixation [34]. Moreover, coal particles produced from HTC act like hydrogen storage, used in lithium-ion batteries and as a starting material for fuel cells [30].

6.6 HYDROTHERMAL CARBONIZATION OF RICE HUSK

The biomass sample that is used for the hydrothermal carbonization process is rice husk. Rice husk is collected from a local rice mill and chopped into less than 1 mm size. The rice husk is stored in bags at room temperature.

$$\text{hydrochar yield}(\%) = \frac{\text{mass of hydrochar}}{\text{initial mass of biomass}} \times 100$$

Hydrochar yield can be calculated by using the above formula. The values of hydrochar yield at different process conditions are shown in Table 6.2.

The ultimate analysis of the hydrochar explains that the hydrochar's carbon content increases with operating conditions. The increase of carbon content is because of the lowering of the hydrogen and oxygen content of hydrochar. This reduction takes place due to dehydration and the decarboxylation of the biomass during the HTC process. As process parameter increases, the dehydration and decarboxylation of the biomass also increases. Table 6.3 shows that fixed

TABLE 6.2
Hydrochar Yield (%) at Different Process Conditions

Temperature (°C)	Time (h)	Hydrochar Yield
170	1.5	63.1
170	3	51.2
170	6.5	48.7
190	1.5	51
190	3	41.2
190	6.5	37.3
210	1.5	43
210	3	36.1
210	6.5	31.2

carbon percentage in hydrochar is increased with increments in temperature and residence time.

Carbon storage factor (CSF) is defined as the carbon quantity left in the hydrochar after decomposition of the feedstock takes place, and it helps to identify the amount of carbon that remains within the solid material after the biomass decomposition. CSF is calculated by [35]

$$CSF = \frac{\text{Carbon remaining in char after HTC}}{\text{Initial mass of feedstock}} = \frac{\%C_{hydrochar} \times yield_{hydrochar}}{m_o}$$

The amount of carbon in the hydrochar as a fraction of the untreated rice husk is called CSF. The carbon content in the hydrochar varies with process conditions. As the carbon was transforming into the liquid phase and gaseous phase, the hydrochar's CSF during HTC ranged between 0.34–0.20 and decreased as the residence time and

TABLE 6.3
Ultimate Analysis of Hydrochar

Temperature (°C)	Time (hour)	Carbon (%)	Hydrogen (%)	Nitrogen (%)	Oxygen (%)	Ash
Rice husk	0	48.14	5.89	0.22	39.65	2.75
170	1.5	55.83	5.63	0.24	31.57	2.9
170	3	57.51	5.28	0.2	29.28	3.04
170	6.5	57.95	5.12	0.23	28.94	3.19
190	1.5	60.01	5.54	0.22	28.74	3.12
190	3	61.54	5.31	0.24	25.73	3.33
190	6.5	64.28	5.08	0.24	24.22	3.67
210	1.5	61.23	5.48	0.21	25.08	3.69
210	3	63.88	5.27	0.25	22.61	3.82
210	6.5	66.01	5.04	0.2	21.93	3.94

reaction temperature were increased from 170–210°C respectively. The carbon storage factors of hydrochar have the greatest value at lower reaction temperatures, and as the temperature and residence time increase, CSF decreases.

Carbon recovery is the amount of carbon present in the hydrochar as a fraction or percentage of the feedstock that is unreacted. Carbon recovery in hydrochar, H_{crec}, is calculated by using the following equation [35]:

$$H_{crec} = \frac{\%C_{hydrochar} \times \text{yield hydrochar}}{\%C_{foedstock} \times m_o} \times 100$$

The carbon recovery value decreases from 72.67 to 41.28 with an increase in process parameters.

High heating value (HHV) is the amount of heat released during the combustion of fuel, which is also known as gross calorific value of the fuel. The high heating value of rice husk and hydrochar material was determined using a bomb calorimeter for some of the conditions. The HHV value of rice husk was 18.32 MJ/Kg, and the maximum heating value of hydrochar was found to be 26.27 MJ/Kg for 210°C and 6.5 h.

6.7 THERMOGRAVIMETRIC ANALYSIS OF HYDROCHAR

Thermogravimetric analysis helps us to understand the degradation process in the presence of heat of the rice husk under pyrolysis conditions. The thermal properties can be studied using a thermogravimetric analyzer, which helps in investigating the kinetics of thermal decomposition of rice husk. There are several models which help in calculating the kinetic parameters.

The application of thermogravimetric methods helps in knowing the physical and chemical processes that take place during the solid mass degradation. The process can be described as follows:

$$\text{Reactant(solid)} \Rightarrow \text{Product 1 (solid)} + \text{Product 2}$$

The conversion rate can be expressed as

$$\frac{dx}{dt} = \beta \frac{dx}{dT} = K(T)f(x)$$

Where
 x is degree of conversion
 $K(T)$ is a function of temperature; $K(T)$ is expressed as:

$$K(T) = A \exp\left(-\frac{E}{RT}\right)$$

Where
 E is the activation energy
 A is pre-exponential factor
 R is the gas constant

6.7.1 KISSINGER-AKAHIRA-SUNOSE METHOD (KAS)

The KAS standard equation is given as [36, 37],

$$\frac{dx}{f(x)} = \frac{A}{\beta} \exp\left(-\frac{E}{RT}\right) dT$$

It is based on the Coats-Redfern method [38], which is given by the following equation,

$$\ln \frac{\beta}{T^2} = \ln \frac{AR}{Eg(x)} - \frac{E}{RT}$$

By plotting the graph for the above equation, we can calculate the pre-exponential factor and apparent activation energy.

6.7.2 FRIEDMAN METHOD

This method is a differential isoconversional model and is given as [39],

$$\ln\left(\frac{dx}{dt}\right) = \ln\left(\beta \frac{dx}{dT}\right) = \ln[Af(x)] - \frac{E}{RT}$$

We can evaluate the values of E and A by plotting the above equation.

6.7.3 FLYNN-WALL-OZAWA METHOD

The Flynn-Wal-Ozawa (FWO) method is based on the integral isoconversional model given as [40, 41]

$$\ln \beta = \ln \frac{AE}{Rg(x)} - 5.331 - 1.052 \frac{E}{RT}$$

For the value of x to be constant, by plotting the graph for the above equation we can calculate both pre-exponential factor and activation energy.

6.7.4 COATS-REDFERN METHOD

This method involves thermal degradation mechanism, which is an integral method, and pre-exponential factor and activation energy values are calculated by plotting the following equation [38]:

$$\ln \frac{g(x)}{T^2} = \ln \frac{AR}{\beta E} - \frac{E}{RT}$$

6.8 MODELING OF HTC PROCESS

In HTC processes, biomass feedstock converts into solid, liquid and gaseous products. To understand the kinetic reaction mechanism, some kinetic models have been already made. There are two kinds of kinetic models available; the first is "one-step kinetic model" and the second is "two-step kinetic model". The one-step kinetic model assumes that the biomass is directly transformed into its final product in one step; no intermediate product is formed during the HTC process. The two-step kinetic model suggests that first biomass is converted into an intermediate product, then it is converted into the final product. Some kinetic models based on the one-step kinetic model have been made already for various feed stocks [42–45]. But Prins et al. [46] suggested a two-step kinetic model that gives an exact description of decomposition kinetics of feedstock.

6.8.1 Two-Step Kinetic Model for the HTC Process

For the decomposition of biomass, the one-step reaction model is used, but this model provides a very rough explanation about the reaction mechanism. To get the accurate results, the two-step kinetic model has been investigated. This model estimates that in the primary step, biomass forms an intermediate product, and then in the second step it converts into the final product. During the decomposition of biomass, volatile product is also formed parallel in the two-step reaction [47].

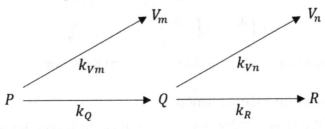

Where
P = initial biomass samples
Q = intermediate product
R = final product
Vm = volatile product formed in the first step
Vn = volatile product formed in the second step

Each biomass feedstock has a different composition, thereupon they also have different kinetic parameters. We developed a simple kinetic model that describes the decomposition kinetics of the biomass for the hydrothermal carbonization process. The values of activation energy and pre-exponential factors can be estimated using the kinetic model. The change in mass with time at 210°C is shown in Figure 6.9.

6.9 CHALLENGES AND FUTURE SCOPE

In the modern world, power consumption and utilization of fossil fuels are two major problems faced by advancing nations. Biomass usage can be a solution for producing the desired fuel and eliminating the fossil fuels application. Hydrothermal carbonization is preferable over other conversion processes, and the properties and applications of hydrochar are quite convincing. However, the process of HTC has not yet been commercialized. The carbon content of the hydrochar produced from rice husk

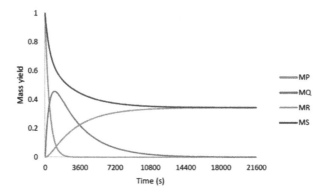

FIGURE 6.9 Evolution of mass with time at 210°C.

is greater, and a good yield is observed. So, rice husk can be more advantageous for the purpose of biofuel production than other agro waste residues.

There are many challenges in the commercialization of the process.

A challenge to environmental impact is the uncertainty of current knowledge. There are products of the reaction still under exploration. Slurry handling is the major challenge, and it requires a long reaction time. Kinetic and energy modeling are challenging due to unknown chemical reactions. The energy requirement for this process must be minimized. The emission gases are not explored, as well as the capture of carbon dioxide.

REFERENCES

1. Dudley B., "BP Statistical Review of World Energy", 2016. http://www.bp.com/statisticalreview.
2. Ozbay N., Putun A.E., Uzun B.B., Putun E., "Bio-crude from biomass: Pyrolysis of cottonseed cake", *Renewable Energy*, 24(3–4), 2001, 615–625.
3. Singh Kambo Harpreet, Dutta Animesh, "A comparative review of biochar and hydrochar in terms of production, physico-chemical properties and applications", *Renewable and Sustainable Energy Reviews*, 45, 2015, 359–378.
4. Salman Zafar, "Importance of Biomass Energy", Bioenergy Consultant, 2014. http://www.bioenergyconsult.com/a-glance-at-biomass-energy.
5. Puig-Arnavat Maria, Bruno Joan Carles, Coronas Alberto, "Review and analysis of biomass gasification models", *Renewable and Sustainable Energy Reviews*, 14(9), 2010, 2841–2851.
6. Mohan Dinesh, Pittman Charles U., Jr., Steele Philip H., "Pyrolysis of wood/biomass for bio-oil: A critical review", *Energy & Fuels*, 20(3), 2006, 848–889.
7. Bridgeman T. G., Jones J. M., Shield I., Williams P. T., "Torrefaction of reed canary grass, wheat straw and willow to enhance solid fuel qualities and combustion properties", *Fuel*, 87(6), 2008, 844–856.
8. Kronholm J., Hartonen K., Riekkola M. L., "Analytical extractions with water at elevated temperatures and pressures", *TrAC Trends in Analytical Chemistry*, 26(5), 2007, 396–412.
9. Schuhmacher J.P., Huntjens F.J., Van Krevelen D.W., "Chemical structure and properties of coal XXVI – Studies on artificial coalification", *Fuel*, 39, 1969, 223–234.

10. Benavente Verónica, Calabuig Emilio, Fullana Andres, "Upgrading of moist agroindustrial wastes by hydrothermal carbonization", *Journal of Analytical and Applied Pyrolysis*, 113, 2015, 89–98.
11. Libra Judy A., Ro Kyoung S., Kammann Claudia, Funke Axel, Berge Nicole D., Neubauer York, Titirici Maria-Magdalena, Fühner Christoph, Bens Oliver, Kern Jürgen, Emmerich Karl-Heinz, "Hydrothermal carbonization of biomass residuals: a comparative review of the chemistry, processes and applications of wet and dry pyrolysis", *Biofuels*, 2, 2011, 71–106.
12. Mumme Jan, Eckervogt Lion, Pielert Judith, Diakité Mamadou, Rupp Fabian, Kern Jürgen, "Hydrothermal carbonization of anaerobically digested maize silage", *Bioresource Technology*, 102, 2011.
13. Funke A., Ziegler F., "Hydrothermal carbonization of biomass: A summary and discussion of chemical mechanisms for process engineering", *Biofuels, Bioproducts & Biorefining*, 4, 2010.
14. Krusea Andrea, Nicolaus Dahmena, "Water – A magic solvent for biomass conversion", *Journal of Supercritical Fluids*, 96, 2015, 36–45.
15. Luijkx G.C.A., *Hydrothermal Conversion of Carbohydrates and Related Compounds*, Technical University of Delft, pp. 31–45, 1994.
16. Sugimoto Y., Miki Y., 1997, "Chemical structure of artificial coals obtained from cellulose, wood and peat". In: A. Ziegler, K. H. van Heek, J. Klein and W. Wanzi (Editors), *Proceedings of the 9th International Conference on Coal Science ICCS '97*, DGMK, vol. 1, pp. 187–190.
17. Brunner G., "Near critical and supercritical water. Part I. Hydrolytic and hydrothermal processes", *The Journal of Supercritical Fluids*, 47(3), 2009, 373–381.
18. Xiao L., Shi Z., Xu F., Sun R., "Hydrothermal carbonization of lignocellulosic biomass", *Bioresource Technology*, 118, 2012, 619–623.
19. Erlach B., Harder B., Tsatsaronis G., "Combined hydrothermal carbonization and gasification of biomass with carbon capture", *Energy*, 45(1), 2012, 329–338.
20. Bergius F., "Beiträgezurtheorie der kohleentstehung", *Naturwissenschaften*, 1, 1928, 1–10.
21. Hashaikeh R., Fang Z., Butler I.S., Hawari J., Kozinski J.A., "Hydrothermal dissolution of willow in hot compressed water as a model for biomass conversion", *Fuel*, 86(10–11), 2007, 1614–1622.
22. Afonso C.A.M., Cerspo J.G. eds., *Green Separation Processes*, Weinheim, Wiley VCH, p. 323–337, 2005.
23. Müller J.B., Vogel F., "Tar and coke formation during hydrothermal processing of glycerol and glucose. Influence of temperature, residence time and feed concentration", *The Journal of Supercritical Fluids*, 70, 2012, 126–136.
24. Erlach B., Tsatsaronis G., "Upgrading of biomass by hydrothermal carbonisation, Analysis of an industrial-scale plant design", 2010, In: *Proceedings of the ECOS – 23rd International Conference, 2010 Jun 14–17*, Lausanne, Switzerland.
25. Antal M.J., Mok W.S.L., Richards G.N., "Mechanism of formation of 5-(hydroxymethyl)2-furaldehyde from D-fructose and sucrose", *Carbohydrate Research*, 199(1), 1990, 91–109.
26. Wallman H., Laboratory Studies of a Hydrothermal Pretreatment Process for Municipal Solid Waste, US Department of Energy, W-7405-Eng-7448, 1995.
27. Titirici M.M., Thomas A., Antonietti M., "Back in the black: Hydrothermal carbonization of plant material as an efficient chemical process to treat the CO_2 problem?", *New Journal of Chemistry*, 31(6), 2007, 787–789.
28. Khemchandani G.V., Ray T.B., Sarkar S., "Studies on artificial coal. 1. Caking power and chloroform extracts", *Fuel*, 53(3), 1974, 163–167.

29. Blazsó M., Jakab E., Vargha A., Székely T., Zoebel H., Klare H., Keil G., "The effect of hydrothermal treatment on a Merseburg lignite", *Fuel*, 65(3), 1986, 337–341.
30. Libra Judy A., Ro Kyoung S., Kammann Claudia, Kammann Claudia, Funke Axel, Berge Nicole D., Neubauer York, Titirici Maria-Magdalena, Fühner Christoph, Bens Oliver, Kern Jürgen, Emmerich Karl-Heinz, "Hydrothermal carbonization of biomass residuals: A comparative review of the chemistry, processes and applications of wet and dry pyrolysis", *Biofuels* 2011 2(1), 89–124.
31. Kläusli Thomas, *Ava, CO_2 Presentation*, Karlsruhe, 16.10.2012.
32. D. Kołodyn'ska, Wnetrzak R., Leahy J.J., Hayes M.H.B., W. Kwapinski, Z. Hubicki, "Kinetic and adsorptive characterization of biochar in metal ions removal", *Chemical Engineering Journal*, 197, 2012, 295–305.
33. Glaser Bruno, Lehmann Johannes, Zech Wolfgang, "Ameliorating physical and chemical properties of highly weathered soils in the tropics with charcoal – A review", *Biology and Fertility of Soils*, 35(4), 2002, 219–230.
34. Hu Bo, Shu-Hong Yu, Wang Kan, Liu Lei, Xu Xue-Wei, "Functional carbonaceous materials from hydrothermal carbonization of biomass: An effective chemical process", *Dalton Transactions*, 2008, 5414–5423.
35. Luo, G., James Strong P., Wang H., Ni W., Shi W., "Kinetics of the pyrolytic and hydrothermal decomposition of water hyacinth", *Bioresource Technology*, 102, 2011, 6990–6994.
36. Liu Zhengang, Balasubramanian R., "Hydrothermal carbonization of waste biomass for energy generation", *Procedia Environmental Sciences*, 16, 2012, 159–166.
37. Danso-Boateng E., Holdich R.G., Shama G., Wheatley A.D., Sohail M., Martin S.J., "Kinetics of faecal biomass hydrothermal carbonization for hydrochar production", *Applied Energy*, 111, 2013, 351–357.
38. Alvarez-Murillo E., Sabio B., Ledesma S. Roman, Gonzalez García C.M., "Generation of biofuel from hydrothermal carbonization of cellulose", *Kinetics Modelling Energy*, 94, 2016, 600–608.
39. Prins Mark J., Ptasinski Krzysztof J., Janssen Frans J.J.G., "Torrefaction of Wood Part 1. Weight loss kinetics", *Journal of Analytical and Applied Pyrolysis*, 77(1), 2006, 28–34.
40. Di Blasi C., Lanzetta M., "Intrinsic kinetics of isothermal xylan degradation in inert atmosphere", *Journal of Analytical and Applied Pyrolysis*, 40–41, 1997, 287–303.
41. Barlaz M.A., "Carbon storage factors during biodegradation of municipal solid waste components in laboratory-scale landfills", *Global Biochemical Cycles*, 12, 1998, 373–380.
42. Kissinger H.E., "Reaction kinetics in differential thermal analysis", *Analytical Chemistry*, 29(11), 1957, 1702–1706.
43. Akahira T., Sunose T. Trans., 1969, Joint convention of four electrical institutes, Paper No. 246, 1969 Research Report, Chiba, Institute of Technology Sci. Technol. 1971, Vol. 16, pp. 22–31.
44. Coats A.W., Redfern J.P., "Kinetic parameters from thermogravimetric data", *Nature*, 201(4914), 1964, 68–69.
45. Friedman H., "Kinetics of thermal degradation of char-forming plastics from thermogravimetry. Application to a phenolic plastic", *Journal Polymer Science, Part C.*, 6, 1964, 183–195.
46. Flynn J.H., Wall L.A., "A quick, direct method for the determination of activation energy from thermogravimetric data", *Polym Lett*, 4, 1966, 323–328.
47. Ozawa T., "A new method of analyzing thermogravimetric data", *Bulletin of the Chemical Society of Japan*, 38(11), 1965, 1881–1886.

7 Production of Biofuels from Algal Biomass

Murali Mohan Seepana, M. Jerold and Rajmohan K. S.

CONTENTS

7.1 Introduction 123
7.2 Biomass Feedstock 124
    7.2.1 Various Feedstocks 124
    7.2.2 Algae as Biomass Feedstock... 124
    7.2.3 Advantage of Macro-Algae as Biomass Feedstock 126
7.3 Cultivation and Nutrients of Algae 127
    7.3.1 Open Pond 127
    7.3.2 Closed-Loop System.. 128
7.4 Methods for Biofuel Production 130
    7.4.1 Chemical Conversion 130
    7.4.2 Thermochemical Conversion... 131
        7.4.2.1 Pyrolysis... 131
        7.4.2.2 Gasification 132
        7.4.2.3 Liquefaction 132
        7.4.2.4 Torrefaction 132
    7.4.3 Biochemical Conversion (BCC) 132
        7.4.3.1 Anaerobic Digestion 133
        7.4.3.2 Fermentation 134
7.5 Scope for Biorefinery... 134
7.6 Summary 135
References.. 135

7.1 INTRODUCTION

The global usage of energy derived from non-renewable resources like fossil fuels has amplified the demand to produce products from low-quality feedstocks such as bitumen, which requires high severity process conditions. Imposing stringent environmental regulations to limit the greenhouse gas emissions and the toxic chemicals emitted from the industries into the atmosphere has restricted refiners to resort to alternate energy resources, preferably those from bio-resources. Any type of fuel produced from biomass is called biofuel. Bioethanol and biodiesel are the two most common biofuels that have the potential to compete with conventional fuels. The US, Brazil, and Canada are the leading bioethanol producing countries (Ranalli, 2007).

Various biomass feedstocks have been researched extensively over the last few decades using various feeds, such as those from agricultural residue, forestry, and municipal residues to produce sustainable products (Naik et al., 2010). These are termed as first- and second-generation feedstocks and may require a vast land area for biofuel production. Algae, however, have been found in abundance in freshwater and marine water possessing high lipid and carbohydrate content, which makes them viable for biofuel production (Demirbas, 2007). They are known to produce superior oil content as opposed to seed crops such as canola, palm oil, and soybean. The high photosynthetic rate of algae makes it an attractive option as a feedstock to produce biofuels such as bioethanol and biodiesel. In this chapter, the importance of algae as feed for biofuel production, with an emphasis on process conditions, important parameters, production techniques for biohydrogen, biomethane, bioethanol, and biodiesel production, is detailed. Finally, the challenges faced in the production of biofuels using algal feedstock and future perspective is discussed.

7.2 BIOMASS FEEDSTOCK

Organic matter that has not become fossilized and has the potential to get converted into fuel is referred to as biomass. Biomass is a promising alternative feedstock which can cater to energy requirements while checking CO_2 emission levels (Zeng et al., 2011).

7.2.1 VARIOUS FEEDSTOCKS

All potential feedstocks which can be used for biomass conversion are categorized as first-, second-, and third-generation feedstock based on their nature as shown in Figure 7.1. (Sims et al., 2010; Singh and Gu, 2010a). Edible seeds (triglycerides) and starch-containing biomasses constitute first-generation feedstock. Grains, cereals, beans, nuts, and starchy vegetables are first-generation feedstock (FGF). Nevertheless, using edible food crops for biofuels production is not encouraged in view of the shortage of food supply to feed the populace. The pursuit of alternative non-edible feedstocks has been encouraged. Non-edible lignocellulosic biomass is a promising alternative for FGF. Lignocellulosic biomass such as forestry residues, virgin wood, energy crops, and agricultural residues are second-generation feedstock (SGF) (Lin and Shuzo, 2006). However, the challenge in converting lignocellulosic biomass into low molecular weight sugar monomers needs to be addressed by developing more effective chemical and physical treatment methods (Zakzeski et al., 2010).

When algae are introduced into SGF, it has the capability to survive in diverse conditions and convert CO_2 into biofuel and useful chemicals (Mondal et al., 2017). This constitutes third-generation feedstock and serves a dual role of converting biomass into biofuel and mitigation of CO_2 emissions (Zeng et al., 2011; Wang et al., 2008).

7.2.2 ALGAE AS BIOMASS FEEDSTOCK

Algae or thallophytes, which were earlier considered aquatic plants, are actually microorganisms which can be unicellular or multicellular. They are photosynthetic microorganisms which grow hydroponically, capable of assimilating carbon

Production of Biofuels from Algal Biomass

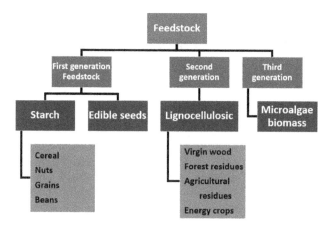

FIGURE 7.1 Classification of biomass feedstock.

heterotrophically as well as mixotrophically (Schmidt et al., 2013). Microalgae photosynthetically convert CO_2 into generating lipids, carbohydrates, and proteins.

Research focus has shifted towards production of alternative fuels using one of the oldest living creatures, microalgae (John, 2011). Algae present in freshwater and marine environments have plenty of beneficial roles, including the consumption of carbon dioxide and production of oxygen. Algae are known for their role as the base for the aquatic food chain, removal of nutrients and pollutants from water, and stabilization of sediments.

More than 50,000 microalgae species have been reported in the literature (Hu et al., 2008). Microalgae can be broadly categorized as red algae, brown algae, yellow-green algae, golden algae, green algae, diatoms, and dinoflagellates, as shown in Figure 7.2.

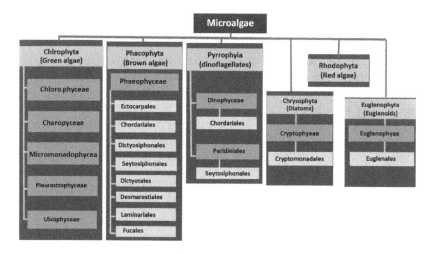

FIGURE 7.2 Family tree of microalgae.

TABLE 7.1
Advantages and Disadvantages of Pretreatment Methods of Algae Biomass

Pretreatment Method	Advantages	Disadvantages
Physical pretreatment	No inhibitory and toxic compounds. Easy to operate and commercializable.	High energy cost. High capital cost.
Thermal pretreatment	No inhibitory and toxic compounds. Commercializable.	Recalcitrant compounds at high temperatures. Energy cost for high temperatures.
Chemical pretreatment	Low energy demand.	Toxicity and inhibition. High chemical cost.
Biological pretreatment	No inhibitory compounds. Low energy demand. High selectivity.	Contamination (onsite enzyme production). High energy cost.

Microalgae biomass can be processed using thermo-chemical, biochemical, or direct combustion technologies (Table 7.1). The biochemical conversion technology route produces hydrogen and methane as primary products (Saxena et al., 2009; Ni et al., 2006). A thermochemical process such as gasification yields syngas, whereas liquefaction and pyrolysis yields bio-oil.

Microalgae biomass is a prospective basis to produce bioplastics, both using the biomass directly as well as using it as a feedstock for secondary processes (Rahman and Miller, 2017). Bioplastics from solid wastes have numerous advantages over petroleum-based plastics, such as usage of renewable substrates, less biodegradable, and fewer GHG emissions (Rajmohan et al., 2019). For instance, poly-3-hydroxybutyrate production is feasible in a microalgal system using the diatom *P. tricornutum* (Hempel et al., 2011).

7.2.3 Advantage of Macro-Algae as Biomass Feedstock

The following advantages make algae biomass feedstock the frontrunner for alternative energy. Microalgae have an immense capability for photosynthesis in a diverse range of light and thermal conditions (Nascimento et al., 2013). Moreover, microalgae can survive and grow under severe conditions by adjusting their nutrient uptake ability according to the change in the nutrient level in the water. Hence, microalgae are found plentifully in freshwater and marine environments.

Algae have a high ratio of surface to volume, which facilitates their efficient growth. However, their large-scale cultivation, harvesting, and conversion technologies still remain a challenge with respect to economy and efficiency. Various applications of microbial algae are illustrated in Figure 7.3. Algae biomass that was grown photoautotrophically has approximately 50% carbon by weight derived from dissolved carbonate or CO_2. According to the stoichiometry, around 1.83 tons of CO_2 is consumed to produce a ton of algal biomass (Chisti, 2007).

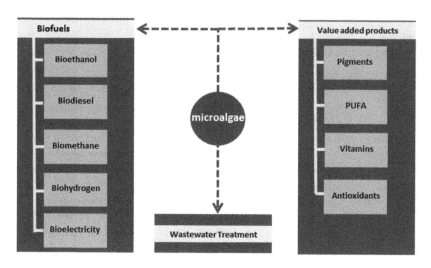

FIGURE 7.3 Applications of microbial algae.

7.3 CULTIVATION AND NUTRIENTS OF ALGAE

Microalgae can be classified as heterotrophic, autotrophic, and mixotrophic depending on their way of growth (Liang et al., 2009). Heterotrophic algae are grown in the absence of sunlight (dark) with a continuous supply of organic carbon source, whereas, in autotrophic cultivation, daylight from sun acts as the energy source while CO_2 is utilized as a carbon source in addition to essential minerals and vitamins. A third type of algae, which grows in the presence of light and requires an organic carbon source, is known as mixotrophic algae, and the growth can be heterotrophic or phototrophic according to the culture conditions. Photobioreactors (PBRs) can be employed for three types of cultivation conditions, and the cost involved and challenges are summarized in Table 7.2 (Chen et al., 2011). Among the three, mixotrophic cultivation is known to result in enhanced biomass yield as it offers the combined benefits of heterotrophic and autotrophic cultivation methods. Currently, two major microalgal cultivation techniques are used, i.e. open pond and closed-loop systems, which are described as follows.

7.3.1 OPEN POND

The most common algae cultivation method is an open pond system. Shallow ponds referred to as raceway ponds or high-rate algal ponds have a depth of 1 foot and size varying from 1 acre to 10 acres. Paddle wheels are used to keep the algae circulation in the pond. A top view of a raceway pond is shown in Figure 7.4a. A minimum length/breadth ratio of 10 is maintained to ensure the flow is not affected by the curves at the semicircular channel ends. Dead zones severely disturb mixing, permit solids to settle, and lead to unnecessary energy losses. To avoid the formation of dead zones and ensure a uniform flow in the semicircular ends of the pond, deflector baffles are usually provided.

TABLE 7.2
Characteristics of Cultivation Conditions

Cultivation Condition	Energy Source	Carbon Source	Reactor Setup	Cost	Challenges in Scaling-Up
Phototrophic	Light	Inorganic	Open pond or PBR	Low	Low cell density, high condensation cost
Heterotrophic	Organic	Organic	Conventional fermenter	Medium	High substrate cost, contamination
Mixotrophic	Light and organic	Inorganic and organic	PBR	High	High equipment cost, high substrate cost
Photo-heterotrophic	Light	Organic	PBR	High	Contamination, high equipment cost, high substrate cost

Source: Chen et al., 2011.

Raceway ponds were first employed for wastewater treatment in the 1950s and commercial production of cyanobacteria and microalgae in the 1960s. The two-stage harvesting method is followed based on the characteristics of algae and process requirements. Every day, a fraction of the pond is generally harvested and processed further to extract fuel.

Residues or dried biomass are used as animal feed. Sapphire Energy aims to produce a million gallons of fuel using an open pond facility in New Mexico. Because of its simple construction, the production and operating costs are low. However, the environment in and around the pond is open, and bad weather can upset algae growth. Contamination with outside organisms may lead to the growth of undesirable species, suppressing the desired algae growth in the pond. The challenge of uneven temperature and light distribution has to be addressed. Moreover, it is not economical with respect to the unit of biomass produced.

7.3.2 Closed-Loop System

To have control over the environment, closed-loop systems are employed wherein invasion by weed algae, zooplankton grazers, and other microorganisms that can affect the algae growth are addressed. Thereby, ideal temperature and light conditions are maintained to cultivate algae that could not survive in open ponds. Photobioreactors are closed-loop culturing systems wherein the light passes through the transparent reactor walls (UV-resistant material) to reach all the cultivated cells (Tredici, 2004). PBRs can be oriented vertically or horizontally depending upon space availability. PBRs can be operated either with sunlight, artificial light, or both.

The current common closed PBRs generally include a horizontal tube, vertical tube (bubble column and airlift), flat panel, stirred tank, and their modified configurations as shown in Figure 7.4. The advantages of PBRs include enhanced areal productivity (g/m^2/day), since operation costs depend on the plant size and enhanced volumetric productivity (g/L/day), since operation cost depends on the water consumption as summarized in Figure 7.5 (Ting et al., 2017).

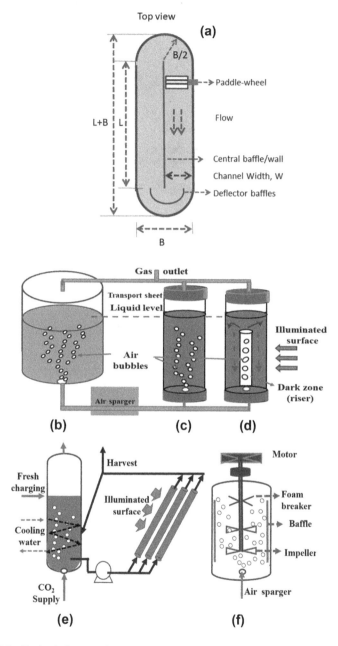

FIGURE 7.4 Typical algae cultivation systems: (a) Open pond system (b) Aerobic digestor (c) Bubble column PBR (d) Airlift PBR (e) Horizontal tube PBR (f) Stirred tank for microalgae cultivation.

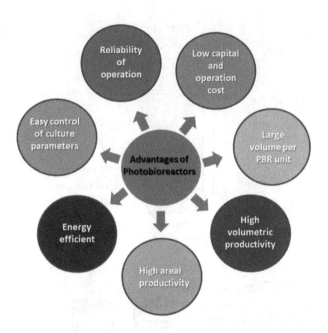

FIGURE 7.5 Advantages of photobioreactors for algae biomass cultivation.

7.4 METHODS FOR BIOFUEL PRODUCTION

Microalgae biomass can be converted into biofuel by using different routes such as thermochemical, chemical, and biochemical, as illustrated in Figure 7.6. In addition, it is also used to generate electricity by direct combustion. Under the above-listed routes, the most commonly used approaches for energy production from microalgae biomass are hydrothermal liquefaction, transesterification, methane production by anaerobic digestion, production of bioethanol, biohydrogen, biomethane, and biobutanol by fermentation of carbohydrates (Singh and Gu, 2010b).

7.4.1 Chemical Conversion

Chemical conversion of biomass into biodiesel is carried out by the transesterification process. Generally, in this technique, the chemical reaction between lipid and alcohol using a catalyst (acid, alkali, or enzyme) takes place to produce fatty acid methyl ester (FAME), referred to as biodiesel, as well as glycerol as a byproduct (Ayyasamy et al., 2018). Base catalysts such as sodium hydroxide, potassium hydroxide, and sodium methoxide offer low costs and higher reaction rate, whereas heterogeneous solid base catalysts such as calcium oxides, ETS-10 (Na and K), anionic exchange resins, and quaternary ammonium silica gels offer easier product recovery and have the potential for bulk capacity processing and regeneration (Lopez et al., 2005). Esterification is a reversible reaction process wherein free fatty acids are converted into alkyl esters via an acid catalyst. Thus, simultaneous esterification and

Production of Biofuels from Algal Biomass

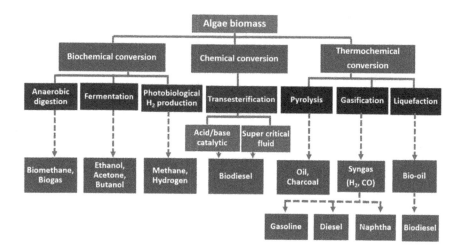

FIGURE 7.6 Biofuel production routes from algae biomass.

transesterification by acid catalysts enhance conversion of biodiesel while suppressing the saponification (Dinesh et al., 2016).

Two approaches are followed to produce biodiesel from the microalgae biomass, viz., the two-stage method lipid extraction process followed by the transesterification of the extracted lipids, and direct transesterification without oil extraction. The former method is widely practiced. Lipid extraction from the microalgae can be done by using a suitable solvent such as alcohol, hexane, or via physical extraction methods such as expeller press, ultrasonic extraction, Soxhlet extractors, and electromechanical methods (Suali and Rosalam, 2012). The advantages of esterified biodiesel include less viscosity, lower density, higher oxygen content, and it is rich in monoglyceride compounds. Identification of an optimum esterification method could enable widespread acceptance of biodiesel in the automobile sector (Dinesh et al., 2016).

7.4.2 Thermochemical Conversion

Gasification, pyrolysis, hydrothermal liquefaction, and torrefaction are commonly used thermochemical pathways for conversion of microalgae biomass. In these processes, thermal decomposition of biomass at an accelerated temperature takes place.

7.4.2.1 Pyrolysis

The word 'pyrolysis' is a combination of two Greek words, 'pyro' and 'lysis'; pyro means 'fire' and lysis means 'separation'. It is a process in which the biomass matter is subjected to high temperatures (around 500°C) in the absence (or low amount) of oxygen under a pressurized environment. It results in partial combustion of biomass, producing liquid fuels and solid residue known as biochar. Liquid fuels known as bio-fuel or bio-crude are produced predominantly with an efficiency of around 80% when subjected to flash pyrolysis at low temperatures. Whereas biochar is basically a charcoal rich in carbon content and it can be used to enhance soil properties among other things. These products are denser in energy than the initial biomass feedstock, which significantly reduces the cost of transportation (Saradhadevi et al., 2019).

7.4.2.2 Gasification

As the name implies, gasification converts biomass feedstock into the flammable gas mixture when subjected to high temperatures (around 800–900°C) in a controlled environment. This takes place in two stages:

(i) The producer gas and biochar are produced as a result of partial combustion of the feedstock.
(ii) Chemical reduction.

Both stages are performed in a gasifier with appropriate separation. The gas (with low CV) produced from a gasifier can be directly used in gas engines and turbines as fuel. The product gas can also be used as feedstock to produce value-added chemicals such as methanol, which can be used as future fuel for transportation. Fixed-bed gasification, fluidized-bed gasification, and their subdivisions are elaborately reported in the literature (Saradhadevi et al., 2019). Moreover, the integrated gasification combined cycle is considered to be a green power generation option.

7.4.2.3 Liquefaction

Liquefaction is the process of producing bio-oil or biocrude from a wet biomass feedstock. It is operated at high-pressure and low-temperature conditions. The bio-oil produced by this process has high energy density and a low heating value (around 35 MJ/Kg), with 5–20 wt.% O_2 (Demirbas, 2011). The advantage of liquefaction is its ability to process any biomass source to produce bio-oil without considering the moisture content of the feedstock. This process is less popular due to its exorbitant costs and complexity associated in processing, as well as the huge equipment cost as compared to pyrolysis (Saradhadevi et al., 2019).

7.4.2.4 Torrefaction

Torrefaction is carried out to increase the energy of algae biomass by decreasing hydrogen and oxygen content while increasing the carbon content. Torrefaction produces high-energy biomass, whereas carbonization produces a low-energy end product. The torrefaction process can be divided into two subcategories based on their pretreatment of feedstock. The detailed description of process conditions for wet and dry torrefaction processes are summarized in Table 7.3.

In addition, value-added chemicals are produced during torrefaction. Moreover, it eliminates the agglomeration-causing elements from the biomass and reduces the corrosion during further processing.

7.4.3 BIOCHEMICAL CONVERSION (BCC)

Biochemical conversion uses enzymes, bacteria, and microorganisms to explore and transform the biomass feedstock into liquid and gaseous fuels. The most

TABLE 7.3
Characteristics of Torrefaction Methods

Thermal Pretreatment Method	Operating Parameters
Wet torrefaction	Temperature: 200–265°C
	Pressure: 200–750 psi
	Residence time: 5 min
	Sample size: 2 g biomass
	Media: hot compressed water
	Cooling procedure: Quick immersion in ice bath
Dry torrefaction	Temperature ranges: 250–310°C
	Pressure: Atmospheric pressure
	Residence time: 80 min
	Biomass sample: 5 g biomass
	Media: Inert gas
	Cooling procedure: flowing nitrogen/indirect water cooling

Source: Yan et al., 2009.

effective and widely used BCC processes are fermentation and anaerobic digestion (bio-methanation). Fermentation is a group of chemical reactions used to transform glucose in plants into acid or alcohol. Whereas anaerobic digestion is defined as a group of biochemical processes in which naturally occurring existing microorganisms are used to convert the biodegradable materials into fuel in an oxygen-free environment. A brief discussion of these topics is elaborated in the following section (Saradhadevi et al., 2019).

7.4.3.1 Anaerobic Digestion

Anaerobic digestion (AD) is a process in which naturally existing microorganisms stabilize the feedstock in an oxygen-free environment and transform them into biogas and bio-fertilizers. It is widely practiced and a reliable process for the treatment of biomass having high moisture content (wet biomass). The degradation of the biodegradable/organic matter in a highly supervised, oxygen-free environment that results in biogas production can be effectively utilized to generate both electricity and heat. This process is mainly used to produce carbon-rich gas and methane from suitable biomass feedstock (Huber, 2006) and can also be used for the treatment of wastewater (Rajmohan et al., 2016, 2018).

Anaerobic digestion (AD) can be categorized as a three-stage process:

(i) In the first stage, biodegradable matter like plants and animal residue containing carbohydrates are decomposed into smaller sizes and digestible forms by a set of bacteria using the appropriate bacteria for this purpose.
(ii) In the second stage, the decomposed matter like amino acids and sugars are converted into CO_2, NH_3, H_2, and organic acids by the second set of bacteria.
(iii) In the last stage, products like organic acids and other compounds are finally converted into methane and carbon dioxide.

The temperature during the process plays a vital role as it directly affects the rate of digestion of bacteria. The material left after the digestion process can be used as fertilizers and low-grade building material such as fiberboard.

7.4.3.2 Fermentation

Fermentation is a process in which the glucose contained in the plants is converted into bio-alcohol (ethanol) and carbon dioxide in the absence of oxygen by the application of yeast. The primary feedstock used for fermentation comprises sugarcane, sweet potato, and corn, as all of them are rich in sugar. Wheat, rice, agriculture waste, woody waste, etc. can also be used, but lignocellulosic biomass like wood and grass have complex conversions due to the presence of long chains (Kumar et al., 2009; Naik et al., 2010). The fermentation process consists of several stages. Firstly, the feedstock is pulverized, and water is added to form a slurry of the feedstock. The slurry is then further converted into a finer form by the breakdown action of heat and enzymes. Other enzymes are also added to transform the starch into sugar. This sugary-rich slurry is transferred into a fermentation chamber followed by the addition of yeasts to the slurry. After 48 hours, the slurry generates bio-alcohol, which can be recovered from the chamber and used as fuel.

7.5 SCOPE FOR BIOREFINERY

'Biorefinery' was coined to represent the simultaneous bioproducts production ranging from electricity, fuels, value-added chemicals, and materials in an economic and eco-friendly approach. Microalgae biorefinery includes cultivation, harvest, conversion, and separation of intermittent and final products (Sankaran et al., 2018; Menon and Rao, 2012). The objective of this approach is mainly to maximize the yield of desired products and minimize the energy requirements for process and waste generation. The high cost of the products formed by microalgae biorefinery is a major challenge to be addressed immediately. This challenge can be addressed by adopting the following strategies:

(i) Cost-effective cultivation of microalgae biomass.
(ii) Adoption of competent harvesting and drying techniques.
(iii) Increasing the yield of biochemicals for nutraceuticals, foods, cosmetics, and aquaculture feed.
(iv) Integration with CO_2 alleviation and recovery of nutrients and heavy metals from wastewater.
(v) Improving the condition for microalgae biomass cultivation.

The above approaches can enhance the simultaneous carbohydrate, lipid, and protein recovery and make the valorization of microalgae biomass cost-effective. The early-harvest algae, mid-harvest algae, and late-harvest algae were rich in protein, carbohydrates, and lipids, respectively. Thus, sufficient time for growth is essential to maximize lipid content and henceforth high gasoline gallon equivalent (GGE) energy. Hybrid wastewater treatment and microalgae production in aquaculture systems is a promising technique which addresses two purposes, i.e. environmental remediation and biomass growth. Moreover, it offers several advantages.

Moreover, usage of microalgae for effluent treatment has various advantages:

(i) Recovery and recycling of nutrients present in the wastewater in the form of fertilizer will reduce the treatment cost considerably.
(ii) Reduction of biological oxygen demand (BOD) in water bodies.
(iii) Enrichment of CO_2 in wastewater balances the ratio of nutrients and leads to the reduction of the cost of harvesting and accelerating the lipid production.

The moderate rise in the production of biodiesel from microalgae is promising. However, this technology takes a backseat due to higher operating costs and uneven availability of biomass. Development of novel cultivation and conversion technologies is essential to contest and replace fossil fuels. The intermittent weather conditions affect the natural light source for photobioreactors, which are crucial for microalgae growth.

Hence, microalgae biorefinery has the potential to cater to future energy demands and value-added biochemicals. However, suitable unit operations and processes need to be developed to make it economically feasible. The concept of the biorefinery is yet to be commercialized and holds potential recovery of sustainable green energy from microalgae biomass.

Upon addressing these challenges, alternative fuel to conventional fossil fuels will become a reality. In turn, it will reduce the dependency on crude oil import. A more isolated analysis can demonstrate that biorefineries are a cost-effective process to invest in. Furthermore, the kinetic studies on the microalgae biomass conversion equally remain an indispensable factor for process design purposes.

7.6 SUMMARY

Microalgae biomass conversion technology offers a potentially authoritative method to the management of various biomasses and wastewater. Microbial technology offers diverse approaches for the cultivation, harvesting, utilization, and production of biofuels. This chapter, while not extensive, has discussed the various cultivation techniques in which microalgae can play a significant role. The area of biorefineries is still being examined because bulk availability of specific biomass is a challenge. While the most technological advancements are being achieved in the thermochemical conversion methods, biochemical conversion routes are looked upon as completely green alternatives. The areas of microalgae biomass conversion continue to pose substantial challenges and offer opportunities for new research and discovery in the future.

REFERENCES

Ayyasamy, Tamilvanan, Kulendran Balamurugan, and Saravanan Duraisamy. "Production, performance and emission analysis of Tamanu oil-diesel blends along with biogas in a diesel engine in dual cycle mode." *International Journal of Energy Technology and Policy* 14(1) (2018): 4–19.

Chen, Chun-Yen, Kuei-Ling Yeh, Rifka Aisyah Duu-Jong Lee, and Jo-Shu Chang. "Cultivation, photobioreactor design and harvesting of microalgae for biodiesel production: A critical review." *Bioresource Technology* 102(1) (2011): 71–81.

Chisti, Yusuf. "Biodiesel from microalgae." *Biotechnology Advances* 25(3) (2007): 294–306.
Demirbas, Ayhan. "Progress and recent trends in biofuels." *Progress in Energy and Combustion Science* 33(1) (2007): 1–18.
Demirbas, Ayhan. "Competitive liquid biofuels from biomass." *Applied Energy* 88(1) (2011): 17–28.
Dinesh, K., A. Tamilvanan, S. Vaishnavi, M. Gopinath, and K. S. Rajmohan. "Biodiesel production using Calophyllum inophyllum (Tamanu) seed oil and its compatibility test in a CI engine." *Biofuels* (2016): 1–7.
Hempel, Franziska, Andrew S. Bozarth, Nicole Lindenkamp, Andreas Klingl, Stefan Zauner, Uwe Linne, Alexander Steinbüchel, and Uwe G. Maier. "Microalgae as bioreactors for bioplastic production." *Microbial Cell Factories* 10(1) (2011): 81.
Hu, Qiang, Milton Sommerfeld, Eric Jarvis, Maria Ghirardi, Matthew Posewitz, Michael Seibert, and Al Darzins. "Microalgaltriacylglycerols as feedstocks for biofuel production: Perspectives and advances." *The Plant Journal* 54(4) (2008): 621–639.
Huber, George W., Sara Iborra, and Avelino Corma. "Synthesis of transportation fuels from biomass: Chemistry, catalysts, and engineering." *Chemical Reviews* 106(9) (2006): 4044–4098.
John, Rojan P., G. S. Anisha, K. Madhavan Nampoothiri, and Ashok Pandey. "Micro and macroalgal biomass: A renewable source for bioethanol." *Bioresource Technology* 102(1) (2011): 186–193.
Kumar, Parveen, Diane M. Barrett, Michael J. Delwiche, and Pieter Stroeve. "Methods for pretreatment of lignocellulosic biomass for efficient hydrolysis and biofuel production." *Industrial and Engineering Chemistry Research* 48(8) (2009): 3713–3729.
Liang, Yanna, Nicolas Sarkany, and Yi Cui. "Biomass and lipid productivities of Chlorella vulgaris under autotrophic, heterotrophic and mixotrophic growth conditions." *Biotechnology Letters* 31(7) (2009): 1043–1049.
Lin, Yan, and Shuzo Tanaka. "Ethanol fermentation from biomass resources: Current state and prospects." *Applied Microbiology and Biotechnology* 69(6) (2006): 627–642.
Lopez, Dora E., James G. Goodwin Jr., David A. Bruce, and Edgar Lotero. "Transesterification of triacetin with methanol on solid acid and base catalysts." *Applied Catalysis. Part A: General* 295(2) (2005): 97–105.
Menon, V., and M. Rao. "Trends in bioconversion of lignocellulose: Biofuels, platform chemicals & biorefinery concept." *Progress in Energy and Combustion Science* 38(4) (2012): 522–550.
Mondal, Madhumanti, Shrayanti Goswami, Ashmita Ghosh, Gunapati Oinam, O. N. Tiwari, Papita Das, K. Gayen, M. K. Mandal, and G. N. Halder. "Production of biodiesel from microalgae through biological carbon capture: A review." *3 Biotech* 7(2) (2017): 99.
Naik, S. N., Vaibhav V. Goud, Prasant K. Rout, and Ajay K. Dalai. "Production of first and second generation biofuels: A comprehensive review." *Renewable and Sustainable Energy Reviews* 14(2) (2010): 578–597.
Nascimento, Iracema Andrade, Sheyla Santa Izabel Marques, Iago Teles Dominguez Cabanelas, Solange Andrade Pereira, Janice Isabel Druzian, Carolina Oliveira de Souza, Daniele Vital Vich, Gilson Correia de Carvalho, and Maurício Andrade Nascimento. "Screening microalgae strains for biodiesel production: Lipid productivity and estimation of fuel quality based on fatty acids profiles as selective criteria." *BioEnergy Research* 6(1) (2013): 1–13.
Ni, Meng, Dennis Y. C. Leung, Michael K. H. Leung, and K. Sumathy. "An overview of hydrogen production from biomass." *Fuel Processing Technology* 87(5) (2006): 461–472.
Rahman, A., and C. D. Miller. "Microalgae as a source of bioplastics." In: Rajesh Prasad Rastogi, Datta Madamwar, and Ashok Pandey (eds.) *Algal Green Chemistry*. Elsevier, Amsterdam, 2017, 121–138.
Rajmohan, K. S., M. Gopinath, and Raghuram Chetty. "Review on challenges and opportunities in the removal of nitrate from wastewater using electrochemical method." *Journal of Environmental Biology* 37(6) (2016): 1519.

Rajmohan, K. S., Margavelu Gopinath, and Raghuram Chetty. "Bioremediation of nitrate-contaminated wastewater and soil." In: *Bioremediation: Applications for Environmental Protection and Management*. Springer: Singapore, 2018, 387–409.

Rajmohan, K. S., Harshit Yadav, S. Vaishnavi, M. Gopinath, and Sunitha Varjani. "Perspectives on bio-oil recovery from plastic waste." In: Sunita Varjani, Ashok Pandey, Edgard Gnansounou, Sindhu Raveendran, Samir Kumar Khanal (eds.) *Current Developments in Biotechnology and Bioengineering: Resource Recovery from Wastes*. Elsevier, Amsterdam, 2019. In press.

Ranalli, P. *Improvement of Crop Plants for Industrial End Uses*. Edited by Paolo Ranalli. Springer: The Netherlands, 2007.

Sankaran, Revathy, Pau L. Show, Dillirani Nagarajan, and Jo-Shu Chang. "Exploitation and biorefinery of microalgae." In: *Waste Biorefinery*. Elsevier, Amsterdam, 2018, 571–601.

Saradhadevi, G., S. Vaishnavi, S. Srinath, Brahm Dutt, and K. S. Rajmohan. "Energy recovery from biomass using gasification. In: Sunita Varjani, Ashok Pandey, Edgard Gnansounou, Sindhu Raveendran, Samir Kumar Khanal (eds.) *Current Developments in Biotechnology and Bioengineering: Resource Recovery from Wastes*. Elsevier, Amsterdam, 2019.

Saxena, R. C., D. K. Adhikari, and H. B. Goyal. "Biomass-based energy fuel through biochemical routes: A review." *Renewable and Sustainable Energy Reviews* 13(1) (2009): 167–178.

Schmidt, Susanne, John A. Raven, C. Paungfoo-Lonhienne, and Chanyarat Paungfoo-Lonhienne. "The mixotrophic nature of photosynthetic plants." *Functional Plant Biology* 40(5) (2013): 425–438.

Sims, Ralph E. H., Warren Mabee, Jack N. Saddler, and Michael Taylor. "An overview of second generation biofuel technologies." *Bioresource Technology* 101(6) (2010): 1570–1580.

Singh, Jasvinder, and Sai Gu. "Biomass conversion to energy in India—A critique." *Renewable and Sustainable Energy Reviews* 14(5) (2010a): 1367–1378.

Singh, Jasvinder, and Sai Gu. "Commercialization potential of microalgae for biofuels production." *Renewable and Sustainable Energy Reviews* 14(9) (2010b): 2596–2610.

Suali, Emma, and Rosalam Sarbatly. "Conversion of microalgae to biofuel." *Renewable and Sustainable Energy Reviews* 16(6) (2012): 4316–4342.

Ting, Han, Lu Haifeng, Ma Shanshan, Yuanhui Zhang, Liu Zhidan, and Duan Na. "Progress in microalgae cultivation photobioreactors and applications in wastewater treatment: A review." *International Journal of Agricultural and Biological Engineering* 10(1) (2017): 1–29.

Tredici, Mario R. "Mass production of microalgae: Photobioreactors." In: Amos Richmond (ed.) *Handbook of Microalgal Culture: Biotechnology and Applied Phycology*, Vol. 1. 2004, 178–214.

Wang, Bei, Yanqun Li, Nan Wu, and Christopher Q. Lan. "CO_2 bio-mitigation using microalgae." *Applied Microbiology and Biotechnology* 79(5) (2008): 707–718.

Yan, Wei, Tapas C. Acharjee, Charles J. Coronella, and Victor R. Vasquez. "Thermal pretreatment of lignocellulosic biomass." Environmental Progress & Sustainable Energy: *An Official Publication of the American Institute of Chemical Engineers* 28(3) (2009): 435–440.

Zakzeski, Joseph, Pieter C. A. Bruijnincx, Anna L. Jongerius, and Bert M. Weckhuysen. "The catalytic valorization of lignin for the production of renewable chemicals." *Chemical Reviews* 110(6) (2010): 3552–3599.

Zeng, Xianhai, Michael K. Danquah, Xiao Dong Chen, and Yinghua Lu. "Microalgae bioengineering: From CO_2 fixation to biofuel production." *Renewable and Sustainable Energy Reviews* 15(6) (2011): 3252–3260.

8 Diffusion Limitations in Biocatalytic Reactions
Challenges and Solutions

Carlin Geor Malar, Muthulingam Seenuvasan and Kannaiyan Sathish Kumar

CONTENTS

8.1 Introduction 139
    8.1.1 Types of Catalysis 140
        8.1.1.1 Homogeneous Catalysis 140
        8.1.1.2 Heterogeneous Catalysis 140
        8.1.1.3 Electrocatalysis 141
        8.1.1.4 Nanocatalysis 141
        8.1.1.5 Photocatalysis 141
        8.1.1.6 Autocatalysis 141
        8.1.1.7 Enzymatic Catalysis (Biocatalysis) 141
        8.1.1.8 Acid-Base Catalysis 142
8.2 Factors Influencing Biocatalytic Action 142
    8.2.1 Substrate Concentration 142
    8.2.2 Enzyme Concentration 143
    8.2.3 Surface Area 143
    8.2.4 Diffusion 143
        8.2.4.1 External Diffusion 144
        8.2.4.2 Internal Diffusion 144
8.3 Approaches to Overcome Diffusional Limitations 145
    8.3.1 Hydrogels 145
    8.3.2 Sensitive Matrices 145
    8.3.3 Non-Porous Supports 146
8.4 Summary 147
References 147

8.1 INTRODUCTION

Catalysis is an accelerating process that takes place with the aid of a substance called a *catalyst*, which increases the reaction rate by lowering the activation energy in an alternative reaction mechanism. Since only a small amount of catalyst is required to exhibit an efficient reaction, it is preferred in most chemical reactions and ensures minimum energy usage [Aguila et al. 2018]. The formation of a temporary intermediate and/or stabilization of the transition states helps in regenerating the catalyst at the end of the reaction.

8.1.1 Types of Catalysis

The catalytic reactions can be classified on various basis, such as based on the physical state of the catalyst, i.e., gas, liquid, solid; based on the catalyst material, i.e., inorganic and organic; based on the mode of catalyst, i.e., homogeneous and heterogeneous; and finally based on the catalytic action, i.e., acid-base catalysis, enzymatic catalysis, photocatalysis or electrocatalysis [Pirousmand and Mahdevi 2018]. Figure 8.1 illustrates the detailed classification of catalytic reactions. Each type of reaction is discussed in detail.

8.1.1.1 Homogeneous Catalysis

In homogeneous catalysis, the catalyst is in the same phase as that of the reactants. High homogeneity can be seen, which results in increased reactivity and selectivity caused by more interactions between them, causing the reaction to occur. Carbonylation, hydrocyanation, hydrogenation, oxidation and metathesis are a few of the most commonly occurring chemical processes in homogeneous catalysis [Hermann et al. 2018]. Most significant homogeneous catalysts include Lewis acids, organometallic complexes, transition metals and organocatalysts.

8.1.1.2 Heterogeneous Catalysis

In heterogeneous catalysis, the phase of the catalyst is different from that of the reactants. This reaction makes the separation process easier and is preferred when the recovery of catalyst is important [Yue 2018]. Well-known heterogeneous catalysis processes include synthesis of ammonia (Haber–Bosch process [Ertl 1990])

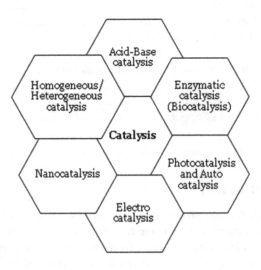

FIGURE 8.1 Types of catalysis.

and synthesis of hydrocarbons (Fischer–Tropsch process). As different phases are involved in this catalytic process, it corresponds to the surface phenomenon, which impose the significance of higher surface area. Recently, the nanoparticles serve as an excellent catalyst due to increased surface area by means of their decreased size. Also, tethering of certain chemical groups prevents agglomeration [Seenuvasan et al. 2013] as well as enhances thermal and mechanical stability in order to increase the catalytic performance.

8.1.1.3 Electrocatalysis

Here the catalysis takes place on the electrode surface where the rate of the electrochemical reaction is modified. The electrocatalytic property of the material vary depending on the electric potential [Wang et al. 2018].

8.1.1.4 Nanocatalysis

This is one of the rapidly growing fields that makes use of the nanomaterials to perform the catalytic activity [Zhou et al. 2018]. Nanoparticles of metals, semi-conductors and polymers are mostly used in the important chemical reactions. Extremely higher activity with longer lifetime and lower energy consumption can be achieved through the nano catalyst. The unique properties and the ability to precisely control shape, size, surface behavior and stability of nanomaterials makes them potential candidates as catalyst.

8.1.1.5 Photocatalysis

The rate of photoreaction that involves the absorption of light is modified by the addition of catalyst and is termed photocatalysis. Depending on the phase of the catalyst with relation to the reactants, it is of two types, namely: homogeneous and heterogeneous photocatalysis.

8.1.1.6 Autocatalysis

Autocatalysis is a catalytic reaction that is easily catalyzed by the products of the same reaction. The rate of catalysis depends on the rate of product formation and thereby increases with the progress of the reaction.

8.1.1.7 Enzymatic Catalysis (Biocatalysis)

Enzymes, usually termed *biocatalysts*, can be obtained from various sources, including plant, animal and microbes. They can be either intracellular or extracellular products of biological substances. Enzymes possess extremely high catalytic property by nature, as they are secondary metabolites, due to which they have wide industrial applications. The major use can be seen in the manufacture of beer, wine, cheese and vinegar [Longo and Sanroman 2006]. Pharmaceutical, food, textile and beverage industries extensively make use of enzymes [Seenuvasan et al. 2018]. More researchers making industrial manufacturing processes are attracted to enzymes because of their versatility [Singh et al. 2016].

Some of the excellent properties of enzymes include their specificity, sensitivity and higher activity. Unlike conventional catalysts, enzyme biocatalysts enhance the rate of the reaction without affecting the thermodynamics of the reaction.

Moreover, high costs, poor recovery, poor stability over wide pH and temperature stops the enzymes from reaching a significant enough level to be used as biocatalysts [Seenuvasan et al. 2014]. Enzyme catalysis is performed under homogeneous conditions which further impose the need of effective enzyme recovery in its active form. Immobilization is the most vibrant technique that can be employed to overcome all such difficulties [Mateo et al. 2007].

Chemical transformations performed by any biological substances, such as enzymes, towards organic components in the reaction are termed biocatalysis. It can neither be considered as a homogeneous nor as a heterogeneous catalysis due to the involvement of microorganisms and living systems. Biocatalysts are of much interest in the pharmaceutical industry due to their high selectivity [Truppo 2017], enabling synthesis of pure compounds. Complete degradation and reduced undesired side reactions are the notable advantages of biocatalysis. In this chapter, biocatalytic reactions will be discussed in detail.

Enzymes are very important molecules that make life possible and accelerate the reactions at the highest rate (10^{17} fold) [Radzicka and Wolfenden 1995]. Their extraordinary behavior of automatic switching on and off based on the presence and absence of signals distinguishes enzymes from other catalysts. Also, metal attachment on the surface enhances the reaction and makes it feasible. Biologically, enzymes are known for their control and regulating activity that emphasize better catalytic property.

8.1.1.8 Acid-Base Catalysis

These are pH-dependent reactions taking place in aqueous medium [Kaufmann et al. 2018]. Potent catalysts used in this process are Brønsted acids/bases, Lewis acids and protonated/deprotonated molecules [H^+/A^-]. The rate of the reaction can be easily modulated by controlling the pH by the mere addition of specific conjugate acids or bases. Acid-base catalysis relies on the Henderson–Hasselbalch equation: $pH = pK_a + \log_{10}([A^-]/[HA])$.

8.2 FACTORS INFLUENCING BIOCATALYTIC ACTION

In general, the rate of the biocatalytic reaction depends mainly on various parameters such as substrate concentration, enzyme concentration, temperature, pH, surface area and diffusion [Maria-Solano et al. 2016]. Enzymes are proteins that are folded to form a binding site with amino acid residues. In some cases, metal ions attach with these amino acids, forming an excellent active site [McCall et al. 2000]. The catalytic reaction with enzymes involves binding and releasing steps of reactant or substrate and product, respectively. The active site of enzymes has an important role in deciding the rate of the reaction, as it should facilitate the compact and tight binding of the substrate for efficient conversion. Simple representation of the Lock and Key theory of enzyme-substrate binding is shown in Figure 8.2.

8.2.1 Substrate Concentration

The reaction rate increases with the increase in the number of collisions between reactant species. It is necessary for the substrate molecules to reach the active sites of

Diffusion Limitations in Biocatalytic Reactions

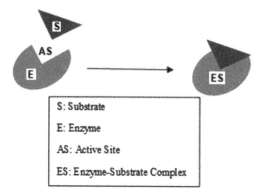

FIGURE 8.2 Lock and key theory.

the enzymes to be converted to product. Therefore, at lower substrate concentration, there may not be sufficient molecules to be occupied in the active sites. Similarly, the presence of a higher concentration of substrate does not affirm high catalytic rate. This is because the active sites of enzymes attain saturation when fully occupied.

8.2.2 Enzyme Concentration

Enzymes are more efficient than any other catalyst because their active sites may be used repeatedly. Catalytic action occurs with the successful binding of substrate molecules on the active site of the enzyme, forming a temporary complex, and the same active site will be occupied by another new substrate molecule with the release of product formed as a result of bio-catalysis. Higher loading is not synonymous with higher reaction rate [Sutormin et al. 2018], and rather it suffers diffusional limitations [Bahamondes et al. 2017]. The relationship between the enzyme concentration and rate of reaction becomes proportional with higher substrate concentration, constant pH and temperature.

8.2.3 Surface Area

It is common that the exposed surfaces take part in the reaction, and these exposed surfaces are considered to be the surface area. This is one of the major factors that determines the rate of the catalytic reaction [Kim 2018]. The surface area of the catalyst or the reactant is increased by decreasing their size, thereby availing numerous surfaces to actively participate in the reaction. The rate of the reaction and surface area of the catalyst are directly proportional to each other. At higher surface area in a heterogeneous catalysis, the number of collisions of the liquid/gas phase with the solid phase increases. The effects of surface area can be clearly understood from Figure 8.3.

8.2.4 Diffusion

In general, molecules frequently collide with each other with conservation of momentum and cause changes in their direction of movement, resulting in random

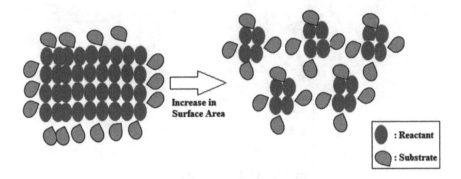

FIGURE 8.3 Effect of surface area of catalyst on reaction rate.

motion. Diffusion is the process by which atoms or molecules move from a region of higher to lower concentration due to random motion. Diffusion is a primary transport mechanism for molecules when the transport distance is small. Hence, for all kinds of chemical reactions to occur, diffusion plays an important role [Viladegut et al. 2017]. It is a process of mass transfer that involves the movement of one molecular species through another. For any successful reaction, either substrate or product should diffuse towards and away from the reaction site, respectively. Diffusion is irreversible as it occurs in only one direction from high to low concentrations.

Reactions in many industries are limited by the rate of diffusion between the catalytic surface and the reactants. Reactions are made less efficient by the diffusional limitations caused by the agglomeration of the catalytic particles [Pang et al. 2015], and hence this is an important factor to be considered.

Two types of diffusion are mainly involved in the catalytic reaction, namely:

- External diffusion
- Internal diffusion

8.2.4.1 External Diffusion

Any insoluble catalyst ought to possess a layer of molecules attached on the surface termed the *Nernst layer* that takes part in diffusing the molecules. This is an unstirred layer and is responsible for the hydrodynamic thickness of the catalyst. If the catalysis is performed under stirring conditions, the thickness of the Nernst layer gets reduced and hence the external diffusion can be neglected.

8.2.4.2 Internal Diffusion

Internal diffusion is also an important factor to be considered for effective catalysis. Here, the diffusion occurs internally within the pores of the catalyst that permit the substrate molecules to be in contact with the catalytic active sites [Dai et al. 2018]. Movement of the products out of the catalyst is also important for easing further catalysis. This is made possible by diffusion of products internally within the catalyst. From this, the importance of surface property of the catalyst can be understood. Figure 8.4 is a schematic representation of external and internal diffusion in a reactor.

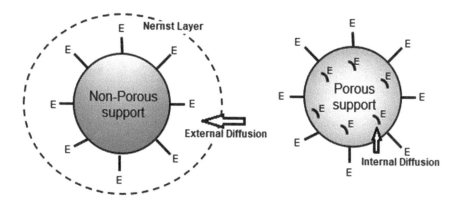

FIGURE 8.4 External and internal diffusion in non-porous and porous support.

8.3 APPROACHES TO OVERCOME DIFFUSIONAL LIMITATIONS

Several methods have been employed to eliminate or overcome the diffusional limitations in a catalytic reaction (Table 8.1). The effect of diffusion on the rate of the catalytic reaction mainly depends on the surface behavior of the catalyst or the support on which the catalyst is attached. Therefore, many approaches have been developed to support the diffusion by modifying the surface porosity. The drawbacks of instability make the enzymes less efficient in performing repeated catalytic reactions. Also, when the enzymes are used in insoluble forms, heterogeneous reactions (solid-liquid) and diffusion limitations hamper the effective catalytic reaction [Majumdar et al. 2016]. Immobilization serves as an excellent technique to bring out extreme use of the biocatalyst. Hence the support used for immobilizing the enzyme has a major role in enabling the diffusion of substrate and product. Some of the most effective strategies adopted to overcome diffusional limitations are discussed here.

8.3.1 Hydrogels

It is important to note that the reaction involving solid substrates/products cannot involve insoluble enzymes. Hence, carriers of other relative phases are used, one of which is a hydrogel. Hydrogel is a gel-like substance made by a network of polymeric chains that can be effectively used as a matrix to attach the enzymes [Haring et al. 2018]. One of the notable properties of hydrogels is that they have the tendency to shrink and re-swell with optimal changes in temperature. Figure 8.5 shows the behavior of the enzyme immobilized on hydrogel. This pumping behavior of hydrogel enhances the diffusion of substrate and product into and out of the gel, overcoming diffusional limitations.

8.3.2 Sensitive Matrices

These are an excellent source for the utmost minimization of diffusional limitations. They include gels/matrices that are sensitive to various parameters like pH, ionic

TABLE 8.1
Various Approaches Applied for Overcoming the Diffusional Limitations

S. No	Strategy	Catalysis	References
1	Using polyelectrolyte complexes	Heterogeneous	[Margolin et al. 1985]
2	Using copolymer	Heterogeneous	[Taniguchi et al. 1989]
3	Using thermo-responsive polymer	Heterogeneous	[Hoshino et al. 1994]
4	Using geometrical transformations	Heterogeneous	[Peters 2012]
5	Using thin-walled photoreactors	Photocatalysis	[Motegh et al. 2014]
6	Using surface-modified nanoparticles	Homogeneous	[Malar et al. 2018]

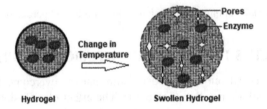

FIGURE 8.5 Schematic representation of hydrogel with change in temperature.

strength, temperature and concentration in reaction medium. Sensitive carriers also solve the problem in solubility of enzymes because here immobilization happens in a reversible manner between an insoluble and soluble state. This is made possible by the change in any of the above-mentioned parameters.

pH-sensitive matrices help in dealing with substrates that are poorly soluble, as it is difficult to maintain good contact with the enzyme. Also, it eases the separation process at the completion of the reaction. Reversible immobilization of the enzymes alters their solubility and helps good recovery. Immobilization of cellulose on a copolymer revealed active recovery with the absence of diffusional limitations [Taniguchi et al. 1989].

Temperature has to be selected carefully such that the catalytic activity of the enzyme is not disturbed as well as favoring the reactants. The response to the temperature change can be favored by the presence of a few bases like NaCl. In general, the support materials are chosen in such a way that the solubility becomes weak with the increase in temperature.

8.3.3 Non-Porous Supports

Recently, attachment of the enzyme on the surface of any non-porous supports [Motegh et al. 2014; Malar et al. 2018; Haring et al. 2018; Martinez-Alejo et al. 2018; Kennedy et al. 1990; Tischer et al. 1999; Haupt et al. 2005] in nanoscale was revealed to eliminate diffusional limitations, whereas the porous structures involve various significant diffusional problems [Martinez-Alejo et al. 2018]. Nanomaterials are used in the form of nanofibers [Kennedy et al. 1990; Tischer and Wedekind 1999;

Haupt et al. 2005; Fernandez-Lorente et al. 2006; Jia et al. 2002; Jia et al. 2003; Gole et al. 2001] and nanoparticles [Tischer and Wedekind 1999; Caruso and Schuler 2000; Daubresse et al. 1996; Liao and Chen 2001; Martins et al. 1996] for effective carriers of biocatalysts. Compatible nanocarriers can be chosen for immobilizing the biocatalyst, and its porous nature can be altered using various techniques [Malar et al. 2018] in order to reduce the diffusional limitations.

8.4 SUMMARY

In this chapter, the diffusional limitations caused by particularly in biocatalytic reactions has been discussed. Strategies related to facilitating the diffusion towards and away from the reaction center are given. One of the major drawbacks of these reactions is their stagnant nature, so it is necessary to know the importance of stirring in any catalytic reaction. So, it is necessary to ensure that the substrate reaches the active site of an enzyme without hindrance to attain a maximum reaction rate. Nanoparticles have extended applications in all areas, with their unique properties, and by further tuning the surface, these properties help to overcome the diffusion-related issues, thereby permitting to relish the complete use of biocatalysts.

REFERENCES

Aguila, B., Sun, Q., Wang, X., O'Rourke, E., Al-Enizi, A.M., Nafady, A., Ma, S. 2018. Lower activation energy for catalytic reactions through host-guest cooperation within metal-organic frameworks. *Angewandte Chemie*. 57: 1–6. doi:10.1002/ange.201803081.

Bahamondes, C., Álvaro, G., Wilson, L., Illanes, A. 2017. Effect of enzyme load and catalyst particle size on the diffusional restrictions in reactions of synthesis and hydrolysis catalyzed by α-chymotrypsin immobilized into glyoxal-agarose. *Process Biochemistry*. 53: 172–179.

Caruso, F., Schüler, C. 2000. Enzyme multilayers on colloid particles: Assembly, stability, and enzymatic activity. *Langmuir*. 16(24): 9595–9603.

Dai, C., Zhang, A., Song, C., Guo, X. 2018. Advances in the synthesis and catalysis of solid and hollow zeolite-encapsulated metal catalysts. *Advances in Catalysis*. 63: 75–115. doi:10.1016/J.bs.acat.2018.10.002.

Daubresse, C., Grandfils, C., Jerome, R., Teyssié, P. 1996. Enzyme immobilization in reactive nanoparticles produced by inverse microemulsion polymerization. *Colloid and Polymer Science*. 274(5): 482–489.

Ertl, G. 1990. Elementary steps in heterogeneous catalysis. *Angewandte Chemie International Edition in English*. 29(11): 1219–1227.

Fernandez-Lorente, G., Palomo, J.M., Mateo, C., Munilla, R., Ortiz, C., Cabrera, Z., Guisan, J.M., Fernandez-Lafuente, R. 2006. Glutaraldehyde cross-linking of lipases adsorbed on aminated supports in the presence of detergents leads to improved performance. *Biomacromolecules*. 7(9): 2610–2615.

Gole, A., Dash, C., Soman, C., et al. 2001. On the preparation, characterization, and enzymatic activity of fungal protease-gold colloid bioconjugates. *Bioconjugate Chemistry*. 12(5): 684–690.

Haring, M., Tautz, M., Alegre-Requena, J.V., et al. 2018. Non-enzyme entrapping biohydrogels in catalysis. *Tetrahedron Letters*. 59(35): 3293–3306.

Haupt, B., Neumann, T.H., Wittemann, A., Ballauff, M. 2005. Activity of enzymes immobilized in colloidal spherical polyelectrolyte brushes. *Biomacromolecules*. 6(2): 948–955.

Herrmann, N., Vogelsang, D., Behr, A., Seidensticker, D. 2018. Homogeneously catalyzed 1, 3-diene functionalization – A success story from laboratory to miniplant scale. *Catalysis.* 10: 5342–5365. doi:10.1002/cctc.201801362.

Hoshino, K., Taniguchi, M., Sasakura, T., et al. 1994. Hydrolysis of starch materials by repeated use of amylase immobilized on a novel thermo-responsive polymer. *Journal of Fermentation and Bioengineering.* 77: 407–412.

Jia, H., Zhu, G., Vugrinovich, B., et al. 2002. Enzyme-carrying polymeric nanofibers prepared via electrospinning for use as unique biocatalysts. *Biotechnology Progress.* 18(5): 1027–1032.

Jia, H., Zhu, G., Wang, P. 2003. Catalytic behaviors of enzymes attached to nanoparticles: The effect of particle mobility. *Biotechnology and Bioengineering.* 84(4): 406–414.

Kaufmann, M., Kruger, S., Mugge, C., Kroh, L.W. 2018. General acid/base catalysis of sugar anomerization. *Food Chemistry.* 265: 216–221.

Kennedy, J.F., Melo, E.H.M., Jumel, K. 1990. Immobilized enzymes and cells. *Chemical Engineering Progress.* 86: 81–89.

Kim, S.H. 2018. Nanoporous gold: Preparation and applications to catalysis and sensors. *Current Applied Physics.* 18(7): 810–818.

Liao, M.H., Chen, D.H. 2001. Immobilization of yeast alcohol dehydrogenase on magnetic nanoparticles. *Biotechnology Letters.* 23(20): 1723–1727.

Longo, M.A., Sanroman, M.A. 2006. Production of food aroma compounds: Microbial and enzymatic methodologies. *Food Technology and Biotechnology.* 44: 335–353.

Majumdar, P., Khan, A.Y., Bandyopadhyaya, R. 2016. Diffusion, adsorption and reaction of glucose in glucose oxidase enzyme immobilized mesoporous silica (SBA-15) particles: Experiments and modelling. *Biochemical Engineering Journal.* 105: 489–496.

Malar, C.G., Seenuvasan, M., Kumar, K.S. 2018. Prominent study on surface properties and diffusion coefficient of urease-conjugated magnetite nanoparticles. *Applied Biochemistry and Biotechnology.* 186(1): 174–185.

Margolin, A.L., Sherstyuk, S.F., Izumrudov, V.A., et al. 1985. Enzymes in polyelectrolyte complexes. The effect of phase transition on thermal stability. *European Journal of Biochemistry.* 146(3): 625–632.

Maria-Solano, M.A., Ortiz-Ruiz, C.V., Munoz-Munoz, J.L., et al. 2016. Further insight into the pH effect on the catalysis of mushroom tyrosinase. *Journal of Molecular Catalysis. Part B: Enzymatic.* 125: 1–16.

Martinez-Alejo, J.M., Benavent-Gil, Y., Rosell, C.M. et al. 2018. Quantifying the surface properties of enzymatically-made porous starches by using a surface energy analyser. *Carbohydrate Polymers.* 200: 543–551.

Martins, F.B., Simoes, I.D., Cruz, E.M., Gaspar, R. 1996. Development of enzyme-loaded nanoparticles: Effect of pH. *Journal of Materials Science: Materials in Medicine.* 7(7): 413–414.

Mateo, C., Palomo, J.M., Fernandez-Lorente, G., et al. 2007. Improvement of enzyme activity, stability and selectivity via immobilization techniques. *Enzyme and Microbial Technology.* 40(6): 1451–1463.

McCall, K.A., Huang, C., Fierke, C.A. 2000. Function and mechanism of zinc metalloenzymes. *The Journal of Nutrition.* 130(5S Suppl): 1437S–1446S.

Motegh, M., Cen, J., Appel, P.W., et al. 2014. Diffusion limitations in stagnant photocatalytic reactors. *Chemical Engineering Journal.* 247: 314–319.

Pang, R., Li, M., Zhang, C. 2015. Degradation of phenolic compounds by laccase immobilized on carbon nanomaterials: Diffusional limitation investigation. *Talanta.* 131: 38–45.

Peters, B. 2012. Headspace diffusion limitations on heterogeneous catalysis in unstirred batch reactors. *Chemical Engineering Science.* 71: 367–374.

Pirouzmand, M., Mahdavi, M. 2018. Biodiesel production from waste cooking oil using Co incorporated MCM-41 as green nano catalyst. *Nanochemistry Letters.* 1: 1–4.

Radzicka, A., Wolfenden, R. 1995. A proficient enzyme. *Science.* 267(5194): 90–93.

Seenuvasan, M., Kumar, K.S., Malar, C.G., et al. 2014. Characterization, analysis and application of fabricated Fe_3O_4–chitosan–pectinase nanobiocatalyst. *Applied Biochemistry and Biotechnology.* 172(5): 2706–2719.

Seenuvasan, M., Malar, C.G., Preethi, S., et al. 2013. Fabrication, characterization and application of pectin degrading Fe_3O_4–SiO_2 nanobiocatalyst. *Materials Science and Engineering. Part C.* 33(4): 2273–2279.

Seenuvasan, M., Vinodhini, G., Malar, C.G., et al. 2018. Magnetic nanoparticles: A versatile carrier for enzymes in bio-processing sectors. *IET Nanobiotechnology.* 12(5): 535–548.

Singh, R., Kumar, M., Mittal, A., Mehta, P.R. 2016. Microbial enzymes: Industrial progress in 21st century. *3 Biotech.* 6(2): 174.

Sutormin, O.S., Sukovataya, I.E., Pande, S., Kratasyuk, V.A. 2018. Effect of viscosity on efficiency of enzyme catalysis of bacterial luciferase coupled with lactate dehydrogenase and NAD(P)H:FMN-Oxidoreductase. *Molecular Catalysis* 458: 60–66.

Taniguchi, M., Kobayashi, M., Fuji, M. 1989. Properties of a reversible soluble-insoluble cellulase and its application to repeated hydrolysis of crystalline cellulose. *Biotechnology and Bioengineering.* 34(8): 1092–1097.

Tischer, W., Wedekind, F. 1999. Immobilized enzymes: Methods and applications. *Topics in Current Chemistry.* 200: 95–126.

Truppo, M.D. 2017. Biocatalysis in the pharmaceutical industry: The need for speed. *ACS Medicinal Chemistry Letters.* 8(5): 476–480.

Viladegut, A., Duzel, U., Chazot, O. 2017. Diffusion effects on the determination of surface catalysis in inductively coupled plasma facility. *Chemical Physics.* 485: 88–97.

Wang, X., Ma, W., Xu, Z., et al. 2018. Metal phosphide catalysts anchored on metal-caged graphitic carbon towards efficient and durable hydrogen evolution electrocatalysis. *Nanoenergy.* 48: 500–509.

Yue, J. 2018. Multiphase flow processing in microreactors combined with heterogeneous catalysis for efficient and sustainable chemical synthesis. *Catalysis Today.* 308: 3–19.

Zhou, Y., Jin, C., Li, Y., Shen, W. 2018. Dynamic behavior of metal nanoparticles for catalysis. *Nanotoday.* 20: 101–120.

9 Recent Advancements and Applications of Nanotechnology in Expelling Heavy Metal Contaminants from Wastewater

Muthulingam Seenuvasan, Venkatachalam Vinothini, Madhava Anil Kumar and Ayyanar Sowmiya

CONTENTS

9.1 Introduction	152
9.2 Characteristics of Wastewater	153
9.3 Conventional Wastewater Treatment	153
9.3.1 Coagulation and Flocculation	153
9.3.2 Precipitation	153
9.3.3 Ion Exchange	154
9.3.4 Electro-Chemical Methods	154
9.3.5 Membrane Separation	154
9.3.6 Adsorption	155
9.4 Pros and Cons of the Conventional Treatment Methods	155
9.5 Nanotechnology for Wastewater Treatment	155
9.5.1 Nanosorbents for Heavy Metal Removal	156
9.5.2 Carbon Nanotube-Based Adsorption of Heavy Metals	157
9.5.3 Nanomembranes for Nanofiltration	157
9.5.4 Nano-Structured Catalyst for Photocatalytic Oxidation	158
9.5.5 Nanomaterials in Water Disinfection	158
9.6 Conclusions	158
References	158

9.1 INTRODUCTION

Water pollution is considered to be one of the most severe environmental problems, an adverse impact of humans and industrial sectors on the planet. Industrial wastewater affects water quality, aquatic flora and fauna, wherein the sources of wastewater are mining, fertilizer, paper and pulp industries. The various contaminants that pollute the water are heavy metals like chromium, mercury, lead, cadmium, zinc, copper and nickel and dye effluents, as listed in Table 9.1. Heavy metals are non-biodegradable and are toxic in nature (Fu and Wang, 2011). At present, the scarcity of freshwater is an ever-increasing problem at a global level. Therefore, there is a need to treat industrial effluent prior to discharging it to water bodies like rivers, lakes, etc. (Kumar et al., 2018). The most commonly used wastewater treatment techniques are coagulation, precipitation, ion exchange, electro-chemical methods, membrane separation and adsorption (Fu and Wang, 2011; Papadopoulos et al., 2004; Nandy et al., 2003). The above-mentioned methods have certain advantages and limitations based on their application. The implementation of nanomaterials in wastewater treatment is widely accepted because of their unique nanoscale dimensions. Carbon-based nanomaterials such as carbon nanotubes (CNTs), graphene-based materials and nano-sized metal oxides of magnesium, aluminum, manganese, cerium and ferric are commonly used (Santhosh et al.,2016).

TABLE 9.1
Heavy Metal Pollutants and Their Effects

Heavy Metals	Sources	Effects
Chromium	Leather tanning, mining, electroplating	Lung cancer, respiratory problems, vomiting
Lead	Lead acid batteries, paints, smelting and thermal power plants	Kidney malfunction
Mercury	Hospital waste (damaged thermometers, barometers), thermal power plants	Spontaneous abortion, congenital malfunctions, gastrointestinal disorders
Arsenic	Smelting and thermal power plants	Respiratory problems, cardiovascular and neurological diseases
Copper	Mining, electroplating and smelting	Cardiovascular disease, liver and kidney damage, gastrointestinal problems
Vanadium	Sulfuric acid plant	Severe eye, nose, throat inflammation, bronchitis and pneumonia
Nickel	Battery industry, smelting and thermal power plants	Dermatitis, headaches and myocarditis
Cadmium	Sludges from paint industry	Renal dysfunction, lung disease and bone defects
Zinc	Smelting and electro-plating	Impairment of growth and reproduction, anemia, liver and kidney failure

9.2 CHARACTERISTICS OF WASTEWATER

Wastewater is polluted water containing organic and inorganic chemicals at concentrations above their carrying capacity. The characteristics of wastewater are broadly classified into physical, chemical and biological characteristics based on the impurities present in the wastewater (Kumar et al., 2017). Physical characteristics of wastewater include color, odor, temperature and turbidity. At its fresh stage, the wastewater is greenish brown or yellowish in color, and over a longer time period it changes into black. The odor of wastewater depends on the presence of impurities, and the odor-producing substances are amines, ammonia, diamines, mercaptens and organic sulfides; its odor quality can be fishy, ammoniacal, rotten egg or decayed. The temperature of wastewater is normally higher than that of the temperature of the water supply, and the average temperature is 20°C. Turbidity depends on the quantity of solids present in a suspended state, and more suspended solids result in high turbidity of wastewater (Merzouket al., 2009).

Chemical characteristics of wastewater are pH, organic and inorganic matter content. Organic matter content is determined by biological oxygen demand (BOD), chemical oxygen demand (COD) and total organic carbon (TOC). BOD is defined as the amount of oxygen required by microorganisms to decompose the biodegradable organic matter under aerobic conditions. High levels of BOD and COD cause greater pollution and also reduce the dissolved oxygen content in water, causing harmful effects to aquatic life systems (Hua et al., 2012). TOC is the amount of organic compounds in the wastewater, and these compounds react with chlorine and produce byproducts which are carcinogenic (Kumar et al., 2012). Microorganisms present in wastewater are usually pathogenic and decide the biological characteristics of wastewater (Chang et al., 2004).

9.3 CONVENTIONAL WASTEWATER TREATMENT

9.3.1 Coagulation and Flocculation

Chemical coagulation and flocculation are widely used in wastewater treatment to eliminate the turbidity and color, especially for treating potable water. The commonly used coagulants are aluminum sulfate, poly aluminum chloride, ferrous sulfate, poly styrene sulfonate, sodium aluminate, silicon derivatives, lime and synthetic organic polymer. For wastewater treatment processes, alum, ferric chloride, poly aluminum chloride and poly styrene sulfonate (Landim et al., 2007; Song et al., 2004; Wang et al., 2007) are commonly used. Flocculants assist the coagulated particles to agglomerate and form clumps of particles, which speed up the settling process. The coagulation-flocculation process is followed by filtration and sedimentation. Factors which affect the coagulation and flocculation process are type and dosage of coagulant/flocculants, pH, mixing speed and time, temperature and retention time (Birjandi et al., 2013).

9.3.2 Precipitation

Precipitation involves the removal of pollutants in the form of precipitates by adding precipitating agents and then removing them by solid separation processes that may include coagulation and flocculation. This method is widely used for the removal of heavy metal and phosphorous from wastewater (Fu and Wang, 2011). The conventional precipitation includes hydroxide and sulfide precipitation. Calcium hydroxide and sodium hydroxide are extensively used for removal of heavy metals like copper and chromium ions (Fu et al., 2012; Mirbagheri and Hosseini, 2005). Coagulants like alum, iron salts and organic polymers are added to increase the removal efficiency of heavy metals. In sulfide precipitation, pyrites and synthetic iron sulfides are used for the removal of copper, cadmium and lead (Özverdi and Erdem, 2006). Precipitation combined with other methods like ion exchange and electrolysis is used for the removal of heavy metals (Kabdaşli et al., 2009; Ghosh et al., 2011).

9.3.3 Ion Exchange

Ion exchange is used to expel metal ions from wastewater by exchanging ions with resin; the parameters affecting the exchange of ions with resin are pH, temperature, metal ion concentration and contact time. Ion exchange resin may be either natural or synthetic resin. Synthetic resins (purolites) and natural resins (zeolites) are widely used to eliminate the wastewater pollutants (Motsi et al., 2011).

9.3.4 Electro-Chemical Methods

Electro-chemical methods are mostly used to remove refractory pollutants from wastewater; they include electro-coagulation, electro-flotation and electro-deposition. In electro-coagulation, the coagulants are produced by electrically dissolving either aluminum or iron ions from aluminum or iron electrodes, which aids in coagulating the pollutants present in the wastewater (Heidmann and Calmano, 2008). Electro-flotation is a solid-liquid separation process which removes pollutants in the form of bubbles generated by electrolysis, and this technique uses suitable metal electrodes (Belkacem et al., 2008).

9.3.5 Membrane Separation

Membrane separation is a pressure-driven process in which the pollutants are removed by membranes which act as a barrier (Mohammad et al., 2015). Membrane separation processes are categorized into ultra-filtration, reverse osmosis, nanofiltration and electro-dialysis; wherein the driving force behind the membrane separation is based on transmembrane pressure for the removal of dissolved and colloidal materials (Samper et al., 2009). Polymer enhanced ultra-filtration is widely used for the expulsion of pollutants. Reverse osmosis is a semi-permeable membrane process that allows the fluid to pass through it, retaining the contaminants, and this technology is widely used as a tertiary treatment. It removes total dissolved solids, heavy metals, organic pollutants, viruses, bacteria and other contaminants. The parameters considered for the removal of pollutants are pretreatment, chemical control and membrane fouling. The nanofiltration process is mostly used for liquid separation, and this process lies between ultra-filtration and reverse osmosis due to the properties of membranes. For removing natural organic matter and inorganic pollutants, nanofiltration is the best potential technology. Nanofiltration mechanisms depend on the charge and size of the particles; the nanofiltration membrane removes nickel, chromium, copper and arsenic (Cséfalvay et al., 2009; Nguyen et al., 2009). Some researchers have carried out integrated reverse osmosis and nanofiltration for the removal of contaminants (Koseoglu and Kitis, 2009). Electro-dialysis is widely used for treating brackish water, seawater and industrial effluents. In the electro-dialysis process, an electric field is the driving force for removing ions through the charged membrane. Ion exchange membranes are broadly classified into cation and anion exchange membranes (Sadrzadeh et al., 2009).

9.3.6 Adsorption

Adsorption is the most effective method for wastewater treatment, as it produces less-contaminated effluent when compared to other processes. Adsorbents used for this process are divided into two categories, namely natural and synthetic adsorbents. Natural adsorbents such as lignocellulosic materials, charcoal, clay minerals, zeolite and ores are used, as well as several biosorbents such as bark, lignin, shrimp, krill, squid, crab shell and algal and microbial biomass (Seenuvasan et al., 2018; Wan and Hanafiah, 2008). The synthetic adsorbents are household waste and industrial waste residues. The most commonly used adsorbent for wastewater treatment is activated carbon due to its porous structure as well as its high surface area, and several low-cost adsorbents are synthesized and used for wastewater treatment (Moradeeya et al., 2017).

9.4 PROS AND CONS OF THE CONVENTIONAL TREATMENT METHODS

The conventional methods have their own advantages and disadvantages, as shown in Table 9.2. Among the conventional methods, precipitation and electrolysis are less efficient, as they produce huge amounts of sludge. Ion exchange and membrane separation are expensive as well as limited to large-scale operations. As a result, there is an emerging need for technological improvements in wastewater treatment.

9.5 NANOTECHNOLOGY FOR WASTEWATER TREATMENT

Nanotechnology-enabled wastewater treatment overcomes the challenges faced by the conventional methods and offers some new treatment opportunities to expand the non-conventional water resources. Nanotechnology is the application of nanomaterials that exhibit significant properties such as high surface area, functionality and remarkable phenomena due to their small dimensions (Seenuvasan et al., 2013). Nanomaterials are small, have tunable pore sizes, are highly reactive, are more accurate and can be produced by eco-friendly techniques. In recent years, the development of various tools and techniques enabled by nanotechnology provides an efficient and effective alternative in wastewater treatment.

TABLE 9.2
Pros and Cons of Conventional Wastewater Treatment

Treatment	Pros	Cons
Coagulation-flocculation	Sludge settling time; easy to remove fine particles, bacteria, viruses and protozoa	High cost, not suitable for heavy metal removal
Precipitation	Simplicity and low cost	Dewatering and disposal problems causes secondary pollution with high capital cost
Ion exchange	High removal efficiency, high treatment capacity	High costs and membrane fouling
Electrolysis	Low sludge formation	High capital cost and expenditure towards electricity
Membrane separation	Flexibility in designing system	Membrane fouling and shorter membrane life and low selectivity
Adsorption	Flexibility in design and operation	Formation of sludge

Nanotechnology is extensively used in wastewater treatment in the removal of various pollutants. Photocatalysis, nanofiltration, nano-adsorption and electro-chemical oxidation are the processes that extensively use the application of nanotechnology in the removal of contaminants from wastewater (Wan et al., 2010). These processes involve the use of titanium dioxide (TiO_2), zinc oxide (ZnO), ceramic membranes, nanomembranes, polymer membranes, carbon nanotubes, submicron nanopowder, metal (oxides) and magnetic nanoparticles for the removal of pollutants (Stafiej and Pyrzynska, 2007; Park et al., 2007). Nanoparticles, when used as catalysts or agents, aid as a functional material in the removal of pollutants by destroying the chemical and functional moieties. Nanomaterials are classified into four groups, namely dendrimers, metal-containing nanoparticles, zeolites and carbonaceous nanoparticles. The environmental promises of nanotechnology include reducing waste production, cleanup of industrial contamination, providing clean drinking water and escalating efficiency.

Nanotechnology plays a vital role in treating wastewater for the removal of toxic heavy metals, natural organic matter, biological contaminants and organic pollutants, nutrients and heavy metals (Kunduru et al., 2017). The increased surface area and specificity of the nanomaterials help in the removal of contaminants even at low concentrations so that the improved efficiency is noticeable in treating wastewater (Amit et al., 2018). Due to their smaller size, nanoparticles have high interaction and reaction capabilities with their counterparts. Energy conservation can be achieved through nanomaterials, which ultimately leads to cost savings. Nanotechnology reduces the number of intervening steps such as energy and materials involved in other conventional treatment techniques.

9.5.1 Nanosorbents for Heavy Metal Removal

Nanomaterials have at least one external dimension that measures 100 nm and exist in different forms such as tubes, particles, rods or fibers (Sheet et al., 2014).

The advanced wastewater treatment technologies exert high efficiencies with the advent of nanotechnology and the use of nanomaterials such as titanium dioxide, zinc oxide and graphene oxides (Peng et al., 2017; Poursani et al., 2016; Karabelli et al., 2008). Iron nanoparticles are used for the removal of copper ions, and the based nanocomposites are known to remove ferric, arsenate and cupric ions from aqueous streams (Esmat et al., 2017; Ghaedi et al., 2012; Karabelli et al., 2008).

Nano-scale zerovalent iron possesses some beneficial effects on environmental protection, as it involves low production costs and high reactivity with contaminants (Wang et al., 2017; Karthikeyan et al., 2014). These materials find their application in treatment of contaminated ground water and also play a major role in adsorption of organic and inorganic pollutants, including solvents, pesticides and other heavy metals. Depending upon contaminant mobility, this technique can be investigated by two treatment methodologies such as mobile contamination and immobile contamination.

9.5.2 Carbon Nanotube-Based Adsorption of Heavy Metals

Carbon nanotubes are graphene sheets rolled up in cylinders, and they are unique in structure and electronic properties of carbon nanomaterials, which have been considered for potential applications in adsorption to uptake a wide range of contaminants due to their larger surface area. Nano-porous carbon, carbon fibers, carbon beads and carbon nanotubes remove specific contaminants such as dichlorobenzene, ethyl benzene and dyes due to their exceptional adsorption facilities (Gupta et al., 2016). Carbon nanotubes are combined with metals to escalate their absorptivity, electrical and thermal properties. The carbon nanotubes are generally functionalized with magnetic iron oxide nanoparticles for the removal of heavy metals from wastewater (Seenuvasan et al., 2014; Chang and Wang, 2002). Carbon nanotubes such as multi-walled carbon nanotubes, alumina coated nanotubes, as well as carbon nanotubes immobilized with calcium alginates, are used to eliminate several divalent heavy metal ions such as Cu^{2+}, Cd^{2+}, Pb^{2+}, Zn^{2+}, Ni^{2+}, Co^{2+} and Mn^{2+} (Gupta et al., 2011; Li et al., 2010).

9.5.3 Nanomembranes for Nanofiltration

Membranes are highly reliable, and the application of nanomembranes in fields of water and wastewater treatment technologies helps in the effective removal of heavy metals, dyes and other contaminants (Murthy and Chaudhari, 2008). Membrane separations are carried out based upon their molecular size. Micro-pollutants and multivalent ions are relatively easily removed using nanomembranes. The membranes are formed by aligning carbon nanotubes vertically together, where the presence of nano-pores of membranes helps to selectively filter the various contaminants (Muthukrishnan and Guha, 2008). Nanofiltration is a high-pressure membrane treatment process utilizing nanomembranes to eliminate various types of contaminants, and the nanofiltration plant consists of a large number of modules with different membrane configuration ranging from 0.9 to 0.55 mm in length and 100 to 300 mm in diameter and arranged either horizontally or vertically. The various membranes such as ceramic, polymeric membranes with anti-fouling coating, composite membranes, metal/metal oxide polymer membrane and zeolite polymer membranes are used.

9.5.4 Nano-Structured Catalyst for Photocatalytic Oxidation

Nano-catalysts, especially inorganic materials, play a vital role in chemical oxidation of organic pollutants. Metal nanoparticles and metal oxides act as a catalyst in oxidation reactions as they exhibit strong catalytic activity. The pollutants are oxidized and get converted into less toxic soluble products, where the nano-catalysts can effectively oxidize the organic and inorganic pollutants. The nano-structured catalysts are particularly used in the advanced oxidation processes, as they generate highly reactive radicals that can easily react with pollutants. Photocatalytic oxidation is the most important tertiary wastewater treatment in which a chemical reaction is induced by the absorption of photons. The various organic pollutants such as volatile organic compounds, pesticides and dyes are easily degraded by photocatalytic oxidation (Lin et al., 2014).

9.5.5 Nanomaterials in Water Disinfection

Nanomaterials such as chitosan, silver nanoparticles, titanium dioxide, fullerene nanoparticles and carbon nanotubes possess great antimicrobial properties in addition to adsorption and catalytic properties. These materials are less reactive, relatively inert in water and do not produce harmful byproducts. These nanomaterials can be applied in water disinfection processes by means of direct action on bacterial cells, breakthrough of the cell membrane and oxidation of cellular components to form dissolved metal ions of reduced toxicity, which can be further removed as precipitates.

9.6 CONCLUSIONS

Nanotechnology offers very efficient, effective and ecofriendly approaches in the treatment of wastewater, and this chapter focuses on the various conventional treatment technologies as well as the role of nanotechnology in the field of wastewater treatment. The development of various nanomaterials and their diverse applications in the form of nano-catalysts, nanomembranes and nano-adsorbents facilitate the removal of various wastewater contaminants. The unique and specific properties of nanomaterials find their application in various sectors, especially in wastewater treatment.

REFERENCES

Amit C, Chandarana H, Kumar MA, and Sunita V. 2018. Nano-technological interventions for the decontamination of water and wastewater. In: X.-T. Bui et al. (eds.), *Water and Wastewater Treatment Technologies, Energy, Environment, and Sustainability.* Springer, Singapore, 487–99.

Belkacem M, Khodir M, and Abdelkrim S. 2008. Treatment characteristics of textile wastewater and removal of heavy metals using the electroflotation technique. *Desalination* 228(1–3): 245–54.

Birjandi N, Younesi H, Bahramifar N, Ghafari S, Zinatizadeh AA, and Sethupathi S. 2013. Optimization of coagulation-flocculation treatment on paper-recycling wastewater: Application of response surface methodology. *J Environ Sci Health A Tox Hazard Subst Environ Eng* 48(12): 1573–82.

Chang IS, Jang JK, Gil GC, Kim M, Kim HJ, Cho BW, and Kim BH. 2004. Continuous determination of biochemical oxygen demand using microbial fuel cell type biosensor. *Biosens Bioelectron* 19(6): 607–13.

Chang YK, Chang JE, Lin TT, and Hsu YM. 2002. Integrated copper-containing wastewater treatment using xanthate process. *J Hazard Mater* 94(1): 89–99.

Cséfalvay E, Pauer V, and Mizsey P. 2009. Recovery of copper from process waters by nanofiltration and reverse osmosis. *Desalination* 240(1–3): 132–42.

Esmat M, Farghali AA, Khedr MH, and El-Sherbiny IM. 2017. Alginate-based nano composites for efficient removal of heavy metal ions. *Int J Biol Macromol* 102: 272–83.

Fu F, and Wang Q. 2011. Removal of heavy metal ions from wastewaters: A review. *J Environ Manage* 92(3): 407–18.

Fu F, Xie L, Tang B, Wang Q, and Jiang S. 2012. Application of a novel strategy-advanced Fenton-chemical precipitation to the treatment of strong stability chelated heavy metal containing wastewater. *Chem Eng J* 189–90: 283–7.

Ghaedi M, Sadeghian B, Amiri Pebdani A, Sahraei R, Daneshfar A, and Duran C. 2012. Kinetics, thermodynamics and equilibrium evaluation of direct yellow 12 removal by adsorption onto silver nanoparticles loaded activated carbon. *Chem Eng J* 187: 133–41.

Ghosh P, Samanta AN, and Ray S. 2011. Reduction of COD and removal of Zn^{2+} from rayon industry wastewater by combined electro-Fenton treatment and chemical precipitation. *Desalination* 266(1–3): 213–17.

Gupta VK, Agarwal S, and Saleh TA. 2011. Synthesis and characterization of alumina-coated carbon nanotubes and their application for lead removal. *J Hazard Mater* 185(1): 17–23.

Gupta VK, Moradi O, Tyagi I, Agarwal S, Sadegh H, Shahryari-Ghoshekandi R, Makhlouf ASH, Goodarzi M, and Garshasbi A. 2016. Study on the removal of heavy metal ions from industry waste by carbon nanotubes: Effect of the surface modification: A review. *Crit Rev Environ Sci Technol* 46(2): 93–118.

Heidmann I, and Calmano W. 2008. Removal of Zn(II), Cu(II), Ni(II), Ag(I) and Cr(VI) present in aqueous solutions by aluminium electrocoagulation. *J Hazard Mater* 152(3): 934–41.

Hua M, Zhang S, Pan B, Zhang W, Lv L, and Zhang Q. 2012. Heavy metal removal from water/wastewater by nanosized metal oxides: A review. *J Hazard Mater* 211–212: 317–31.

Kabdaşli I, Arslan T, Olmez-Hanci T, Arslan-Alaton I, and Tünay O. 2009. Complexing agent and heavy metal removals from metal plating effluent by electrocoagulation with stainless steel electrodes. *J Hazard Mater* 165(1–3): 838–45.

Karabelli D, Çağri U, Shahwan T, Eroğlu AE, Scott TB, Hallam KR, and Lieberwirth I. 2008. Batch removal of aqueous Cu^{2+} ions using nanoparticles of zero-valent iron: A study of the capacity and mechanism of uptake. *Ind Eng Chem Res* 47(14): 4758–64.

Karthikeyan S, Kumar MA, Maharaja P, Rao BP, Sekaran G, and Sekaran G. 2014. Process optimization for the treatment of pharmaceutical wastewater catalyzed by poly sulpha sponge. *J Taiwan Inst Chem E* 45(4): 1739–47.

Koseoglu H, and Kitis M. 2009. The recovery of silver from mining wastewaters using hybrid cyanidation and high-pressure membrane process. *Miner Eng* 22(5): 440–44.

Kumar MA, Baskaralingam P, Aathika ARS, and Sivanesan S. 2018. Role of bacterial consortia in bioremediation of textile recalcitrant compounds. In: S. Varjani, Gnansounou E., Gurunathan B., Pant D., Zakaria Z. (eds.), *Waste Bioremediation. Energy, Environment, and Sustainability*. Springer, Singapore, 165–83.

Kumar MA, Kumar VV, Premkumar MP, Baskaralingam P, Thiruvengadaravi KV, Anuradha D, and Sivanesan S. 2012. Chemometric formulation of bacterial consortium-AVS for improved decolorization of resonance-stabilized and hetero-polyaromatic dyes. *Bioresour Technol* 123: 344–51.

Kumar MA, Poonam S, Kumar VV, Baskar G, Seenuvasan M, Anuradha D, and Sivanesan S. 2017. Mineralization of aromatic amines liberated during the degradation of a sulfonated textile colorant using *Klebsiella pneumoniae* strain AHM. *Proc Biochem* 57: 181–9.

Kunduru KR, Nazarkovsky M, Farah S, Pawar RP, Basu A, and Domb AJ. 2017. Nanotechnology for water purification: Applications of nanotechnology methods in wastewater treatment. *Water Purif*: 33–74.

Landim AS, Filho GR, and Nascimento De Assunção Rosana Maria. 2007. Use of polystyrene sulfonate produced from waste plastic cups as an auxiliary agent of coagulation, flocculation and flotation for water and wastewater treatment in municipal department of water and wastewater in Uberlandia-MG, Brazil. *Polym Bullet* 58(2): 457–63.

Li Y, Liu F, Xia B, Du Q, Zhang P, Wang D, Wang Z, and Xia Y. 2010. Removal of copper from aqueous solution by carbon nanotube/calcium alginate composites. *J Hazard Mater* 177(1–3): 876–80.

Lin ST, Thirumavalavan M, Jiang TY, and Lee JF. 2014. Synthesis of ZnO/Zn nano photocatalyst using modified polysaccharides for photodegradation of dyes. *Carbohydr Polym* 105: 1–9.

Merzouk B, Gourich B, Sekki A, Madani K, and Chibane M. 2009. Removal turbidity and separation of heavy metals using electrocoagulation-electroflotation technique. A case study. *J Hazard Mater* 164(1): 215–22.

Mirbagheri SA, and Hosseini SN. 2005. Pilot plant investigation on petrochemical wastewater treatment for the removal of copper and chromium with the objective of reuse. *Desalination* 171(1): 85–93.

Mohammad AW, Teow YH, Ang WL, Chung YT, Oatley-Radcliffe DL, and Hilal N. 2015. Nanofiltration membranes review: Recent advances and future prospects. *Desalination* 356: 226–54.

Moradeeya P, Kumar MA, Thorat RB, Rathod M, Khambhaty Y, and Basha S. 2017. Screening of nanocellulose for biosorption of chlorpyrifos from water: Chemometric optimization, kinetics and equilibrium. *Cellulose* 24(3): 1319–32.

Motsi T, Rowson NA, and Simmons MJH. 2011. Kinetic studies of the removal of heavy metals from acid mine drainage by natural zeolite. *Int J Miner Proc* 101(1–4): 42–9.

Murthy ZV, and Chaudhari LB. 2008. Application of nanofiltration for the rejection of nickel ions from aqueous solutions and estimation of membrane transport parameters. *J Hazard Mater* 160(1): 70–7.

Muthukrishnan M, and Guha BK. 2008. Effect of pH on rejection of hexavalent chromium by nanofiltration. *Desalination* 219(1–3): 171–78.

Nandy T, Shastry S, and Pathe and Kaul SN. 2003. Pre-treatment of currency printing ink wastewater through coagulation-flocculation process. *Water Air Soil Pollut* 148(1–4): 15–30.

Nguyen CM, Bang S, Cho J, and Kim KW. 2009. Performance and mechanism of arsenic removal from water by a nanofiltration membrane. *Desalination* 245(1–3): 82–94.

Özverdi A, and Erdem M. 2006. Cu^{2+}, Cd^{2+} and Pb^{2+} adsorption from aqueous solutions by pyrite and synthetic iron sulphide. *J Hazard Mater* 137(1): 626–32.

Papadopoulos A, Fatta D, Parperis K, Mentzis A, Haralambous KJ, and Loizidou M. 2004. Nickel uptake from a wastewater stream produced in a metal finishing industry by combination of ion-exchange and precipitation methods. *Sep Purif Technol* 39(3): 181–8.

Park HJ, Jeong SW, Yang JK, Kim BG, and Lee SM. 2007. Removal of heavy metals using waste eggshell. *J Environ Sci* 19(12): 1436–41.

Peng W, Li H, Liu Y, and Song S. 2017. A review on heavy metal ions adsorption from water by graphene oxide and its composites. *J Mol Liq* 230: 496–504.

Poursani AS, Nilchi A, Hassani A, Shariat S, and Nouri J. 2016. The synthesis of nano TiO_2 and its use for removal of lead ions from aqueous solution. *J Water Res Protect* 8(8): 438–48.

Sadrzadeh M, Mohammadi T, Ivakpour J, and Kasiri N. 2009. Neural network modeling of Pb^{2+} removal from wastewater using electrodialysis. *Chem Eng Proc* 48(8): 1371–81.

Samper E, Rodríguez M, De la Rubi MA, and Prats D. 2009. Removal of metal ions at low concentration by micellar-enhanced ultrafiltration (MEUF) using sodium dodecyl sulfate (SDS) and linear alkylbenzene sulfonate (LAS). *Sep Purif Technol* 65(3): 337–42.

Santhosh C, Velmurugan V, Jacob G, Jeong SK, Grace AN, and Bhatnagar A. 2016. Role of nanomaterials in water treatment applications: A review. *Chem Eng J* 306: 1116–37.

Seenuvasan M, Kumar KS, Malar GCG, Preethi S, Kumar MA, and Balaji N. 2014. Characterization, analysis, and application of fabricated Fe_3O_4-chitosan-pectinase nanobiocatalyst. *Appl Biochem Biotechnol* 172(5): 2706–19.

Seenuvasan M, Malar GCG, Preethi S, Balaji N, Iyyappan J, Kumar MA, and Kumar KS. 2013. Fabrication, characterization and application of pectin degrading Fe_3O_4-SiO_2 nanobiocatalyst. *Mat Sci Eng C* 33(4): 2273–9.

Seenuvasan M, Suganthi JRG, Sarojini G, Malar GCG, Priya ME, and Kumar MA. 2018. Effective utilization of crustacean shells for preparing chitosan composite beads: Applications in ameliorating the biosorption of an endocrine disrupting heavy metal. *Desalin Water Treat* 121: 28–35.

Sheet I, Kabbani A, and Holail H. 2014. Removal of heavy metals using nanostructured graphite oxide, silica nanoparticles and silica/graphite oxide composite. *Energy Procedia* 50: 130–8.

Song Z, Williams CJ, and Edyvean RGJ. 2004. Treatment of tannery wastewater by chemical coagulation. *Desalination* 164(3): 249–59.

Stafiej A, and Pyrzynska K. 2007. Adsorption of heavy metal ions with carbon nanotubes. *Sep Purif Technol* 58(1): 49–52.

Wan N, Zhu L, Wang D, Wang M, Lin Z, and Tang H. 2010. Sono-assisted preparation of highly-efficient peroxidase-like Fe_3O_4 magnetic nanoparticles for catalytic removal of organic pollutants with H_2O_2. *Ultrason Sonochem* 17(3): 526–33.

Wan Ngah WS, and Hanafiah MA. 2008. Removal of heavy metal ions from wastewater by chemically modified plant wastes as adsorbents: A review. *Bioresour Technol* 99(10): 3935–48.

Wang C, Gao X, Chen Z, Chen Y, and Chen H. 2017. Preparation, characterization and application of polysaccharide-based metallic nanoparticles: A review. *Polymers* 9(12): 689.

Wang JP, Chen YZ, Ge XW, and Yu HQ. 2007. Optimization of coagulation-flocculation process for a paper-recycling wastewater treatment using response surface methodology. *Colloids Surf A* 302(1–3): 204–10.

10 Organic Flocculation as an Alternative for Wastewater Treatment

Devlina Das

CONTENTS

10.1 Introduction 163
10.2 Flocculants from Natural Sources 165
    10.2.1 Bio-sludge 165
    10.2.2 Microbes as Sources of Bioflocculants 166
    10.2.3 Plants as Sources of Bioflocculants 167
    10.2.4 Biopolymers as Sources of Bioflocculants 167
10.3 Preparation of Flocculant Composites... 169
10.4 Technological Advances 170
    10.4.1 Adsorption-Coupled Flocculation 170
    10.4.2 Ultrasonic-Assisted Flocculation 172
    10.4.3 UV-Coupled Flocculation 172
10.5 Conclusion 174
Acknowledgments.. 174
References.. 174

10.1 INTRODUCTION

The presence of organic and inorganic pollutants in water, released by natural and anthropogenic activities, has become an important concern of this era. The release of harmful wastes like pesticides, textile dyes, chemicals, phenol derivatives and heavy metals has been reported to have harmful environmental impacts on aquatic ecosystems (Frinhani and Carvalho 2010; Orias and Perrodin 2013). These toxic compounds result in an increase in the biological oxygen demand of the receiving waters, thereby diminishing the level of dissolved oxygen and causing disturbance to the entire ecosystem. Changes in pH, formation of suspended solids, as well as the adherence of toxic substances (like heavy metals and pesticides) to the suspended particles have also been reported (Ghaly et al. 2014; Dasgupta et al. 2015). Some of the popular technologies available for decontamination and treatment of wastewater include chemical precipitation, ion exchangers, chemical oxidation/reduction, reverse osmosis, electrodialysis, ultrafiltration and adsorption. However, many of the available processes proposed cannot be used on an industrial scale due to technological drawbacks and other economic reasons. An industrial effluent treatment process must be designed accordingly to satisfy objectives such as lowering the levels of pollution or zero-level discharge. Generally, wastewater treatment involves the following steps, namely: (i) primary treatment step using mechanical, physical and chemical methods; (ii) secondary treatment using chemical or biological methods; and (iii) tertiary treatment to remove the remaining pollutants or the molecules produced during the secondary purification (e.g., the removal of salts produced by the mineralization of organic matter).

Flocculation is a frequently applied process in the primary purification of industrial wastewater (and in some cases in secondary and tertiary treatment) (Mishra et al. 2002; Ahmad et al. 2008). This technology not only removes the suspended matter but also binds ionic moieties, which further reduces the cost and energy requirements in wastewater treatment. This technology involves the dispersal of one or several agents which destabilize the colloidal particles, leading to the formation of micro-flocs. Bonding the micro-floc particles together through the addition of a flocculation additive causes formation of larger, denser flakes, which are more easily separated. So, a simple separation step eliminates the floc.

The coagulants and flocculants frequently used are mineral additives including metal salts such as alum, polyaluminum chloride and synthetic polymers such as polyacrylamide. Alum, an inorganic coagulant, has already been acknowledged for its application in traditional groundwater purification. But residual aluminum has been linked to serious health issues such as the development of Alzheimer's disease (AD) and senile dementia (Martyn et al. 1997; Lee 2000). Using these chemical substances may have environmental consequences, which include: (i) an increase in metal concentration in water affecting human health, (ii) excessive consumption, (iii) pH sensitivity, (iv) inefficiency towards fine particles, (v) production of large volumes of toxic sludge and (vi) dispersion of acrylamide oligomers, which may also create health hazards (neurotoxins). For these reasons, organic coagulants and flocculants have been considered for environmental applications (Swati and Govindan 2005; Usman et al. 2016).

Current research in the area of organic flocculation is mainly being pursued with the biopolymers sourced from (i) crustaceans, (ii) plants and (iii) microbes. Naturally occurring and commonly used bioflocculants include chitosan, dextran, clay, pullulan, psyllium, alginates, starch, plant gum, vegetable extracts, tannins, tamarind seed kernels and Moringa seed kernel extracts and date palm rachis extracts. Rising concerns of environmental pollution problems and health issues make utilization of bioflocculants derived from natural sources an important step towards sustainable environmental technology. This chapter offers deep insights on the preparation of organic flocculants/flocculant composites from natural sources, operational parameters to suit different industrial wastewater conditions and recent advancement in the area of flocculation technology viz. adsorption-flocculation, sonication-flocculation and UV-flocculation. Some of the applications of these flocculant/flocculant composites would include drinking water treatment, wastewater treatment, dye removal, algal cell harvest and recovery of adenovirus.

10.2 FLOCCULANTS FROM NATURAL SOURCES

10.2.1 Bio-sludge

Wastewater sludge is an important environmental issue, which is a matter of concern. Treatment and disposal of wastewater sludge represent approximately 50% of wastewater treatment costs. Recycling of produced sludge can be beneficial to prevent the environmental problems and to improve the value of sludge. However, in bio-sludge, the polymeric substances are tightly attached with microbial cells, and most of the functional groups are occupied by linking with cell membrane, and they exhibit low flocculation activity (Guang-Hui et al. 2009). Polymeric substances can be detached from the cell, making the functional groups free by physico-chemical means, and thereafter they can be collected as active bioflocculants. Research conducted on various extraction techniques has been mainly correlated to the composition, sludge dewatering ability and biofiltration efficiency. However, research is scant in the area of determination of flocculation efficiency. The extraction process includes (i) physical techniques like centrifugation and sonication in order to harvest sludge and reduce the floc size and (ii) chemical techniques including formaldehyde to fix the cells and prevent the cell lysis. The efficiency of extraction was determined by both individual and combined treatment of these methods. Figure 10.1 shows the effects of different extraction methods on flocculation efficiency.

Based on the efficiency of flocculation, a combined effect of formaldehyde with other physical techniques like sonication, heat treatment, NaOH and EDTA was quite significant in enhancing or affecting the flocculation efficiency (Nguyen et al. 2016). Formaldehyde plays a major role in reducing the colloidal effect of sonication, which enhances the overall flocculation activity of the flocculant. In the case of EDTA, a decrease in flocculation efficiency was observed due to the reaction with polymeric proteins on the membrane surface (Figure 10.1).

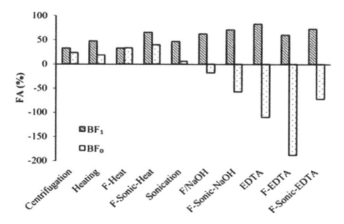

FIGURE 10.1 Effect of various extraction methods on flocculation activity of BFs (BF_1: purified BFs; BF_0: crude BFs) (Source: Nguyen et al. 2016).

10.2.2 Microbes as Sources of Bioflocculants

Microbes have commonly been a source of bioflocculants. In order to make the process more sustainable, major research has been focused on developing low-cost and sustainable enrichment media for growth of bioflocculant-producing microbes. Research was carried out on microbial flocculants, and its applications are given in Table 10.1. The extraction process has been both physical and chemical in nature. The bioflocculation activity is a function of carbon sources, nitrogen sources, duration of fermentation and extraction techniques. Rice straw has been used as a sustainable fermentation media for the growth of bioflocculant-producing microbes (Zhao et al. 2012). Optimal flocculation activity was noted after 100 days. Novelty of this research was the enhancement in flocculation activity (97.35%) by choosing an optimum ratio of bioflocculant-producing media to synthetic flocculant (5:1). The enhancement of flocculation activity was due to the consumption of the synthetic flocculant by microbes. A similar report is available on the use of palm oil mill effluent as a substrate for production of bioflocculant from *Aspergillus niger* (Aljuboori et al. 2014). Corn stover hydrolsates have been reported as a medium for bioflocculant production using seven bacterial species

TABLE 10.1
Bioflocculants from Microbial Sources

S. No.	Microbial Sources	Flocculation Activity (%)	Application	Reference
		Bacteria		
1.	*Cobetia sp. Bacillus sp.*	90.0	Wastewater, brewery, dairy and river water	Ugbenyen et al. (2015)
2.	*Cellulomonas sp. Streptomyces sp.*	91.0	Wastewater treatment	Nwodo et al. (2014)
3	*Rhodococcus erythropolis*	90.0	Remove disperse dye solutions	Peng et al. (2014)
4.	*Methylobacterium sp. Acetobacterium sp.*	95.0	Wastewater treatment	Luvuyo et al. (2013)
5.	*Chryseomonas luteola*	85.0%	Decolorized dye wastewater	Syafalni et al. (2012)
6.	*Klebsiella sp.* TG-1	84.0	Waste residue from food industry	Liu et al. (2013)
7.	*Bacillus sp.* Gilbert	91.0	Wastewater treatment	Nontembioso et al. (2011)
8.	*Azotobacter indicus*	90.0	Treated dairy, wool, starch, and sugar industries wastewater	Patil et al. (2011)
		Fungi		
9.	*Aspergillus niger*	86.2	Wastewater treatment	Aljuboori et al. (2014)
10.	*Aspergillus parasiticus*	92.4	Reactive Blue 4 dye removal	Deng et al. (2005)

isolated from activated sludge (Wang et al. 2013). The nitrogen source plays an important role in bioflocculant production. Acetonitrile has been reported as an important nitrogen source for the production of bioflocculant from *Klebsiella oxytoca*, and a high flocculation activity was noted towards the removal of heavy metals (Fan et al. 2019). Glutamic acid was found to be beneficial for an enhanced bioflocculant production. The purification was brought about by centrifugation and ethanol-based precipitation. The flocculation activity was noted as 76.8%. Food industry waste has also been used as a substrate for production of intracellular bioflocculant by *Klebsiella sp.* (Liu et al. 2013). The extraction process involved both physical and chemical techniques. Primary extraction of microbial bioflocculants was done by centrifugation and ultrasonication. Chemical pretreatment techniques, namely sterilization, alkaline-thermal and acid-thermal treatment of sludge, have been considered for growth of bioflocculant-producing microbes. Suspended solid dosage played an important role for the production of bioflocculant. The nature of bioflocculant was in the forms of broth, capsular and slime. Along with calcium and kaolin, these microbial bioflocculants played an important role towards the removal of COD, ammonium and turbidity via three mechanisms, namely: (i) binding to negatively charged kaolin particles, (ii) reduction of inter-particle distance between kaolin particles and (iii) bridging by bioflocculant to reduce the inter-particle distance between the kaolin-Ca complex. Therefore, microbial bioflocculants could be a sustainable source of wastewater treatment and sludge recycling.

10.2.3 Plants as Sources of Bioflocculants

Bioflocculants from plant sources have also been reported, which may be useful towards cost minimization during wastewater treatment (Table 10.2). A common plant, okra, has been used as a model bioflocculant production source due to the economic feasibility of scaling up to 220 T/year (Lee et al. 2018). The most convenient strategy to extract bioflocculant was reported as follows: use of vegetable washer → vegetable seed remover → vegetable slicer → conventional or microwave extractor → plate and frame filter (batch process) or hydrocyclone (continuous process) → forced circulation evaporator → rotary dryer as shown in Figure 10.2.

Temperature plays an important role in plant-based bioflocculant extraction. The effects of extraction conditions, including solvent types and extraction temperatures on the flocculation activity of the mucilage, water, acid, chelating and alkali fractions of pectin extracted from the cladodes of *Opuntia ficus-indica* (OFI), were evaluated. Among all, the alkali fraction of pectin extracted at a high temperature was found to exhibit a high flocculation efficiency (Belbahloul et al. 2015).

10.2.4 Biopolymers as Sources of Bioflocculants

Biopolymers have been used conventionally as a bridging and binding agent. Rapid research in extraction techniques is focused on a greener method to avoid chemicals, high temperatures and prolonged times. The ultrasonic-assisted extraction method has been a promising and eco-friendly technology for the extraction of bioflocculants

TABLE 10.2
Bioflocculants from Plant Sources

S. No.	Scientific Name	Common Name	Active Ingredients	References
1.	Abelmoschus esculentus	Okra	L-rhamnose, D-galactose, L-galacturonic acid	Lee et al. (2015)
2.	Plantago psilium	Psillium	L-arabinose, D-xylose, L-galacturonic acid	Mishra et al. (2002)
3.	Prosopis velutina	Mesquite seed gum	Galactomannan	Carpinteyro-Urban and Torres (2013)
4.	Moringa oleifera	Sajna	D-galactose	Miller et al. (2008)
5.	Opuntia ficus-indica	Cactus	D-xylose, L-galacturonic acid	Nharingo et al.(2015)
6.	Conicia indica	Kenaf	L-rhamnose, D-galactose	Varsha and Jay (2012)
7.	Cyamopsis tetragonoloba	Guar seed	D-galactose	Carpinteyro-Urban and Torres (2013)

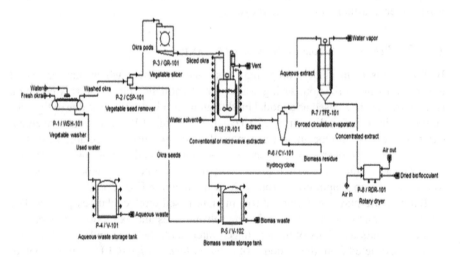

FIGURE 10.2 Process flowsheet for bioflocculant production from okra (Source: Lee et al. 2018).

from biopolymers. Alginate has been used extensively due to the flocculation and decolorization capability (Sand et al. 2010; Diaz-Barrea et al. 2014), owing to the presence of free hydroxyl and carboxyl groups distributed along with the polymer chain (Yang et al. 2014). Alkali is a chemical commonly chosen for extraction of alginate from seaweeds (Fawzy et al. 2017). Ultrasound-assisted extraction of alginate from *Sargassum muticum* was found to affect the structure, molar mass and thermo-rheological properties (Flórez-Fernández et al. 2019). Parameters affecting the ultrasound-assisted extraction process include pH, temperature, algae/water ratio, ultrasound power and duration. A major advantage of this process was the high recovery of biopolymers obtained per unit weight of the feed. Comparative studies on ultrasonic and chemical extraction of alginate from brown and red seaweed confirmed a higher recovery of the same from ultrasonic-assisted treatment as compared to chemical treatment (Mohammed et al. 2018). Formalin pollution and water wastage are two of the main problems in the alginate industry. In this case, application of the Fenton process was found to decrease the viscosity of alginate to a significant extent, thereby eliminating the need of formalin (Gao et al. 2018). This technique was also found to reduce the water consumption by 60%.

The characteristics of certain bioflocculants are affected by the extraction method chosen. In the case of chitosan extraction, chemical techniques increase the degree of deacetylation but result in a decrease in molecular weight. To overcome these problems, an alternative procedure based on enzymatic methods has attracted great attention recently because of its simple and eco-friendly nature. However, enzymatic extraction requires a lot of time, which may vary from several hours up to a few days. As an alternative, microbial fermentation can be considered as a favorable method for chitin preparation. But various parameters like carbon source concentration, pH fluctuations, inoculum dosage, fermentation time and temperature make this method more complex compared to chemical methods.

In recent years, microwave chemistry has received much attention, as it can speed up the reaction. Microwave heating, as an alternative to conventional techniques, has been proved to be more rapid and efficient for several chemical techniques. Microwave technology has been successfully used to convert chitin to chitosan in less time (Knidri et al. 2016). Ultrasound irradiation-assisted technique has resulted in chitosan, which is completely soluble in acid. These technologies are eco-friendly, efficient and would eliminate the need of chemicals for the extraction of bioflocculant.

10.3 PREPARATION OF FLOCCULANT COMPOSITES

The composite flocculation technology was introduced for addressing specific pollutants in wastewater and to decrease the dosage requirements in wastewater treatment. Composite flocculant consists of a synthetic polymer (as a bridging component) and an inorganic coagulant (as a binding component). In this context, a powdered composite composed of polymethacrylate as a coagulant and aluminum sulfate as an inorganic binding agent has been reported to exhibit a high clarification index (Kadooka et al. 2016). Titanium xerogel, along with poly DADMAC, was found to remove *E. coli* and turbidity (Wang et al. 2019). Similarly, a high cadmium removal efficiency was exhibited by a composite of anionic polyacrylamide and inorganic PAC (Zhao et al. 2018). A flocculation process using the composite polyaluminum chloride-epichlorohydrin dimethylamine (PAC-EPI-DMA) was employed for the treatment of an anionic azo dye (Reactive Brilliant Red K-2BP dye) (Yang et al. 2013).

Composite flocculants have recently been studied using bioflocculants. In the case of biopolymeric flocculants, the change in structure was correlated with the pollutant removal efficiencies. Removal of two typical antibiotics (norfloxacin and tylosin) at trace levels in turbid water simultaneously containing natural organic matters (NOMs) was conducted using two phenylalanine(Phe)-modified-chitosan flocculants (CHS-Phe and CHS-PPhe) with different molecular architectures (Du et al. 2018). It was reported that for hydrophilic and flexible CHS-Phe linear molecules, bridging and sweeping flocculation for antibiotics were enhanced by electrostatic attraction, π-electron-containing interaction and H-bond between introduced Phe groups and antibiotic molecules. However, for comb-like CHS-PPhe, excessively hydrophobic polyPhe branches aggregated to the coils. This phenomenon caused fewer exposed flocculation sites outwards, leading to poor flocculation efficiencies.

A novel amphoteric chemically bonded composite flocculant (carboxymethyl chitosan-*graft*-polyacrylamide, denoted as CMC-*g*-PAM) was evaluated for its flocculation efficiency in terms of the floc sizes. In acidic conditions, a flat configuration was favored when the polymeric flocculant was adsorbed onto the particle surface, leading to a slower initial floc growth rate, but larger and denser flocs were produced. Bridging was the dominant mechanism in neutral and alkaline conditions. A faster initial rate of bridging resulted in smaller and more open floc structures. A rearrangement process in neutral pH subsequently led to more compact flocs.

Bioflocculant oil emulsion composite is a new approach to this technology. A composite composed of Moringa protein extract oil has been used to harvest algae (Kandasamy et al. 2018). Under optimum conditions (MPOE dosage of 50 ml/L, pH 8, mixing time 4 min), a flocculation efficiency of 86% was noted. Hence, it can be concluded that flocculant composites can serve as simple, cost-effective and target-specific agents for wastewater treatment and cell harvesting.

10.4 TECHNOLOGICAL ADVANCES

10.4.1 Adsorption-Coupled Flocculation

Wastewater treatment using adsorption and flocculation techniques has been considered two different unit operations. However, the lacunae still exist with respect to (i) dosage requirements, (ii) leaching, (iii) sludge disposal and (iv) binding of pollutants. In the case of adsorbents, problems associated with pollutant leaching and disposal exist. In the case of flocculants, suspended solids are mainly removed, whereas ionic impurities are not addressed.

In order to address the limitations associated with these two technologies, clay as a natural mineral element has been used, owing to its effectiveness towards binding ionic pollutants and coagulating by varying the operational parameters. Application of clay can be a good option from the point of view of natural abundance, renewability and environmental sustainability. Due to their low cost and high sorption properties, clay materials are considered robust adsorbents. Natural clay minerals possess a high surface area, chemical and mechanical stability, and can serve as effective adsorbents and coagulation agents. Commonly reported clays include bentonite, montmorillonite, kaolinite and illite. In montmorillonite, the negative charges mainly arise from the isomorphous substitution of Al^{3+} by Mg^{2+} in the O sheet. Therefore, montmorillonite has good cation exchange capacity (CEC) and layer expansion capacity, thereby having the ability to reduce anionic moieties. Studies conducted using both KSF and K10 varieties modified with polymeric Al/Fe have been reported to reduce both suspended solids and chemical oxygen demand of sewage wastewater (Ramesh et al. 2007). Experiments conducted with clay suggested a relationship between porosity and ion exchange capacity, which was the possible mechanism for adsorption of basic dyes, namely basic green 5 (BG5) and basic violet 10 (BV10) (Sarma et al. 2016). Bentonite is a common variant of montmorillonite and has been used as an adsorbent for many toxic metals and dyes owing to its high cation exchange capacity (80–130 meq/100 g) and surface area (150–200 m^2/g) (Murray 2000). A combination of bentonite and kaolinite has also been reported to reduce the level of leaching.

Experiments conducted on polyaromatic hydrocarbons and heavy metals suggested that the presence of humic acid content enabled clays to exhibit a higher degree of pollutant binding (Zhou et al. 2005). Post adsorption, the coagulation/flocculation of clay suspensions has been brought about by biopolymers (Figure 10.3). Chitosan has primarily been used as a coagulation/flocculation agent (Feng et al. 2017). In the case of bentonite clays, the optimum dosage has been reported as 5 mg/L, which justified the process feasibility. Surface properties of chitosan if altered can be beneficial in the flocculation process. Polyacrylamide grafted chitosan has been reported to exhibit higher surface roughness and water solubility at a diverse pH (2.0–12.0) (Yand et al. 2013). These properties enable these grafted biopolymers to address high turbidity (clay suspensions) and ionic moieties effectively. The mechanism is primarily known as 'bridging'. The floc size formation is dependent on the shear rates. Experiments conducted on kaolin suggested that macro-flocs can be obtained under low-medium shear rates, whereas under high shear rates, micro-flocs could be obtained, which would be difficult to dewater (Wang et al. 2018).

Salinity is an important condition determining the flocculation efficiency of clay. Studies were conducted on kaolinite, bentonite and illite under saline

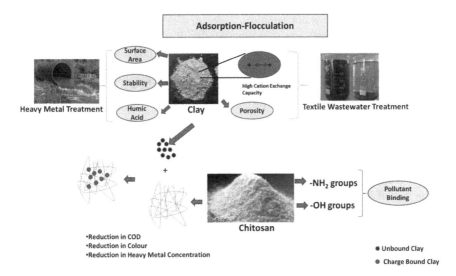

FIGURE 10.3 Clay and chitosan-based adsorption-flocculation.

conditions in the presence of an external polymer which suggested that the dewatering capacity depends on the clay structures and the ability to form the dense floc (water excluding) or loose floc (water trapping) (Ajao et al. 2018). Owing to the diverse structure of clay and the functional group of biopolymers, the remediation of wastewater can be performed using adsorption coupled with flocculation as a single unit operation, which could be beneficial in terms of cost-effective and energy-efficient technology.

10.4.2 Ultrasonic-Assisted Flocculation

Ultrasonic irradiation has been reported to initiate the polymerization of many different vinyl monomers through the generation of free radicals. Polymerization of styrene (Nagatomo et al. 2016), acrylonitrile (Selvaraj et al. 2015), methyl methacrylate (Marimuthu and Murugesan 2017), acrylamide and acrylic acid (Das et al. 2013; Jeon et al. 2018) monomers by ultrasound energy has been thoroughly investigated. Further studies have been carried out on cashew gum chemically grafted under ultrasound-assisted methods. Instrumental methods, namely FTIR, proton NMR and thermogravimetry, confirmed the presence of grafted acrylamide on the cashew gum surface. The flocculation ability of the grafted biopolymer was tested using a kaolin suspension. Based on the anionic character of kaolin particles in suspension and the protonation of the amide nitrogen group ($CONH^{3+}$) under acidic conditions, pH 3.5 was chosen for the flocculation procedure, where the turbidity of the kaolin suspension declined from 432 NTU to less than 10 NTU after 9 min of sedimentation (Klein et al. 2018). Ultrasonication and flocculation have also been used as two separate unit operations for sample preparation with the major aim of determining the level of polyaromatic hydrocarbon in water (Fan et al. 2007). Flocculation was performed using aluminum sulfate and flocculation aid florisil followed by ultrasonic and solid phase extraction. These three-unit operations could recover up to 94% of PAHs from the water samples.

Study on dye removal was performed under ultrasonication onto smectite clay. Dye removal up to 100% could be achieved during this process (Hamza et al. 2018). Thus, there is a scope of ultrasonication for the treatment of wastewater containing various pollutants and also in the preparation of grafted biopolymers, which would serve as natural bioflocculants (Figure 10.4).

10.4.3 UV-Coupled Flocculation

The integrated technology involving flocculation and UV treatment (for generation of radicals) has been mainly used for (i) COD reduction in wastewater and (ii) reduction of harmful microbes/viruses in water (Figure 10.5). Studies conducted on stabilized landfill leachate confirmed 75% reduction in COD by mineral-based flocculation, followed by ultraviolet-based sulfate radical oxidation processes. This integrated system could remove up to 91% COD content from wastewater (Ishak et al. 2018).

Similar work was reported on the treatment of biodiesel effluent using flocculation and UV-based degradation. The organic load drastically reduced after 6 hours

Organic Flocculation

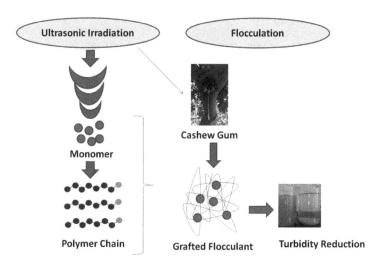

FIGURE 10.4 Ultrasonication-assisted flocculation.

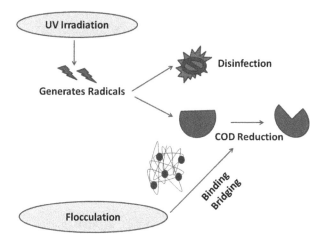

FIGURE 10.5 UV-assisted flocculation.

at an acidic pH, and further reduction in acute toxicity from 89% to 20% was noted, which justified the process as an efficient remediation technology to address organic load and toxicity. A two-step technique for addressing issues related to drinking water has been reported (Guerrero-Latorre et al. 2016).

Ultraviolet treatment followed by use of flocculation-chlorination tablets could successfully reduce the level of the hepatitis virus from drinking water samples. Results obtained were effectively found to follow the EPA norms (Costa et al. 2017). From the data available in the literature, it can be inferred that this ultraviolet treatment coupled with flocculation can not only effectively reduce the organic load but also play an important role in water disinfection and sanitation.

10.5 CONCLUSION

Bio-based flocculants are non-toxic, benign to the environment, fairly shear stable and display effective removal ability for turbidity, suspended solids, colors and dyes in larger dosages compared to organic synthetic flocculants. They can serve as an eco-friendly alternative to inorganic Al-based flocculants or toxic synthetic polymers. A sustainable approach has been adopted to produce and extract these bioflocculants from various natural sources. Sludge recycling for the growth of biopolymer-producing microbes is a strategy to produce bioflocculants and manage solid waste. The extraction process using microwave irradiation and ultrasonication eliminates the need to use toxic chemicals. Composite flocculants developed using natural biopolymers in an emulsion or grafted form have been found to exhibit better efficiency as compared to composites composed of synthetic polymers and inorganic salts. Recent advancement on the use of clay-based adsorption-coupled flocculation, ultrasonication-assisted flocculation and UV-coupled flocculation suggest that bio-based flocculants can play a vital role towards the development of sustainable environmental technologies for wastewater treatment.

ACKNOWLEDGMENTS

The author is grateful for the financial support provided by BIRAC (Biotechnology Ignition Grant, Call 11, partnered with IKP Knowledge Park, Hyderabad), DBT, Govt. of India (2017–2019).

REFERENCES

Ahmad, A.L., Wong, S.S., Teng, T.T., Zuhairi, A. 2008. Improvement of alum and PACl coagulation by polyacrylamides (PAMs) for the treatment of pulp and paper mill wastewater. *Chemical Engineering Journal* 137(3): 510–517.

Ajao, V., Bruning, H., Rijnaarts, H., Temmink, H. 2018. Natural flocculants from fresh and saline wastewater: Comparative properties and flocculation performances. *Chemical Engineering Journal* 349: 622–632.

Aljuboori, A.H.R., Uemura, Y., Osman, N.B., Yusup, S. 2014. Production of a bioflocculant from *Aspergillus niger* using palm oil mill effluent as carbon source. *Bioresource Technology* 171: 66–70.

Belbahloul, M., Zouhri, A., Anouar, A. 2015. Bioflocculants extraction from Cactaceae and their application in treatment of water and wastewater. *Journal of Water Process Engineering* 7: 306–313.

Carpinteyro-Urban, S., Torres, L.G. 2013. Use of response surface methodology in the optimization of coagulation–flocculation of wastewaters employing biopolymers. *International Journal of Environmental Research* 7(3): 717–726.

Costa, N.M., Silva, V.M., Damaceno, G., Sousa, R.M.F., Richter, E.M., Machado, A.E.H., Trovó, A.G. 2017. Integrating coagulation-flocculation and UV-C or H_2O_2/UV-C as alternatives for pre- or complete treatment of biodiesel effluents. *Journal of Environmental Management* 203(1): 229–236.

Das, R., Ghorai, S., Pal, S. 2013. Flocculation characteristics of polyacrylamide grafted hydroxypropyl methyl cellulose: An efficient biodegradable flocculant. *Chemical Engineering Journal* 229: 144–152.

Dasgupta, J., Sikder, J., Chakraborty, S., Curcio, S., Drioli, E. 2015. Remediation of textile effluents by membrane based treatment techniques: A state of the art review. *Journal of Environmental Management* 147: 55–72.

Deng, S., Yu, G., Ting, Y.P. 2005. Production of a bioflocculant by *Aspergillus parasiticus* and its application in dye removal. *Colloids and Surfaces, Part B: Biointerfaces* 44(4): 179–186.

Diaz-Barrera, A., Gutierrez, J., Martinez, F., Altamirano, C. 2014. Production of alginate by *Azotobacter vinelandii* grown at two bioreactor scales under oxygen-limited conditions. *Bioprocess and Biosystems Engineering* 37(6): 1133–1140.

Du, H., Yang, Z., Tian, Z., Huang, M., Yang, W., Zhang, L., Li, A. 2018. Enhanced removal of trace antibiotics from turbid water in the coexistence of natural organic matters using phenylalanine-modified-chitosan flocculants: Effect of flocculants molecular architectures. *Chemical Engineering Journal* 333: 310–319.

Fan, H.C., Yu, J., Chen, R.P., Yu, L. 2019. Preparation of a bioflocculant by using acetonitrile as sole nitrogen source and its application in heavy metals removal. *Journal of Hazardous Materials* 363: 242–247.

Fan, S.L., Zhao, L., Lin, J.M. 2007. Flocculation-ultrasonic assisted extraction and solid phase clean-up for determination of polycyclic aromatic hydrocarbons in water rich in colloidal particulate with high performance liquid chromatography and ultraviolet-fluorescence detection. *Talanta* 72(5): 1618–1624.

Fawzy, M.A., Gomaa, M., Hifney, A.F., Abdel-Gawad, K.M. 2017. Optimization of alginate alkaline extraction technology from *Sargassum latifolium* and its potential antioxidant and emulsifying properties. *Carbohydrate Polymers* 157(10): 1903–1912.

Feng, B., Peng, J., Zhu, X., Huang, W. 2017. The settling behavior of quartz using chitosan as flocculant. *Journal of Materials Research and Technology* 6(1): 71–76.

Flórez-Fernández, N., Domínguez, M., Torres, D. 2019. A green approach for alginate extraction from *Sargassum muticum* brown seaweed using ultrasound-assisted technique. *International Journal of Biological Macromolecules* 124: 451–459.

Frinhani, E.M.D., Carvalho, E.F. 2010. Monitoramento da qualidade das águas do Rio do Tigre, Joaçaba, SC. *Unoesc & Ciência – ACET* 1: 49–58.

Gao, F., Liu, X., Chen, W., Guo, W., Chen, L., Li, D. 2018. Hydroxyl radical pretreatment for low-viscosity sodium alginate production from brown seaweed. *Algal Research* 34: 191–197.

Ghaly, A.E., Ananthashankar, R., Alhattab, M., Ramakrishnan, V.V. 2014. Production, characterization and treatment of textile effluents: A critical review. *Journal of Chemical Engineering and Process Technology* 5: 182.

Guang-Hui, Y., Pin-Jing, H., Li-Ming, S. 2009. Characteristics of extracellular polymeric substances (EPS) fractions from excess sludges and their effects on bioflocculability. *Bioresource Technology* 100(13): 3193–3198.

Guerrero-Latorre, L., Gonzales-Gustavson, E., Hundesa, A., Sommer, R., Rosina, G. 2016. UV disinfection and flocculation-chlorination sachets to reduce hepatitis E virus in drinking water. *International Journal of Hygiene and Environmental Health* 219(4–5): 405–411.

Hamza, W., Dammak, N., Hadjltaief, H.B., Eloussaief, M., Benzina, M. 2018. Sono-assisted adsorption of crystal violet dye onto Tunisian smectite clay: Characterization, kinetics and adsorption isotherms. *Ecotoxicology and Environmental Safety* 163: 365–371.

Ishak, A.R., Hamid, F.S., Mohamad, S., Tay, K.S. 2018. Stabilized landfill leachate treatment by coagulation-flocculation coupled with UV-based sulfate radical oxidation process. *Waste Management* 76: 575–581.

Jeon, W.Y., Choi, Y.B., Kim, H.H. 2018. Ultrasonic synthesis and characterization of poly(acrylamide)-co-poly(vinylimidazole)@MWCNTs composite for use as an electrochemical material. *Ultrasonics Sonochemistry* 43: 73–79.

Kadooka, H., Jami, M.S., Tanaka, T., Iwata, M. 2016. Mechanism of clarification of colloidal suspension using composite dry powdered flocculant. *Journal of Water Process Engineering* 11: 32–38.

Kandasamy, G., Raehanah, S., Shaleh, M. 2018. Flotation removal of the microalga *Nannochloropsis* sp. using *Moringa* protein–oil emulsion: A novel green approach. *Bioresource Technology* 247: 327–331.

Klein, J.M., de Lima, V.S., da Feira, J.M.C., Camassola, M., Brandalise, R.N., Forte, M.M.C. 2018. Preparation of cashew gum-based flocculants by microwave- and ultrasound-assisted methods. *International Journal of Biological Macromolecules* 107(Part B): 1550–1558

Knidri, H.E., Khalfaouy, R.E., Laajeb, A., Addaou, A., Lahsini, A. 2016. Eco-friendly extraction and characterization of chitin and chitosan from the shrimp shell waste via microwave irradiation. *Progress Safety Environmental Protection* 104(A): 395–495.

Lee, C.S., Chong, M.F., Binner, E., Gomes, R., Robinson, J. 2018. Techno-economic assessment of scale-up of bio-flocculant extraction and production by using okra as biomass feedstock. *Chemical Engineering Research and Design* 132: 358–369.

Lee, R. 2000. Coagulation and flocculation in wastewater treatment. *Water Wastewater* 29: 141.

Liu, Z.Y., Hu, Z.Q., Wang, T., Chen, Y.Y., Zhang, J., Yu, J.R., Li, Y.L., Zhang, Y.F., Li, Y.L. 2013. Production of novel microbial flocculants by *Klebsiella* sp. TG-1 using waste residue from the food industry and its use in defecating the trona suspension. *Bioresource Technology* 139: 265–271.

Luvuyo, N., Nwodo, U.U., Mabinya, L.V., Okoh, A.I. 2013. Studies on bioflocculant production by a mixed culture of *Methylobacterium* sp. Obi and *Actinobacterium* sp. Mayor. *BMC Biotechnology* 13(1): 62.

Marimuthu, E., Murugesan, V. 2017. Influence of ultrasonic condition on phase transfer catalyzed radical polymerization of methyl methacrylatein two phase system – A kinetic study. *Ultrasonics Sonochemistry* 38: 560–569.

Martyn, C.N., Coggon, D.N., Inskip, II., Lacey, R.F., Young, W.F. 1997. Aluminum concentrations in drinking water and risk of Alzheimer's disease. *Epidemiology* 8(3): 281–286.

Miller, S., Fugate, E., Craver, V.O., Smith, J.A., Zimmerman, J.B. 2008. Toward understanding the efficacy and mechanism of *Opuntia* spp. as a natural coagulant for potential application in water treatment. *Environmental Science and Technology* 42(12): 4274–4279.

Mishra, M.A., Bajpai, M., Rajani, S., Mishra, R.P. 2002. *Plantago psyllium* mucilage for sewage and tannery effluent treatment. *Iranian Polymer Journal* 11: 381–386.

Mohammed, A., Bissoon, R., Bajnath, E., Mohammed, K., Lee, T., Bissram, M., John, N., Jalsa, N.K., Lee, K.Y., Ward, K. 2018. Multistage extraction and purification of waste *Sargassum natans* to produce sodium alginate: An optimization approach. *Carbohydrate Polymers* 198: 109–118.

Murray, H.H. 2000. Traditional and new applications for kaolin, smectite, and palygorskite: A general overview. *Applied Clay Science* 17(5-6): 207–221.

Nagatomo, D., Horie, T., Hongo, C., Ohmura, N. 2016. Effect of ultrasonic pretreatment on emulsion polymerization of styrene. *Ultrasonics Sonochemistry* 31: 337–341.

Nguyena, V.H., Klai, N., Nguyena, T.D., Tyagi, R.D. 2016. Impact of extraction methods on bio-flocculants recovered from backwashed sludge of bio-filtration unit. *Journal of Environmental Management* 180: 344–350.

Nharingo, T., Zivurawa, M.T., Guyo, U. 2015. Exploring the use of cactus *Opuntia ficus indica* in the biocoagulation–flocculation of Pb(II) ions from wastewaters. *International Journal of Environmental Science and Technology* 12(12): 3791–3802.

Nontembiso, P., Sekelwa, C., Leonard, M.V., Anthony, O.I. 2011. Assessment of bioflocculant production by *Bacillus* sp. Gilbert, a marine bacterium isolated from the bottom sediment of Algoa Bay. *Marine Drugs* 9(7): 1232–1242.

Nwodo, U.U., Green, E., Mabinya, L.V., Okaiyeto, K., Rumbold, K., Obi, L.C., Okoh, A.I. 2014. Bioflocculant production by a consortium of *Streptomyces* and *Cellulomonas* species and media optimization via surface response model. *Colloids Surfaces Biointerfaces* 116: 257–264.

Orias, F., Perrodin, Y. 2013. Characterisation of the ecotoxicity of hospital effluents: A review. *The Science of the Total Environment* 454 : 250–276.

Patil, S.V., Patil, C.D., Salunke, B.K., Rahul, B., Salunkhe, R.B., Bathe, G.A., Patil, D.M. 2011. Studies on characterization of bioflocculant exopolysaccharide of *Azotobacter indicus* and its potential for wastewater treatment. *Applied Biochemistry and Biotechnology* 163(4): 463–472.

Peng, L., Yang, C., Zeng, G., Wang, I., Dai, C., Long, Z. 2014. Characterization and application of bioflocculant produced by *Rhodococcus erythropolis* using sludge and livestock waste-water as a cheap culture media. *Applied Biochemistry and Biotechnology* 98(15): 6847–6858.

Ramesh, A., Hasegawa, H., Maki, T., Ueda, K. 2007. Adsorption of inorganic and organic arsenic from aqueous solutions by polymeric Al/Fe modified montmorillonite. *Separation and Purification Technology* 56: 90–100.

Sand, A., Yadav, M., Mishra, D.K., Behari, K. 2010. Modification of alginate by grafting of N-vinyl-2-pyrrolidone and studies of physicochemical properties in terms of swelling capacity, metal-ion uptake and flocculation. *Carbohydrate Polymers* 80(4): 1147–1154.

Sarma, G.K., Gupta, S.S., Bhattacharyya, K.G. 2016. Adsorption of crystal violet on raw and acid-treated montmorillonite, K10 in aqueous suspension. *Journal of Environmental Management* 171: 1–10.

Selvaraj, V., Sakthivel, P., Rajendran, V. 2015. Effect of ultrasound in the free radical polymerization of acrylonitrile under a new multi-site phase-transfer catalyst – A kinetic study. *Ultrasonics Sonochemistry* 22: 265–271.

Swati, M., Govindan, V.S. 2005. Coagulation studies on natural seed extracts. *Journal of Indian Water Works Association* 37(2): 145–149.

Syafalni, S., Abustan, I., Ismail, N., Kwan, T.S. 2012. Production of bioflocculant by *Chryseomonas luteola* and its application in dye water treatment. *Modern Applied Science* 6(5): 13.

Ugbenyen, A.M., Vine, N., Simonis, J.J., Basson, A.K., Okoh, A.I. 2015. Characterization of a bioflocculant produced from the consortium of three marine bacteria of the genera *Cobetia* and *Bacillus* and its application for wastewater treatment. *Journal of Water, Sanitation and Hygiene for Development* 5(1): 81–88.

Usman, B., Saidat, O.G., Abdulwahab, G. 2016. Enhancement of pumpkin seed coagulation efficiency using a natural polyelectrolyte coagulant aid. *International Journal of Chem Tech Research* 9(5): 781–793.

Varsha, P., Jay, P. 2012. Mucilage extract of *Cocinia indica* fruit as coagulant–flocculant for turbidity water treatment. *Asian Journal of Plant Science and Research* 2(4): 442–445.

Wang, Z., Nan, J., Ji, X., Yang, Y. 2018. Effect of the micro-flocculation stage on the flocculation/sedimentation process: The role of shear rate. *The Science of the Total Environment* 633: 1183–1191.

Wang, L., Ma, F., Lee, D.J., Wang, A., Ren, N. 2013. Bioflocculants from hydrolysates of corn stover using isolated strain *Ochrobactium ciceri* W2. *Bioresource Technology* 145: 259–263.

Wang, X., Gan, Y., Zhang, S. 2019. Improved resistance to organic matter load by compositing a cationic flocculant into the titanium xerogel coagulant. *Separation and Purification Technology* 211: 715–722.

Yang, C.H., Shih, M.C., Chiu, H.C., Huang, K.S. 2014. Magnetic *Pycnoporus sanguineus*-loaded alginate composite beads for removing dye from aqueous solutions. *Molecules* 19(6): 8276–8288.

Yang, Z., Liu, X., Gao, B., Zhao, S., Wang, Y., Yue, Q., Li, Q. 2013. Flocculation kinetics and floc characteristics of dye wastewater by polyferric chloride–poly-epichlorohydrin–dimethylamine composite flocculant. *Separation and Purification Technology* 118: 583–590.

Yang, Z., Yang, H., Jiang, Z., Cai, T., Li, H., Li, H., Li, A., Cheng, R. 2013. Flocculation of both anionic and cationic dyes in aqueous solutions by the amphoteric grafting flocculant carboxymethyl chitosan-*graft*-polyacrylamide. *Journal of Hazardous Materials* 254–245: 36–45.

Zhao, C., Shao, S., Zhou, Y., Yang, Y., Shao, Y., Zhang, L., Zhou, Y., Xie, L., Luo, L. 2018. Optimization of flocculation conditions for soluble cadmium removal using the composite flocculant of green anion polyacrylamide and PAC by response surface methodology. *The Science of the Total Environment* 645: 267–276.

Zhao, G., Ma, F., Wei, L., Hong, C. 2012. Using rice straw fermentation liquor to produce bioflocculants during an anaerobic dry fermentation process. *Bioresource Technology* 113: 83–88.

Zhou, P., Yan, H., Gu, B. 2005. Competitive complexation of metal ions with humic substances. *Chemosphere* 58(10): 1327–1337.

11 Power Production in Microbial Fuel Cells (MFC)
Recent Progress and Future Scope

E. Elakkiya and Subramaniapillai Niju

CONTENTS

11.1	Introduction	180
11.2	Energy Crisis in India	181
11.3	Waste to Energy Options in India	182
11.4	Background of Microbial Fuel Cells (MFC)	182
	11.4.1 Extra Cellular Electron Transfer by Bacteria	183
11.5	Factors Affecting MFC Performance	184
	11.5.1 Operating Factors	185
	11.5.1.1 pH of the System	185
	11.5.1.2 Organic Loading Rate	185
	11.5.1.3 Hydraulic Retention Time	185
	11.5.1.4 Electrochemically Active Biofilm Formation	186
	11.5.1.5 Temperature	186
	11.5.2 MFC Material Properties	186
	11.5.2.1 Anode	186
	11.5.2.2 Cathode	188
	11.5.2.3 Separator	190
	11.5.3 Number of Chambers	190
11.6	Large-Scale MFC Architecture	191
	11.6.1 Tubular MFC	192
	11.6.2 Stacked MFC	193
	11.6.3 Separate Electrode Modules	193
11.7	Power Harvesting in MFC	194
11.8	Real-Time MFC Testing	195
11.9	Future and Scope	198
11.10	Conclusion	198
References		198

11.1 INTRODUCTION

Industrialization has widely contributed to the economic development of independent India. Industrialization has generated employment for the rural population, aiding urbanization. Though these changes have positively influenced the economy of the nation, they have had deleterious effects on the environment. Industries pollute the environment and the urban population produces large amounts of municipal waste. In order to protect the existing natural resources and to curb pollution, the Indian government has formulated various stringent legislatures, forcing industries to treat waste before disposal. It has formed a statutory body, Central Pollution Control Board (CPCB), to monitor and control industries from polluting the environment (CPCB 2017). Municipalities have also set up sewage treatment plants and proper solid waste disposal to protect nearby water bodies. To comply with discharge standards prescribed by the CPCB, the industries have installed wastewater treatment plants (WWTP) with the main focus of reducing pollution without considering the energy input required.

The current global energy crisis has increased crude oil prices. India as a developing nation is finding it difficult to meet the energy demand of the world's second largest population. It meets nearly 33% of energy requirements from the import of raw materials for power generation (MNRE). The current energy scenario has compelled the country to install and operate energy-efficient WWTP without compromising the effluent quality. Energy-efficient WWTP plants necessitate both reductions in energy input and energy production in waste management processes (Gu et al. 2017).

The current energy recovery from wastewater is by anaerobic treatment of wastewater to produce biogas with energy content. The biogas produced is combusted and the energy is utilized in combined heat and power (CHP) turbines (Chinnaraj & Venkoba Rao 2006). The conversion from chemical to thermal and later to electrical energy leads to conversion losses. Microbial fuel cells (MFC) are a robust technology of directly producing electrical energy from wastewater (Rabaey & Verstraete 2005). MFC is based on the principle of bio-electrogenesis: electricity production by bacteria (Venkata Mohan et al. 2014). The increase in publications on MFC in the past decade from India highlights the scope for MFC applications in India (Patel et al. 2017).

Current trends in MFC studies are based on laboratory-scale experiments. Large-scale MFC has not found its ground in India yet. A detailed analysis of operating MFC performance in laboratory-scale reactors helps in developing better designs of large-scale reactors. MFC power production is affected by physical, chemical and biological factors (Venkata Mohan et al. 2014). The MFC configuration and materials used for construction determine the internal resistance of the cells and hence affects the MFC performance (Oliveira et al. 2013). Basically, MFC consists of three important physical components: (i) the anode, which is the site of bacterial growth and substrate oxidation; it also plays a crucial role in electron collection; (ii) the cathode is the region where oxygen reduction reaction occurs; and (iii) the membrane, with the primary function of separating the anode and cathode reactions. Enormous research on material synthesis and modification has been carried out to improve electron transfer to the anode and to improve oxygen reduction reaction in the cathode (S. Li et al. 2017). Efforts have been made to review and consolidate the data available on anode and cathode materials (Zhou et al. 2011), but very limited research and review are available for membranes used in MFC. Contradictorily, membrane is the main component contributing to MFC material cost. Hence, in the following section an attempt has been made to analyze the role of membranes or separators in MFC and the improvements made to reduce cost and increase performance. The recent interest in MFC with capacitive bio-anodes enabling production and storage of electric current (Santoro et al. 2016) has been reviewed briefly. Operating factors, such as pH, which controls both the electrochemical and biological reactions, and other factors, such as concentration of organics in wastewater, shear stress on biofilm formed and ambient temperature, play important roles in determining power production in MFC has been highlighted. Figure 11.1 illustrates the factors affecting MFC performance. In a detailed analysis of these factors is the initial step to understanding MFC functioning and the key to developing efficient large-scale MFC reactors.

11.2 ENERGY CRISIS IN INDIA

Energy is the basis for economic development of a nation; the self-sufficiency in energy resources and generation capacity are critical factors influencing a nation's overall growth. Coal is the major source of energy in India. Nearly 57.3% of power production is from coal, and the rest is met by oil, gas, hydro, nuclear and renewable energy. The Government of India has formulated numerous policies to increase the share of renewable sources to grid power. Currently, 20.1% of power is produced from renewable energy sources (Ministry of New and Renewable Energy

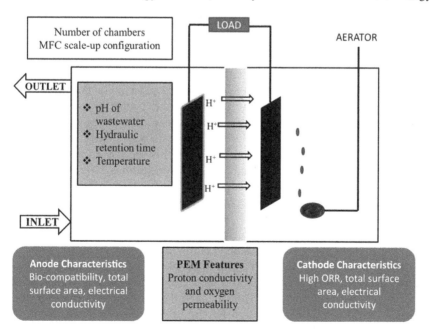

FIGURE 11.1 Factors influencing MFC performance.

Annual Report 2015–2016). The second major issue with regard to power in India is transmission and distribution losses. The power production units are centralized, and hence electrical energy is transmitted over larger distances to the end users. Decentralization of power production could considerably reduce power losses (Decentralised Systems n.d.). Biomass and waste-based power is one of the viable opportunities available for decentralizing power production.

11.3 WASTE TO ENERGY OPTIONS IN INDIA

As the second largest populated country in the world with rising urbanization, India generates enormous amounts of waste. Energy generation from waste is a sustainable solution, as it addresses two important issues of environmental protection and the energy crisis (Waste to Energy Potential in India n.d.). Indian cities produce large amounts of wastewater, of which only 10% is treated. The rest is let untreated into the environment (Kazmi et al. 2015). Biogas production via anaerobic treatment of wastewater is the current available option for energy recovery. The biogas generated has to be further converted to electrical energy before it can be utilized. The conversion of biogas to electricity requires infrastructure installation, which further increases the cost of energy production. MFC converts the organic compounds in wastewater to electricity, directly reducing the conversion losses, and reduces the infrastructure installation. In the following section, we will further look into the workings of MFC and its application in power recovery from waste.

11.4 BACKGROUND OF MICROBIAL FUEL CELLS (MFC)

In 1911, Potter first demonstrated that bacteria are capable of producing electric current (Venkata Mohan et al. 2014). Though practical developments have been made since then, only after synthetic mediators for electron transfer were used was analytical MFC developed. Later, the usage of synthetic mediators was discouraged due to its toxicity to microbial growth. In the initial years, MFC studies were carried out using pure substrate; later it was diverted to use wastewater as substrate, as it could contain waste disposal along with power production (Pandey et al. 2016). Bio-electrogenesis is the background of power production in MFC, which has been briefed in this section. In nature, bacteria metabolize the substrate to produce electrons and protons. The electrons produced pass through a cascade of redox molecules to reach a terminal electron acceptor (TEA), by which it derives energy in the form of adenosine triphosphate (ATP). Oxygen is usually used as a terminal electron acceptor by aerobic bacteria. Anaerobic bacteria are capable of utilizing other electron acceptors, such as nitrates and sulfates (Venkata Mohan et al. 2014). Most TEA are soluble and can diffuse bacterial cells. In MFC, these electrons are to be transferred to an electron acceptor outside the bacterial cell. Exoelectrogens are groups of bacteria capable of transferring electrons outside the cell, and this phenomenon is known as extra-cellular electron transfer (EET). Exoelectrogens are used in the anode chambers of MFC; these bacteria utilize the anode as a TEA and transfer the electrons to the anode (Logan 2009).

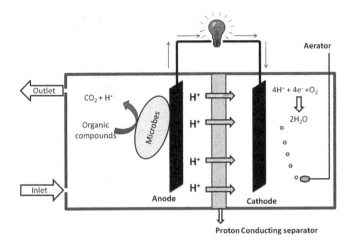

FIGURE 11.2 MFC operation and reactions occurring in anode and cathode chambers.

In a typical MFC operating with wastewater as a substrate, the anode chamber is filled with wastewater and is maintained in anaerobic conditions. Refer to Figure 11.2. Exoelectrogens utilize the substrate in wastewater to produce electrons and protons. The electrons are directly transferred to anode by exoelectrogens. The electrons then pass through external electrical circuits to reach the cathode, producing electric current. The cathode chamber is aerated, and a separator is used to separate anode and cathode reactions. The protons pass through the separator to reach the cathode chamber. In the cathode, the electrons, protons and oxygen combine to produce water. In the anode compartment, oxidation of substrate occurs, and in the cathode chamber, oxygen is reduced (Logan et al. 2006; Aelterman et al. 2006; Rabaey & Verstraete 2005).

11.4.1 Extra Cellular Electron Transfer by Bacteria

Exoelectrogens are capable of extracellular electron transfer without the addition of artificial mediators. Direct electron transfer by conductive pili and redox active proteins present on the cell surface, and mediated transfer via redox shuttles present in the wastewater or metabolites produced by bacteria, are two mechanisms by which bacteria transfer electrons to anodes (Kumar et al. 2015).

Direct electron transfer happens by passage of electrons through conductive pili and membrane-bound proteins without involving any diffused redox mediators. Figure 11.3 shows the direct electron transfer between bacteria and anode (Logan 2009). Electron transfer by membrane-bound proteins occurs only in bacteria having direct physical contact with the anode. In thicker biofilm, as the distance between bacteria and the anode surface is high, the conductive pili plays a prominent role in electron transfer (Mohan, Mohanakrishna et al. 2008).

Mediated electron transfer occurs via metabolites produced by bacteria which act as redox shuttles. The electron shuttles may be redox compounds present in the wastewater. Redox shuttles are present in the oxidized state inside the bacterial

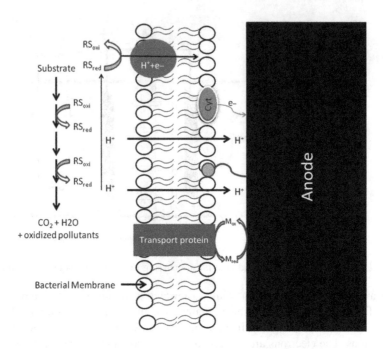

FIGURE 11.3 Bio-electrogenesis and EET in the anode chamber of MFC.

membrane, which get readily reduced and move across the membrane to reach the anode, transferring the electrons to anodes returning to the oxidized state. In the initial studies, artificial mediators were used, which was later discouraged due to their toxicity to microbial growth (Kumar et al. 2015). Phenazines and quinines are groups of molecules segregated by bacteria and are capable of mediating electron transfer (Jayaprakash et al. 2016). The redox mediators secreted by one organism can be utilized by other organisms (Venkata Mohan et al. 2014).

11.5 FACTORS AFFECTING MFC PERFORMANCE

MFC is a complicated system where electrochemical and biological reactions occur at the same time in a single reactor. An intensive study on the reactor design, operation, long-term performance and material stability are essential for maintaining sustainability and efficiency of large-scale systems (Oliveira et al. 2013; Janicek et al. 2014). Studies on lab-scale MFC concentrate on optimizing the factors affecting MFC power production and material modifications to improve power production. Factors affecting power generation have been widely studied, and they include biological factors (biofilm formed, the substrate utilization rate) (Venkata Mohan et al. 2014), operating factors (pH of anolyte and catholyte, temperature and hydraulic retention time), mass transfer phenomenon occurring inside the reactor, materials used in MFC construction and its stability and performance on long-term operations (Oliveira et al. 2013). An elaborate study of these factors is essential to design MFC with improved power production and enhance treatment of wastewater.

In the following section we will be reviewing the factors affecting power generation improvements made and the future scope of MFC in wastewater treatment facilities.

11.5.1 Operating Factors

11.5.1.1 pH of the System

pH plays a crucial role in determining the anodic and cathodic reactions. Microbial metabolism and growth is highly dependent on the anolyte/wastewater pH. It has been widely accepted that near neutral pH is favorable for microbial growth in the anode chamber, but at this pH, competitive growth of methanogens reduces coulombic efficiency and power production (Oliveira et al. 2013). Operation at acidic pH has increased the coulumbic efficiency (CE), as it reduces the growth of methanogens (Raghavulu et al. 2009). In the cathode chamber, oxygen reduction reactions occur constantly, hence H^+ ions are continuously consumed and the pH of catholyte increases, whereas in the anode compartment, the microbes are continuously producing protons, reducing the pH of wastewater (Oliveira et al. 2013). Hence, a pH gradient is found to occur between the two compartments. To reduce the effect of pH gradient, a variety of buffer systems have been tested. In particular, phosphate buffer has been widely used to maintain the pH of wastewater and catholyte in MFC (Mohan, Saravanan et al. 2008); the cost of phosphate buffer reduces the feasibility of implementing it in large-scale systems. Bicarbonate buffer systems (Fan et al. 2015) and zwitterion (Nam et al. 2010) have been experimented with to maintain pH in MFC, but the usage of buffer systems increased the cost of operation of MFC. Hence, CO_2 addition to the cathode was made to produce carbonic acid. This process was economically favorable, as CO_2 produced in various industrial processes could be used. Another approach was by inverting cell polarity; switching between anodes and cathodes can reduce the pH gradient, and the anodic microbes were capable of carrying out oxygen reduction reaction (ORR), but more analysis on long-term stability is required (Cheng et al. 2010).

11.5.1.2 Organic Loading Rate

The concentration of the substrate or pollutants is a key factor influencing microbial growth and decides the microbial metabolism occurring inside the MFC reactor. In a MFC system utilizing wastewater as substrate, the organic loading rate (OLR) determines the substrate concentration. At high loading the power production is high, but reduction in coulombic efficiency and waste treatment was found (Oliveira et al. 2013). The reduction in coulombic efficiency was due to the competitive growth of methanogens at high substrate concentrations. Hence it is important to optimize OLR of a reactor to attain higher power and waste treatment (Martin et al. 2010; Velvizhi & Mohan 2012).

11.5.1.3 Hydraulic Retention Time

In a continuous system, the hydraulic retention time (HRT) determines the substrate concentration and shear stress provided to anode biofilm. At higher HRT the power generation increased but the chemical oxygen demand (COD) removal efficiency and coulombic efficiency was found to be lower (Oliveira et al. 2013). At higher flow rate, there is higher shear stress to the biofilm formed over the anode, and hence it results in thinner and homogenous biofilm, capable of direct electron transfer (Juang et al. 2011). It is important to optimize the HRT based on both required effluent quality and power production.

11.5.1.4 Electrochemically Active Biofilm Formation

The biofilm formed over the anode surface is crucial in determining the rate of electron transfer to anode. Various experiments have been carried out to increase exoelectrogens on anode surfaces. Usually biofilm formation is made by supplying anolyte/wastewater in batch, fed-batch or continuous modes. In batch and fed-batch modes, fresh anolyte is supplied once the electrical output falls below the threshold level. In a continuous mode, the electrolyte environment is more stable, and constant flow of anolyte provides shear stress for biofilm formation (Santoro et al. 2017). As stated earlier, at higher shear stress the film formed is thinner and consists of highly exoelectrogenic bacteria (Rickard et al. 2004). Anode potential plays an important role in the bio-catalyst formed and the electron transfer mechanism. Poised anode potential or whole cell potential enhances exoelectrogenic bacteria populations in the biofilm. Positive anode potential increases attachment of negatively charged bacteria, and negatively charged anodes are used in the production of products (Venkata Mohan et al. 2014).

11.5.1.5 Temperature

Temperature plays an important role in determining MFC performance, as it affects the system kinetics, mass transfer, thermodynamic factors and nature of the microbial community formed. Temperature determines the microbial growth and hence substrate utilization, hence it is important to maintain optimum temperature for maximum power generation (Oliveira et al. 2013). As a tropical country, the temperature in India is moderate, and hence MFC operation throughout the year is feasible.

11.5.2 MFC Material Properties

11.5.2.1 Anode

The anode surface is the point of contact for exoelectrogens and anode/MFC. Anode materials must pose certain features to improve interactions between the bacterial biofilm and material surface (Santoro et al. 2017). The most important features are (i) high electrical conductivity, (ii) good mechanical strength and stability on long-term operations, (iii) high surface area, aiding bacteria attachment, (iv) biocompatibility, and (v) low-cost and environmentally friendly (S. Li et al. 2017).

Carbonaceous and metallic materials have been tested and studied in MFC. The carbonaceous electrode used in MFC includes graphite rod, graphite plate, carbon rod, carbon mesh, carbon veil, carbon paper, carbon cloth, carbon brush, granular graphite, granular activated carbon and reticulated vitreous carbon, which are commercially available carbon materials (Wei et al. 2011; Hernández-Fernández et al. 2015). Metal-based materials include stainless steel plate, stainless steel mesh, stainless steel scrubber, silver sheet, nickel and copper sheet (Santoro et al. 2017). Carbon cloth is a widely-used carbon material in MFC due its high surface area, porosity, electrical conductivity and flexibility. Carbon brush is of high interest due to reduced inter-electrode spacing in MFC. The drawback in carbon brush is the cost of the titanium wire present in the center, and recent work is being aimed at lowering the overall cost. Granular activated carbon (GAC) is used due to its bio-compatibility and is highly porous in nature; the important drawback is the reduction in electrical conductivity. In most of the situations, GAC is used as packing material and not as standalone material; usually carbon rods are used as current collectors (S. Li et al. 2017). Though metal anodes pose high electrical conductivity, they fail in biocompatibility. The usage of copper and nickel anodes in MFC releases ions which are toxic to microbes in the anode chamber (Santoro et al. 2017). Hence, carbonaceous anodes are preferred over metallic ones.

Various surface modifications have been made, aiming to (i) increase bacterial attachment by increasing surface charge, (ii) increase the presence of oxygen and nitrogen functional groups to facilitate bacterial attachment (Wei et al. 2011), (iii) improve hydrophilicity (Yang et al. 2016) and (iv) immobilize mediators. The surface modifications include chemical treatments, coating doping, thermal and electrochemical treatment.

Recently, researchers have shifted their focus on improving capacitive behavior of anode used in MFC in order to convert MFC to power producing and storing devices. Most of the materials used in improving anode performance, such as carbon nano-tubes, graphene, graphite particles, carbon-nanofibers, conductive polymers, metallic oxides and composite materials, also have wide applications in supercapacitors. Deeke et al. (2012) initially experimented with the power performance and capacitive behavior of anode materials coated with capacitive materials in MFC. In their study, they used carbon black mixed with NMP (n-methyl-2-pyrrolidone) and PVDF (polyvinylidene fluoride) as capacitive materials, and the materials were coated in graphite plate anode to 0.5 mm thickness. The capacitive behavior of this kind of MFC is illustrated in Figure 11.4. It was found that the MFC with capacitive anode performed better, and it was due to the rough surface of the anodes coated with capacitive material; the charge storage was also better.

The research was extended in their next study (Deeke et al. 2013) of the influence of thickness of the capacitive layer on MFC power production and capacitance. It was found that as the thickness was reduced, the performance improved due to the increase in specific surface area and reduced charge transfer resistance. Later, Feng et al. (2014) used graphene and polypyrrole (PPy) as capacitive materials with the base material of graphite felt. A remarkable increase in power was produced, and this type of MFC could be the future of energy storage technologies. Y. Wang et al. (2016) used electrodeposited MnO_2 on carbon felt and studied the capacitive behavior, along with its influence on cathodic reaction. The hexa-valent chromium reduction was studied in a cathode chamber of MFC. It was found that the MnO_2 electrodeposited anodes performed better and also improved the cathodic reaction of MFC. The improved cathodic reaction was due to increased electrons available for chromium reduction. A summary of these studies is available in Table 11.1. Santoro, Soavi, Serov et al. (2016) have used enzyme-based cathodes and additional capacitive cathodes where the MFC system could perform on par with supercapacitors, opening up a new field of MFC based supercapacitors. Most of these studies are on lab-scale reactors; the scale-up and its effects on performance has to be further studied.

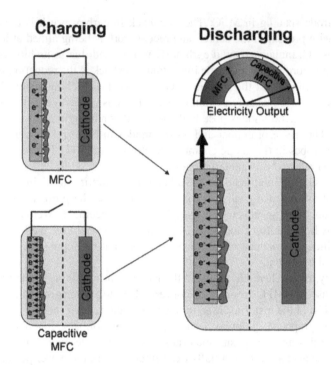

FIGURE 11.4 Internal capacitor in MFC (adapted from Deeke, Sleutels, Hamelers & Buisman, 2012).

11.5.2.2 Cathode

Cathode is one of the major components determining the MFC performance; the oxygen reduction reaction (ORR) occurring at the cathode is considered to be the rate limiting step in power production of MFC. In the cathode, at acidic condition, $2e^-$ combines with $2H^+$ and O_2 to produce H_2O_2, which is later reduced to H_2O, whereas in alkaline conditions the OH^- ions are reduced by $4\ e^-$ to produce H_2O (Santoro et al. 2017). Though in most chemical fuel cells the acid or alkaline conditions are preferred, in biotic fuel cells or MFC, neutral conditions are exploited. As the ORR reaction occurs at a low rate, a catalyst is required to improve the reaction rate. Commonly used cathode catalysts are platinum, platinum group metals, earth-abundant metals supported on carbon materials, or carbon material itself. The commonly used cathode carbon materials are graphite rod, plate, carbon cloth, paper and activated granular carbon. Utilization of platinum-free catalyst reduced the overall cost, and it includes utilization of metals such as manganese, cobalt, nickel, iron and copper (S. Li et al. 2017). Though Pt-based catalyst performed well in the initial period, its performance subsided in long-term operations. Reduction in performance was due to fouling of the cathode. Fouling was also found on other carbon cathodes, but cleaning could restore 90% of MFC performance in carbon-based material, whereas Pt-based catalyst could not be revived (Santoro et al. 2017). Granular activated carbon is regarded as the best cathode material.

TABLE 11.1
Capacitive Anodes in MFC

S. No	MFC Type	Electrode Material	Capacitive Material	Substrate	Cycle Time	Power Performance	Inferences
1	Dual chambered	Plain graphite plate	Carbon black + polymer (NMP + PVDF)	Anolyte 10mM acetate and phosphate buffer Catholyte 100mM ferricyanide	10 mins charging 20 mins discharge	0.79 A/m^2 for non-capacitive anode and 1.02 A/m^2 for capacitive anode	Higher time for discharge High surface roughness Internal capacitor improved performance
2	Dual chambered	Plain graphite plate	Carbon black + polymer (NMP + PVDF) 0.2, 0.5 and 1.5 mm of polymer thickness studied	Anolyte 10mM acetate and phosphate buffer Catholyte 100mM ferricyanide	—	40% higher performance than non-capacitive cathodes	Lower thickness, better performance Due to high specific surface area and reduced charge transfer resistance
3	Dual Chambered	Graphite felt	Graphene + PPy	Anolyte lactate growth medium Catholyte phosphate buffer	—	—	Remarkable increase in power performance High energy storage
4	Dual chambered	Carbon felt	Electrodeposition MnO$_2$	Anolyte – growth medium Catholyte- potassium dichromate	40 min discharge	35 times better	MnO$_2$ showed capacitive behavior Improved reduction reaction in cathode

It is commonly mixed with a polymer such as PTFE and rolled or pressed against the current collectors. External pressure provided to press GAC is more effective, and thermal treatment further increased the MFC performance (Wei et al. 2011; Santoro et al. 2017).

11.5.2.3 Separator

The separator placed between the anode and cathode plays a crucial role in MFC performance. It influences the mass transfer phenomenon, ohmic losses, and power generation. Functions of the separator in MFC include the separation of anodic oxidation and the cathodic reduction reaction. It should minimize the passage of substrate from the anode to cathode chamber and should poses low oxygen permeability (W. Li et al. 2011). Nafion is a persulfonated membrane which has been extensively used in MFC, has the highest proton exchange capacity and low oxygen permeability, confers low resistance to the cell and increases the total power output. Major disadvantages of the utilization of Nafion for MFC are the cost of materials and bio-fouling. Cation exchange membranes (CEMs) are popularly known as proton exchange membranes in MFC, and Nafion is widely used in MFC. CEM usually suffers from pH splitting caused by competition of other cations with protons. The pH of the anode reduces gradually, affecting the electrochemical and biological reactions on the anode side. This phenomenon is known as pH splitting and is a major drawback in power generation. AEM are anion exchange membranes and can be used in proton transfer using phosphate or carbonate buffer as a carrier. The cost of phosphate buffer is high, and hence its application in practical MFC operations is difficult; though carbonate buffer is cheap it has not been studied widely. Bipolar membrane is widely used in microbial desalination rather than MFC. Size selective separators including micro-filtration membranes, ultrafiltration membranes, porous glass fiber, nylon mesh and glass fiber have been used as separators. Micro porous membranes are efficient in separating anodic and cathodic reactions but are not efficient in reducing diffusion of oxygen and substrate between chambers. J-cloth can be effectively used to improve proton exchange, but the growth of bacteria drastically reduced the performance as a separator (W. Li et al. 2011).

Ceramic membranes are the new arena in MFC separators. The cost of membranes used in MFC separators contributes to the installation of MFC; ceramic membranes are cheap and provide natural environments for bacteria in MFC. Studies reveal that ceramic membranes show efficiency on par with ionic membranes. The common ceramic materials used in MFC studies are earthen pots, titanium oxide terracotta, earthenwares, clayware, red soil, black soil, goethite, pyrophyllite and mullite. Pasternak et al. (2016) and Winfield et al. (2016) provide a broader review on ceramic membranes used in MFC and their performance.

11.5.3 NUMBER OF CHAMBERS

A typical MFC consist of two chambers separated by independent membranes. Initial work with MFC exploited dual-chambered MFC with water as a catholyte, as ORR occurring at the cathode was very low aeration of water; utilization of ferricyanide, permanganate and phosphate buffers were tested (Venkata Mohan et al. 2014). Though the power improved, the chemical catholytes used were highly toxic and not sustainable. Hence, research shifted to air-cathode MFC, where metal catalyst was used to improve ORR. Air cathodes without synthetic mediators are widely used in MFC studies due to easy scale-up and sustainability. The use of air cathodes removes the requirement of costly CEM, and cheap separators could be used. The recent development of utilization of algae in the bio-cathode has given dual-chambered MFC a rebirth. The algal MFC is regarded as biorefineries (Gajda et al. 2015). Triple-chambered MFC with single a cathode and two anodes is less explored and has the potential to generate higher power than dual-chambered MFC (Samsudeen et al. 2014).

11.6 LARGE-SCALE MFC ARCHITECTURE

In the past decade, enormous research has been made on lab-scale MFC, studying the factors influencing power generation and the type of wastewater that can be used as substrate. These studies highlight that MFC is capable of utilizing a wide range of wastewater from easily biodegradable food processing industry to complex pharmaceutical and crude oil refining wastewater (Pandey et al. 2016). The combination of biological and electrochemical processes enhances the treatment efficiency in MFC (W.-W. Li et al. 2013). Though MFC poses numerous advantages over other existing wastewater treatment technologies, the low power production and high material cost are the limiting factors in regard to its large-scale application.

It is essential for researchers to gradually scale-up MFC from the lab-scale to larger systems. In studies pertaining to the direct scaling up of lab-scale compartment-type reactors to larger volumes, low power output was observed. The reduction was due to the reduced electrode area to volume of ratio (Oliveira et al. 2013; Janicek et al. 2014). Hence it is important to maintain the electrode area to volume of reactor optimum while scaling-up the system. MFC reactor design is crucial in large-scale operations, and scientists have been trying various configurations to improve energy recovery.

Large-scale MFC design includes: (i) tubular MFC, consisting of concentric tubes of anode and cathode; (ii) stacked MFC, consisting of plate and frame type anode and cathode chambers; and (iii) modularized separate electrodes in a single compartment. In the following section, we will have a brief discussion of these designs and the large-scale system studied until now (Janicek et al. 2014).

The basic requirements of large-scale MFC for high efficiency and performance are:

  (i) Capability to handle large volume of wastewater
 (ii) Low internal resistance to attain high power generation
(iii) Low HRT without compromising final effluent quality
(iv) Stability and consistent performance in long-term operation
 (v) Low installation, maintenance and operating costs
(vi) Operating at ambient temperatures and environmental conditions

11.6.1 Tubular MFC

Tubular MFCs are the most common scale-up design used and studied. The simplicity of the design and near optimal cross-section reducing the dead space is the most attractive feature of tubular MFC. It consists of a tubular cathode wrapped by a layer of separator and then anode. The reversed configuration, where a tubular anode is wrapped by a separator and then by cathode material, is also available. The supporting materials used in tubular MFC include polyvinyl-chloride, poly propylene, cylindrical bottles, measuring cylinders, nylon tubing or CEM in rolled form. Graphite or carbon brush, granular carbon material and sometimes flat electrodes have been employed as anodes in these reactors (Janicek et al. 2014). The cathode in most cases is carbon cloth, fiber or veil coated with platinum catalyst. Most of the studies have used separators to isolate the anode and cathode reactions, and very few reports have used separator-less MFC designs. The reactor must be very narrow to increase the electrode area to the volume of the reactor. The drawback in these systems is that the entire system has to be dismantled for cleaning up any of the electrode or membrane (Zhang et al. 2013; Zhuang et al. 2012) (Figure 11.5).

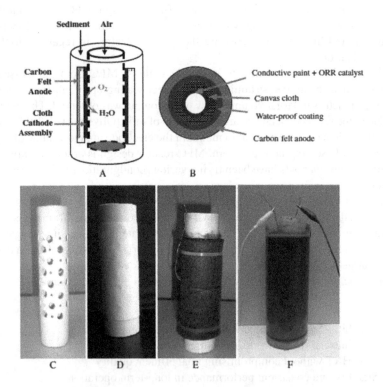

FIGURE 11.5 Tubular MFC: (A) complete configuration, (B) anode and cathode, (C)–(F) the tubular MFC construction (adapted from Yuan, Zhou, & Zhuang, 2010).

11.6.2 Stacked MFC

Stacked MFC consists of plate and frame arranged in stacks, and the electrical connection between individual cells can be made in a series or parallel (Figure 11.6). This design aids the reactor in handling higher amounts of wastewater without compromising the electrode area-to-volume ratio. The important advantage of this system is reduction in inter-electrode distance to a minimum, lowering the internal resistance of the system (Janicek et al. 2014). The voltage reversal phenomenon due to load variation in the end of the system where individual cells are connected in series reduces the final power production (Ledezma et al. 2013). The practical issue faced while operating these systems is the dead volumes created during the wastewater channelization; it paves the way to a reduction in total operating volume. Arrangements and modifications in MFC stacks have been made to minimize voltage reversal and dead volumes (Wu et al. 2016).

11.6.3 Separate Electrode Modules

Biofouling of membrane and fouling of cathodes are predominant factors reducing power production on long-term operations of MFC. It is essential to clean the MFC components at frequent intervals to restore their power performance. Tubular and

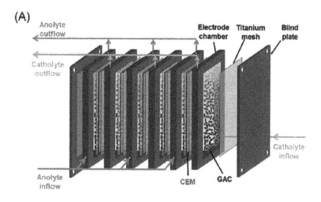
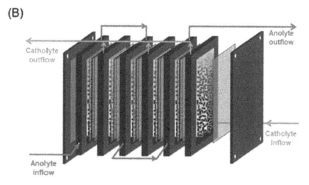

FIGURE 11.6 Plate and frame MFC (adapted from Wu et al., 2016).

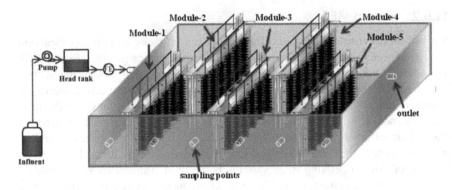

FIGURE 11.7 Separate anode and cathode modules (adapted from Dong et al. 2015).

stacked architecture consists of anode, cathode and membrane integrated or attached to each other; for cleaning of any one of the system components, complete dismantling of the reactor is required (Kim et al. 2015). This additional exercise contributes to the maintenance of MFC. Hence, an improved configuration consisting of separate modules of anode and cathode would make cleaning easy (Dong et al. 2015) (Figure 11.7).

11.7 POWER HARVESTING IN MFC

MFC has been operated and current collection was made continuously until Dewan and Beyenal (2009) demonstrated that energy production from MFC can be improved if current collection is made intermittently. The study demonstrates the usage of the same MFC in continuous energy harvesting (CEH) and in intermittent energy harvesting (IEH) modes. The resistors were used to discharge current in CEH, and capacitors were used for IEH. It was observed that a higher power could be produced in intermittent energy harvesting mode. The continuous withdrawal of high currents led to mass transport limitations to electro-active species near the anode. This phenomenon is known as limiting current in electrochemistry. In the IEH cycle, the electro-active species had enough time to replenish while the capacitor was discharging current. Developments in large-scale systems mainly focus on improving reactor characteristics rather than the energy harvesting mechanisms. Recent interest in improving energy generation in MFC have paved the way to develop efficient energy harvesting systems from MFC. Deeke et al. (2012) have worked with MFC with internal capacitor, in which the anode of the capacitor is coated with capacitive material, whereas the experiments have been performed with MFC consisting of non-capacitive and capacitive anodes. Systems with capacitive anodes outperformed the MFC with non-capacitive anodes. Further studies into MFC capacitive materials are essential to improve power production and storage. Santoro, Soavi, Arbizzani et al. (2016) have illustrated an integrated system using novel hydrogen-evolving reaction electrodes for H_2 production. The system could generate pulsed current and H_2. The combined system aided synergistic systems, improving anodic reactions, which in turn produced a high current pulse. It is important to focus on energy harvesting systems designed for MFC applications to minimize power losses.

11.8 REAL-TIME MFC TESTING

The environmental conditions in wastewater treatment plants vary widely with the laboratory conditions. The wastewater differs in composition on a daily basis; changing seasons affect the wastewater temperature. The final product requirement determines the production process employed in industries, and hence the wastewater composition varies constantly. It is not wise to conclude MFC efficiency by studying the scaled-up systems in laboratories or controlled conditions. It is essential to study MFC performance in actual WWTP environments before installation. Very limited work is available for MFC working in real-field conditions.

In this section, we will briefly review the studies on MFC performed in actual wastewater treatment plants. Zhang et al. (2013) constructed a tubular MFC which was installed in WWTP in the USA. The system could attain high treatment and power efficiency, as the power generation reduced as the temperature of its surroundings dropped. Refer to Figure 11.8 for the MFC design. The system was operated for 400 days, after which there was a gradual decrease in performance; the decrease was due to biofouling of the cathode and membranes.

Martinucci et al. (2015) have installed floating MFC of various sizes in denitrification tanks of domestic wastewater treatment plants. The study was performed in Milano-Nosedo WWTP (Figure 11.9c), the world's largest one, in the city of Milan. MFC was operated for six months continuously. Figure 11.9a,b shows floating MFC design and vegetative growth on MFC operated at the wastewater treatment plant. The change in climate and water flow patterns influenced the power performance in MFC. Particularly during the month of November, precipitation caused reduced COD loading and power production.

FIGURE 11.8 The tubular MFC operating in actual WWTP (adapted from Zhang et al. 2013).

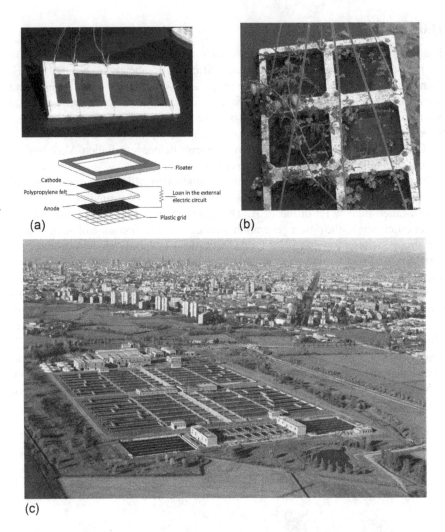

FIGURE 11.9 (a) Floating MFC construction; (b) vegetative growth on MFC operated in denitrification tank; (c) WWTP in Milano-Nosedo, Milan City, Italy (adapted from Martinucci et al. 2015).

Hiegemann et al. (2016) have studied the energy balance in integrating MFC into existing wastewater treatment plants. Energy saving could be made, though there was a reduction in biogas produced. The lower sludge production led to lower energy required for sludge treatment, and the energy required for aeration was reduced. Table 11.2 summarizes the performance of MFC operated at a real WWTP. All the above-mentioned work highlights the fact that MFC performance was not only based on its internal configuration; the external environment widely influenced the power production. Hence, results obtained from studies carried out in controlled conditions may vary to actual reactor performance operating in natural WWTP.

Power Production in Microbial Fuel Cells (MFC)

TABLE 11.2
Performance of MFC in Real Wastewater Treatment Facility

S. No	MFC Architecture	Place of Installation	Treatment Capacity	Type of Wastewater	Treatment Efficiency	Power	Inference	Reference
1	Tubular MFC Carbon brush anode Cathode 1-Activated Carbon Cathode 2-Platinum catalyst	South Shore water reclamation facility	4 L	Effluent from primary treatment	COD 65–70%	0.0255 Kwh/m^3	Fouling caused reduced performance on long-term operation.	(Zhang et al. 2013)
2	Floating type MFC anode and cathode of carbon cloth Separator-polypropylene	Denitrification tank of wastewater treatment plant in Milano-Nosedo, Italy	Small – 20x15cmx10 Medium – 30x20cmx10 Large – 40x30cmx10	Sewage wastewater	-	S – 750mA/m^2 M – 500mA/m^2 L – 150mA/m^2	Smaller size produced high power As COD loading decreased in rainy season the performance reduced	(Martinucci et al. 2015)
3	Single chamber MFC anode-graphite brush Cathode-carbon cloth loaded with Pt catalyst	After primary clarifier at WWTP in Bottorp, Germany	45L each 11.4Lx4	Sewage + cold mine discharge	COD 24% TSS 40% Nitrates 28%	0.36Kwh/Kg COD	Integration of MFC into WWTP is energy-saving option Reduced sludge production led to less energy requirement for sludge treatment	(Hiegemann et al. 2016)

11.9 FUTURE AND SCOPE

MFC technology is still in the infant stage. Further work in MFC requires testing of scaled-up systems in actual environments rather than controlled atmospheres. The development of cheaper materials with higher stability for long-term operations, without compromising the efficiency in power production and wastewater treatment efficiency, would aid in scaling-up and installation of MFC in wastewater treatment plants. Unlike other treatment technologies, MFC integrates and exploits biological and electrochemical treatment in a single reactor, conferring higher treatment efficiency. MFC is capable of handling variations in wastewater composition and is capable of degrading complex waste, which is usually difficult to degrade by existing biological treatments. The power harvesting methods have to be optimized to obtain higher power conversions. H. Wang et al. (2015) has consolidated and reviewed the energy harvesting system available to recover high energy from MFC. Optimizing MFC power generation and harvesting systems with low-cost materials would make MFC a reliable option for wastewater treatment.

11.10 CONCLUSION

The global energy scenario, depletion of fossil fuels and increasing energy demands have played a pivotal role in raising the value of energy as a commodity. Energy generation from waste is a sustainable approach to produce power and provides economic support to the industries installing waste treatment plants. Wastewater is a store of chemical energy, but in a dilute form; bio-methanation is one of the viable options for recovering energy, but the conversion of biogas to electrical energy suffers from various conversion losses. MFC is a robust technology with the potential to convert energy in wastewater to electrical energy, directly reducing the conversion loses. The added advantage of a low environmental footprint by reducing CO_2 production from WWTP is the most attractive feature of MFC. Though MFC cannot be used as a standalone technology in WWTP, its role is important in attaining self-sufficiency and energy in waste treatment. Current research trends show the rising interest among those in the scientific community towards MFC and other bio-electrochemical systems. Intensive research for fabrication of low-cost materials with high stability and efficiency in long-term operations is required. Studies on MFC working in WWTP highlights the adaptability of MFC to uncontrolled atmosphere, providing a positive note in scaling-up MFC. Further studies on large-scale MFC and design have to be made to bring wide acceptability among the industrial community.

REFERENCES

Aelterman, P; Rabaey, K; Clauwaert, P & Verstraete, W 2006. "Microbial fuel cells for wastewater treatment". *Water Science and Technology: A Journal of the International Association on Water Pollution Research* 54(8): 9–15.

Cheng, Ka Yu; Ho, Goen & Cord-Ruwisch, Ralf 2010. "Anodophilic biofilm catalyzes cathodic oxygen reduction". *Environmental Science and Technology* 44(1): 518–525.

Chinnaraj, S & Venkoba Rao, G 2006. "Implementation of an UASB anaerobic digester at bagasse-based pulp and paper industry". *Biomass and Bioenergy* 30(3): 273–277.

CPCB. 2017. http://cpcb.nic.in/Introduction/.
Decentralised Systems n.d. Ministry of New and Renewable Energy. https://mnre.gov.in/decentralized-systems.
Deeke, Alexandra; Sleutels, Tom H.J.A.; Ter Heijne, Annemiek; Hamelers, Hubertus VM & Buisma, Cees JN 2013. "Influence of the thickness of the capacitive layer on the performance of bioanodes in Microbial Fuel Cells Alexandra". *Journal of Power Sources* 243(2013): 611–616.
Deeke, Alexandra; Sleutels, TH; Hamelers, HV & Buisman, CJ 2012. "Capacitive bioanodes enable renewable energy storage in microbial fuel cells". *Environmental Science and Technology* 46(6): 3554–3560.
Dewan, Alim; Beyenal, Haluk & Lewandowski, Z 2009. "Intermittent energy harvesting improves the performance of microbial fuel cells". *Environmental Science & Technology* 43(12): 4600–4605.
Dong, Yue; Qu, Youpeng; He, Weihua; Du, Yue; Liu, Jia; Han, Xiaoyu & Feng, Yujie 2015. "A 90-liter stackable baffled microbial fuel cell for brewery wastewater treatment based on energy self-sufficient mode". *Bioresource Technology* 195: 66–72.
Fan, Yanzhen; Hu, Hongqiang; Liu, Hong; Qian, Fang; Morse, Daniel E.; Mink, Justine E; Qaisi, Ramy M; Logan, Bruce E; Hussain, Muhammad M & Choi, Seokheun 2015. "Sustainable power generation in microbial fuel cells using bicarbonate buffer and proton transfer mechanisms". *Biosensors and Bioelectronic* 69(3): 8154–8158.
Feng, Chunhua; Lv, Z; Yang, X & Wei, C 2014. "Anode modification with capacitive materials for a microbial fuel cell: an increase in transient". *Physical Chemistry Chemical Physics* 2014(16): 10464–10472.
Gajda, Iwona; Greenman, J; Melhuish, C & Ieropoulos, I 2015. "Self-sustainable electricity production from algae grown in a microbial fuel cell system". *Biomass and Bioenergy* 82: 87–93.
Gu, Yifan; Li, Yue; Li, Xuyao; Luo, Pengzhou; Wang, Hongtao; Robinson, Zoe P; Wang, Xin & Wu, Jiang 2017. "The feasibility and challenges of energy self-sufficient wastewater treatment plants". *Applied Energy* 204: 1463–1475.
Hernández-Fernández, FJ; De Los Ríos, A Pérez; Salar-García, MJ & Ortiz-Martínez, VM 2015. "Recent progress and perspectives in microbial fuel cells for bioenergy generation and wastewater treatment". *Fuel Processing Technology* 138: 284–297.
Hiegemann, Heinz; Herzera, Daniel; Nettmanna, Edith; Lübkena, Manfred; Schulteb, Patrick; Schmelzb, Karl-Georg; Gredigk-Hoffmannc, Sylvia & Wicherna, Marc 2016. "An integrated 45 L pilot microbial fuel cell system at a full-scale wastewater treatment plant". *Bioresource Technology* 218: 115–122.
Janicek, Anthony; Fan, Yanzen & Liu, Hong 2014. "Design of microbial fuel cells for practical application: A review and analysis of scale-up studies". *Biofuels* 5(1): 79–92.
Jayaprakash, Jayapriya; Parthasarathy, Abinaya & Viraraghavan, Ramamurthy 2016. "Decolorization and degradation of monoazo and diazo dyes in pseudomonas catalyzed microbial fuel cell". *Environmental Progress and Sustainable Energy* 35(6): 1623–1628.
Juang; Yang, PC; Chou, HY & Chiu, LJ, Der Fong et al. 2011. "Effects of microbial species, organic loading and substrate degradation rate on the power generation capability of microbial fuel cells". *Biotechnology Letters* 33(11): 2147–2160.
Kazmi, A; Singh A; Bawa, M; Starkl, I; Patil, P & Khale, V 2015. "Sewage treatment and management in Goa, India: A case study". *Water Quality Exposure and Health* 8(1): 1–11.
Kim, Kyoung Yeol et al. 2015. "Assessment of microbial fuel cell configurations and power densities". *Environmental Science and Technology Letters* 2(8): 206–214.
Kumar, Ravinder; Singh, L; Wahid, ZA & Din, MFM 2015. "Exoelectrogens in microbial fuel cells toward bioelectricity generation: A review". *International Journal of Energy Research* 39(8): 1048–1067.

Ledezma, Pablo; Greenman, John & Ieropoulos, Ioannis 2013. "MFC-cascade stacks maximise COD reduction and avoid voltage reversal under adverse conditions". *Bioresource Technology* 134: 158–165.

Li, Shuang; Cheng, Chong & Thomas, Arne 2017. "Carbon-based microbial-fuel-cell electrodes: From conductive supports to active catalysts". *Advanced Materials* 29(8): 1–30.

Li, Wen-wei; Sheng, GP; Liu, XW & Yu, HQ 2011. "Recent advances in the separators for microbial fuel cells". *Bioresource Technology* 102(1): 244–252.

Li, Wen-Wei; Yu, Han-Qing & He, Zhen 2013. "Towards sustainable wastewater treatment by using microbial fuel cells-centered technologies". *Energy and Environmental Science* 7(3): 911–924.

Logan, BE 2009. "Exoelectrogenic bacteria that power microbial fuel cells". *Nature Reviews in Microbiology* 7(5): 375–381.

Logan, Bruce E; Hamelers, Bert; Rozendal, René; Schröder, Uwe; Keller, Jürg; Freguia, Stefano; Aelterman, Peter; Verstraete, Willy & Rabaey, Korneel 2006. "Microbial fuel cells: Methodology and technology". *Environmental Science and Technology* 40(17): 5181–5192.

Martin, E; Savadogo, O; Guiot, SR & Tartakovsky, B 2010. "The influence of operational conditions on the performance of a microbial fuel cell seeded with mesophilic anaerobic sludge". *Biochemical Engineering Journal* 51(3): 132–139.

Martinucci, E; Pizza, F; Perrino, D; Colombo, A; Trasatti, SPM; Lazzarini Barnabei, A; Liberale, A & Cristiani, P 2015. "Energy balance and microbial fuel cells experimentation at wastewater treatment plant Milano-Nosedo". *International Journal of Hydrogen Energy* 40(42): 1–7.

Ministry of New and Renewable Energy Annual Report 2015–2016. https://mnre.gov.in/annual-report.

Mohan, S Venkata; Mohanakrishna, G; Reddy, B Purushotham; Saravanan, R & Sarma, PN 2008. "Bioelectricity generation from chemical wastewater treatment in mediator-less (anode) microbial fuel cell (MFC) using selectively enriched hydrogen producing mixed culture under acidophilic microenvironment". *Biochemical Engineering Journal* 39(1): 121–130.

Mohan, S et al. 2008. "Bioelectricity production from wastewater treatment in dual chambered microbial fuel cell (MFC) using selectively enriched mixed microflora: Effect of catholyte". *Bioresource Technology* 99: 596–603.

Nam, Joo Youn; Kim, Hyun Woo; Lim, Kyeong Ho; Shin, Hang Sik & Logan, Bruce E 2010. "Variation of power generation at different buffer types and conductivities in single chamber microbial fuel cells". *Biosensors and Bioelectronics* 25(5): 1155–1159.

Oliveira, VB; Simões, M; Melo, LF & Pinto, AMFR 2013. "Overview on the developments of microbial fuel cells". *Biochemical Engineering Journal* 73: 53–64.

Pandey, Prashant; Shinde, Vikas N; Deopurkar, Rajendra L; Kale, Sharad P; Patil, Sunil A & Pant, Deepak 2016. "Recent advances in the use of different substrates in microbial fuel cells toward wastewater treatment and simultaneous energy recovery". *Applied Energy* 168: 706–723.

Pasternak, Grzegorz; Greenman, John & Ieropoulos, Ioannis 2016. "Comprehensive study on ceramic membranes for low-cost microbial fuel cells". *ChemSusChem* 9(1): 88–96.

Patel, Rushika; Zaveri, Purvi & Munshi, Nasreen S 2017. "Microbial fuel cell, the Indian scenario: Developments and scopes". *Biofuels* 7269(January): 1–8.

Rabaey, Korneel & Verstraete, Willy 2005. "Microbial fuel cells: Novel biotechnology for energy generation". *Trends in Biotechnology* 23(6): 291–298.

Raghavulu, S; Veer; Goud, RK & Sarma, PN 2009. "Effect of anodic pH microenvironment on microbial fuel cell (MFC) performance in concurrence with aerated and ferricyanide catholytes". *Electrochemistry Communications* 11(2): 371–375.

Rickard, Alexander H; McBain, AJ; Stead, AT & Gilbert, P 2004. "Shear rate moderates community diversity in freshwater biofilms". *Applied and Environmental Microbiology* 70(12): 7426–7435.

Samsudeen, Nainamohamed; Radhakrishnan, TK & Matheswaran, Manickam 2014. "Performance comparison of triple and dual chamber microbial fuel cell using distillery wastewater as a substrate". *Environmental Progress & Sustainable Energy* 34(2): 589–594.

Santoro, Carlo; Arbizzani, C; Erable, B & Ieropoulos, I 2017. "Microbial fuel cells: From fundamentals to applications. A review". *Journal of Power Sources* 356: 225–244.

Santoro, Carlo; Soavi, Francesca; Serov, Alexey & Arbizzani, Catia 2016. "Self-powered supercapacitive microbial fuel cell: The ultimate way of boosting and harvesting power". *Biosensors and Bioelectronics* 78: 229–235.

Santoro, Carlo; Soavi, Francesca; Arbizzani, Catia; Serov, Alexey; Kabir, Sadia; Carpenter, Kayla; Bretschger, Orianna & Atanassov, Plamen 2016. "Co-generation of hydrogen and power/current pulses from supercapacitive MFCs using novel HER iron-based catalysts". *Electrochimica Acta* 220: 672–682.

Velvizhi, G; Mohan, S & Venkata 2012. "Electrogenic activity and electron losses under increasing organic load of recalcitrant pharmaceutical wastewater". *International Journal of Hydrogen Energy* 37(7): 5969–5978.

Venkata Mohan, S; Velvizhi, G; Annie Modestra, J & Srikanth, S 2014. "Microbial fuel cell: Critical factors regulating bio-catalyzed electrochemical process and recent advancements". *Renewable and Sustainable Energy Reviews* 40: 779–797.

Wang, Heming; Park, Jae-do & Ren, Zhiyong Jason 2015. "Practical energy harvesting for microbial fuel cells: A review". *Environmental Science & Technology* 49(6): 3267–3277.

Wang, Yuyang; Wen, Qing; Chen, Ye & Yin, Jinling 2016. "Enhanced performance of a microbial fuel cell with a capacitive bioanode and removal of Cr (VI) using the intermittent operation". *Applied Biochemistry and Biotechnology* 180(7): 1372–1385.

Waste to Energy Potential in India n.d. Ministry of New and Renewable Energy. https://mnre.gov.in/waste-energy.

Wei, Jincheng; Liang, Peng & Huang, Xia 2011. "Recent progress in electrodes for microbial fuel cells". *Bioresource Technology* 102(20): 9335–9344.

Winfield, Jonathan; Gajda, I; Greenman, J & Ieropoulos, I 2016. "A review into the use of ceramics in microbial fuel cells". *Bioresource Technology* 215: 296–303.

Wu, Shijia; Li, Hui; Zhou, Xuechen; Liang, Peng; Zhang, Xiaoyuan; Jiang, Yong & Huang, Xia 2016. "A novel pilot-scale stacked microbial fuel cell for efficient electricity generation and wastewater treatment". *Water Research* 98: 396–403.

Yang, Na; Ren, Y; Li, X & Wang, X 2016. "Effect of graphene-graphene oxide modified anode on the performance of microbial fuel cell". *Nanomaterials (Basel)* 6(9): 174.

Zhang, Fei; Ge, Zheng; Grimaud, Julien; Hurst, Jim & He, Zhen 2013. "Long-term performance of liter-scale microbial fuel cells treating primary effluent installed in a municipal wastewater treatment facility". *Environmental Science and Technology* 47(9): 4941–4948.

Zhou, Minghua; Chi, Meiling; Luo, Jianmei; He, Huanhuan & Jin, Tao 2011. "An overview of electrode materials in microbial fuel cells". *Journal of Power Sources* 196(10): 4427–4435.

Zhuang, Li; Yuan, Y; Wang, Y & Zhou, S 2012. "Long-term evaluation of a 10-liter serpentine-type microbial fuel cell stack treating brewery wastewater". *Bioresource Technology* 123: 406–412.

12 Synthesis, Characterization and Antimicrobial Properties of CuO-Loaded Hydrophobically Modified Chitosans

P. Uma Maheswari, K. Sriram and K. M. Meera Sheriffa Begum

CONTENTS

12.1 Introduction	204
12.1.1 Chitosan as a Nanocarrier for Drug Delivery	204
12.1.2 Surface Functionalization of Chitosan	204
12.1.3 Copper and Copper Oxide Nanoparticles	204
12.1.4 Antimicrobial Studies – Chitosan Materials	205
12.1.5 Need of a Hybrid Functionalized Drug Carrier	206
12.2 Preparation of Hydrophobically-Modified Chitosans	206
12.2.1 ^{1}H NMR Analysis	207
12.2.2 FTIR Analysis	207
12.2.3 AFM Analysis	207
12.2.4 SEM Analysis	209
12.3 Preparation of CuO Nanoparticles	210
12.4 Preparation of CuO-Loaded Chitosan Nanoparticles	212
12.5 Biocompatibility	215
12.6 Antimicrobial Assay	215
12.7 Conclusion	217
Acknowledgment	217
References	217

12.1 INTRODUCTION

12.1.1 CHITOSAN AS A NANOCARRIER FOR DRUG DELIVERY

Chitosan is a natural polysaccharide and a well-proven biocompatible material that has numerous applications in various fields. Recently, chitosan application in drug delivery is tremendously increasing due to its excellent biocompatibility, biodegradability, low allergenic and mucoadhesive property compared with other natural polymers (Michael et al. 2013). Chitosan nanoparticle is the most preferable carrier for drug and gene delivery as it forms stable ionic complexes with them. It is also the most suitable carrier for organ-specific drug delivery applications like colon cancer, lung cancer, liver cancer, etc.

12.1.2 SURFACE FUNCTIONALIZATION OF CHITOSAN

Chitosan has been modified with a wide range of functional groups to tune its physico-chemical properties towards drug delivery challenges. The chemical modifications are done by attaching either hydrophilic or hydrophobic groups with the amine group of the chitosan (Wan Ngah and Liang 1999; Sashiwa et al. 2000; Holappa et al. 2004; Villaa et al. 2013; Zhou et al. 2013; Rong et al. 2013). These modifications have improved chitosan towards solubility, structural stability and antimicrobial characteristics (Cai et al. 2015; Xu et al. 2010; Muzzarelli et al. 2001). The hydrophobic modifications of chitosan, such as alkylation, acylation, cholic acid etc., have gained more importance in carrying lipid-soluble drugs and in their sustained delivery (Kiparissides and Kammona 2012; Leceta et al. 2013; Wu et al. 2005; Balan et al. 2015; Zhang et al. 2016). Also, the resultant outcome of the modifications changes the surface nature, nanoparticle-forming ability of chitosan and its interaction with other bio macromolecules like protein and DNA (Gonil et al. 2014; Hu et al. 2014). The chitosan nanocomposite has wide applications in the biomedical field, such as tissue engineering (Shavandi et al. 2015). The hydrophobic modification in chitosan leads to the sustained rate of drug release, protects the drug from gastric fluid degradation and enhances the mucoadhesive properties of the carrier.

12.1.3 COPPER AND COPPER OXIDE NANOPARTICLES

Copper is a micro-nutrient for humans, and it is also a structural component of almost all microorganisms. However, at high concentrations, copper generates free radicals that disrupt the biomolecule's synthesis (Sivaraj et al. 2014). Copper oxide also shows antimicrobial activities equivalent to silver oxide nanoparticles. Copper oxide is inexpensive and easy to synthesize in bulk quantities, which makes it more cost-effective than silver nanoparticles (Yurderi et al. 2015). CuO nanoparticles are effective against drug-resistant pathogens such as *S. aureus* and *E. coli*. These features of CuO nanoparticles have been used in the development of antibacterial and antifungal composite materials to improve their biomedical applications (Huang et al. 2015; Bondarenko et al. 2013; Pourbeyram and Mohammadi 2014; Ruparelia et al. 2013).

12.1.4 ANTIMICROBIAL STUDIES – CHITOSAN MATERIALS

Friedman et al. (2013) investigated the antimicrobial and immunological properties and the feasibility of using nanoparticles (NPs) to deliver antimicrobial agents to treat a cutaneous pathogen. In vitro studies were performed using chitosan and alginate against *Propionibacterium acnes*. By electron microscopy imaging it was observed that the chitosan-alginate NPs were found to induce the disruption of the *P. acnes* cell membrane, providing a mechanism for the bactericidal effect. The chitosan-alginate NPs also exhibited anti-inflammatory properties and have the potential utility of drug therapy for the treatment of dermatologic conditions with infectious and inflammatory components.

Zain et al. (2014) produced silver and copper nanoparticles by chemical reduction of their respective nitrates by ascorbic acid in the presence of chitosan using microwave heating. Particle size was increased by increasing the concentration of nitrates and decreasing the chitosan concentration. Surface zeta potentials were positive for all the nanoparticles produced, and these varied from 27.8 to 33.8 mV. Antibacterial activities of Ag, Cu, mixtures of Ag and Cu and Ag-Cu bimetallic nanoparticles were tested using *Bacillus subtilis* and *Escherichia coli*. Of the two, *B. subtilis* was proved more susceptible under all the process conditions investigated. Silver nanoparticles displayed higher activity than the copper and mixtures of Ag and Cu nanoparticles which had been compared with the same mean particle size. However, when compared on an equal concentration basis, Cu nanoparticles proved more lethal to bacteria due to a higher surface area. The highest antibacterial activity was obtained with bimetallic Ag-Cu nanoparticles with minimum inhibitory concentrations (MIC) of 0.054 and 0.076 mg/L against *B. subtilis and E. coli*, respectively.

Saharan et al. (2015) synthesized Cu-chitosan nanoparticles and evaluated them for their growth promotory and antifungal efficacy using tomato (*Solanum lycopersicum* Mill.). Physico-chemical characterization was performed. Their study highlighted the stability and porous nature of Cu-chitosan nanoparticles which established a substantial growth promotory effect on tomato seed germination, seedling length and fresh and dry weight at 0.08, 0.10 and 0.12% level. These nanoparticles caused 70.5 and 73.5% inhibition of mycelia growth and 61.5 and 83.0% inhibition of spore germination in *Alternaria solani* and *Fusarium oxysporum*, respectively, at 0.12% concentration in an in vitro model. The overall results confirmed the significant growth promotory as well as antifungal capabilities of Cu-chitosan nanoparticles.

Vo et al. (2017) applied magnetite-carboxymethyl chitosan (MNP-CMCS) nanoparticles to obtain magnetically retrievable and deliverable carriers with antimicrobial effect. The presence of carboxylate groups in CMCS not only enhanced the antimicrobial activity but also enabled Ag ions' chelating ability to induce the in situ formation of Ag nanoparticles. The deposition of Ag nanoparticles on the surface of MNP-CMCS enhanced its antimicrobial activity against planktonic cells due to the dual action of CMCS and Ag NPs. The prepared MNP-CMCS-Ag was efficiently delivered into an existing biofilm under the guidance of an applied magnetic field.

Ong et al. (2008) investigated enhancing chitosan wound dressing activity by incorporating a procoagulant (polyphosphate) and an antimicrobial agent (silver). Silver-loaded chitosan-polyphosphate exhibited a significantly greater bactericidal activity than chitosan-polyphosphate by in vitro studies, achieving a complete kill of *Pseudomonas aeruginosa* and *Staphylococcus aureus* consistently. The silver dressing also significantly reduced mortality from 90% to 14.3% in *P. aeruginosa* wound infection models in mice.

Yoksan and Chirachanchai (2009) prepared silver nanoparticles by γ-ray irradiation-reduction under simple conditions, using chitosan as a stabilizer. The nanoparticles were spherical, with an average size of 7–30 nm, as observed from TEM. The silver nanoparticles exhibited antimicrobial activities against *Escherichia coli* and *Staphylococcus aureus*. Their results suggested that silver nanoparticles dispersed in chitosan solution can be directly applied in antimicrobial fields, including antimicrobial food packaging and biomedical applications.

Wang et al. (2012) combined chitosan with nano-ZnO to increase its antimicrobial activity, using polyvinyl alcohol as a support to form composite nanofibers by electro spinning. *Escherichia coli* and *Candida albicans* were used to test the antimicrobial efficacy of the newly synthesized chitosan-nano-ZnO composite. The minimum inhibitory concentration of chitosan/nano-ZnO against *C. albicans* was 160 μg/ml, which was closer to the concentration of the treated composite with the lowest fluorescence intensity. The cell damage was also observed by SEM, which indicated that nano-ZnO in the nanofibrous membranes played a cooperative role in the antimicrobial process of chitosan.

12.1.5 NEED OF A HYBRID FUNCTIONALIZED DRUG CARRIER

The metal nanoparticles with hydrophobically functionalized polymer coating form hybrid nanocarriers, which are gaining more attraction towards targeted drug delivery. These hybrid nanocarriers offer improved bioavailability and biocompatibility of polymers in addition to the enhanced drug loading by the large surface area of metal nanoparticles. The effect of pure chitosan and hydrophobically modified chitosans with CuO nano particles towards antimicrobial activity was investigated. Chitosan was first successfully modified by vanillin, naphthalene acidic acid (NAA), tyrosine, phendione and phenyl benzimidazole (phbenzim) ligands. The modified chitosan materials were characterized to ensure the modifications and their surface morphology. The modified chitosans were checked for biocompatibility with L929 fibroblast normal cell lines through MTT assay. The modified chitosans were added with CuO nano particles and expected to reasonably increase the antimicrobial property of chitosan and modified chitosans towards the strains *Staphylococcus epidermis* and *Candida albicans*.

12.2 PREPARATION OF HYDROPHOBICALLY-MODIFIED CHITOSANS

The various hydrophobic ligands such as vanillin, NAA, tyrosine, phendione and phenyl benzimidazole (phbenzim) were employed to functionalize chitosan for enhancement of biomedical properties. The deacetylated chitosan was dissolved in acid medium at 40°C. The chitosan solution was slowly mixed with pure ligands, and the reaction mixture was stirred continuously for 72 h in the presence of the deprotonating base, triethylamine, to react ligands with the amino group. The dusty white chitosan solution completely changed its color during the amino group modification (Figure 12.1a–e). Then the reaction mixture was neutralized. The product was centrifuged, washed repeatedly and dried for characterization through analytical techniques.

12.2.1 ^{1}H NMR Analysis

The ^{1}H NMR spectrum of chitosan, pure ligands and functionalized chitosan are shown in Figure 12.2a–e. The vanillin represented the signals of methoxyl protons at 3.85 ppm, the phenolic proton at 6.5 ppm, the aromatic protons at 7–7.4 ppm and the aldehyde protons at 9.7–10.3 ppm. The chitosan spectrum indicated the broad signals of methyl, methylene, hydroxyl and amino protons at 2, 3.2, 4.3 to 4.7 and 4.8 ppm, respectively. In chitosan-vanillin, the presence of broadened aromatic protons of vanillin signals in the downfield region of 8 to 8.4 ppm (Sajomsang et al. 2008), together with the broad chitosan signals, was observed. The NAA ligand showed the sharp aromatic peaks in the region of 8.5 to 9 ppm (Kowapradit et al. 2008). On chemical modification of chitosan with NAA, the presence of several aromatic proton peaks of NAA signals, together with the chitosan signals in the downfield region, was observed. The presence of broadened aromatic protons of tyrosine signals in the downfield region of 6.8 to 7.3 ppm (Gasper et al. 2015), together with the broad chitosan signals of methyl and amino protons (2–4.8 ppm), was observed. In the chitosan-phendione spectrum, the presence of broadened aromatic protons of phendione signals in the downfield region of 7.5 to 9.4 ppm (Rabea et al. 2009), together with the broad chitosan signals, was observed. In the modified chitosan-benzimidazole, aromatic protons of phenylbenzimidazole in the region of 7 to 8 ppm, together with the broad chitosan signals, were observed. This confirms the incorporation of aromatic groups of ligands to the amino group of chitosan polymer chain.

12.2.2 FTIR Analysis

Figure 12.3a–e represented the FTIR spectra of chitosan, ligands of (a) vanillin, (b) NAA, (c) phendione, (d)tyrosine, (e) benzimidazole and their modified chitosan to identify the occurrence of hydrophobic modifications in the drug carrier materials. It was clearly observed that the characteristic peaks of chitosan and ligands were indicated in all the IR spectra of modified chitosans (Sriram et al. 2017, 2018a,b).

12.2.3 AFM Analysis

The surface topology changes of chitosan and functionalized chitosan were understood by AFM analysis. Figure 12.4(I–V) shows the 2D and 3D structures and surface profiles of both chitosan and modified chitosans, respectively. The surface roughness data were collected at different places to identify the surface smoothness of the carrier materials. The surface of the modified chitosan has improved its smoothness due to the hydrophobic substitution on the NH_2 group of chitosan. This was clearly observed by fewer peaks and valleys in the modified chitosan images.

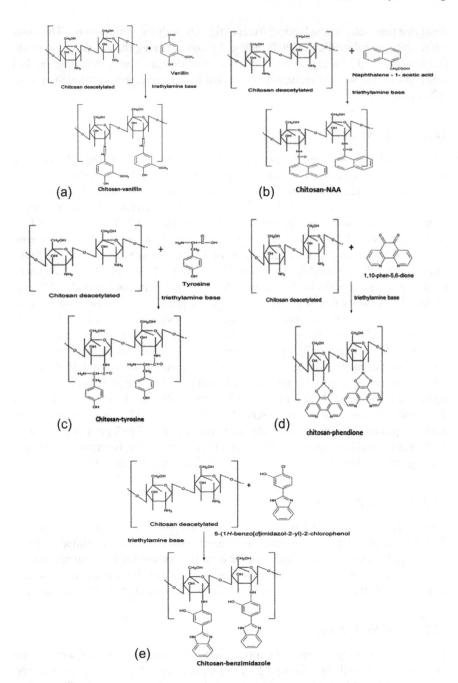

FIGURE 12.1 Reaction mechanism of chitosan with (a) vanillin, (b) NAA, (c) tyrosine, (d) phendione and (e) benzimidazole.

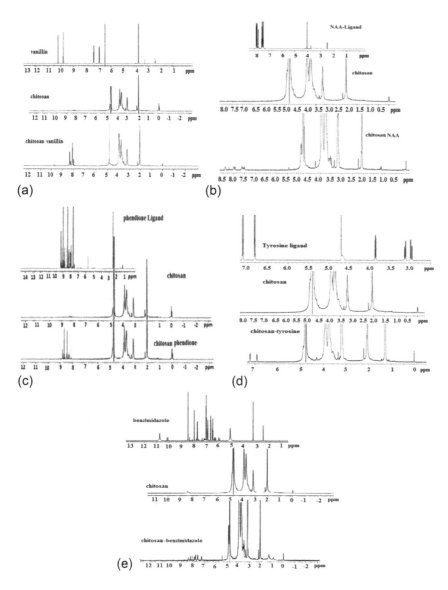

FIGURE 12.2 ¹H NMR spectra of chitosan, ligands and functionalized chitosan.

12.2.4 SEM Analysis

The SEM images of chitosan as well as modified chitosan nanoparticles are shown in Figure 12.5. The chitosan-vanillin formed a distinct nanostructure without any aggregation in comparison with chitosan nanoparticles. SEM images of NAA showed small nanospheres of about 30 to 40 nm in size. The chitosan-phendione was obtained as a cluster during the nanoparticle formation; this may be due to the inclusion of a bulky planar aromatic group, which induces the hydrophobic

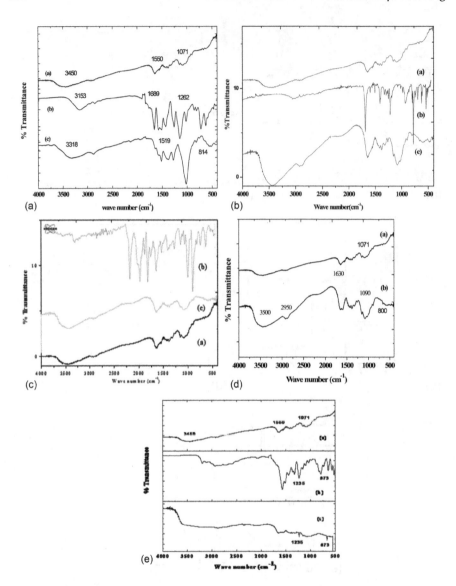

FIGURE 12.3 FTIR spectra of chitosan, ligands and functionalized chitosan.

intermolecular interactions. The phenyl benzimidazole-modified chitosan indicated nanosphere structure in a more clustered manner of about 40 to 110 nm in size. Chitosan-tyrosine had formed more aggregated nanostructures in comparison with chitosan nanoparticles.

12.3 PREPARATION OF CUO NANOPARTICLES

The CuO nanoparticles were prepared by the chemical reduction method. Copper (II) chloride solution of 0.01M was added in 0.01M $NaBH_4$ solution under vigorous

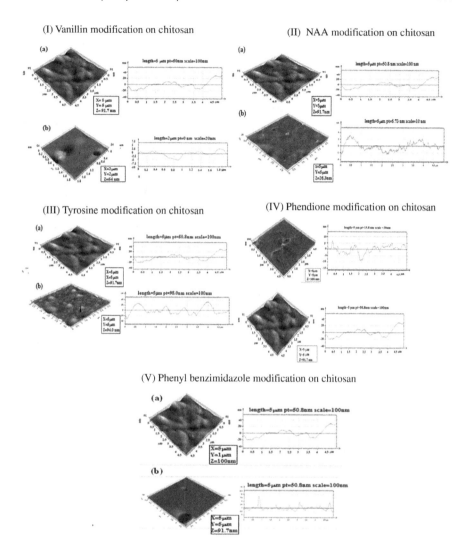

FIGURE 12.4 (I–V) AFM images of (a) chitosan and (b) modified chitosans.

stirring using polyvinylpyrrolidone (pvp) solution as a stabilizing agent. The mixture was stirred for 2 h and the resultant black precipitate was separated. The particle was washed repeatedly and dried in a vacuum.

The XRD pattern of the CuO nanoparticle is shown in Figure 12.6a, which matches exactly with the standard code of CuO nanoparticle (JCPDS 72-0629). The crystal size (dc) was measured by the Debye-Scherrer formula by substituting the values of β and θ, which were calculated from the diffraction peak at (2 0 0). The Debye-Scherrer formula is given in equation (12.1):

$$dc = \frac{k\lambda}{\beta \cos\theta} \tag{12.1}$$

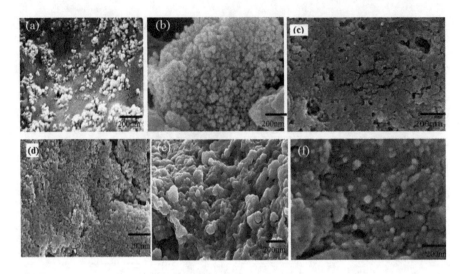

FIGURE 12.5 SEM images of (a) chitosan, (b) chitosan-vanillin, (c) chitosan-NAA, (d) chitosan-phendione, (e) chitosan-benzimidazole and (f) chitosan-tyrosine.

where k is the Scherrer constant, λ is wavelength of the X-ray, β is the full width at half maximum and θ is the angle measured. The calculated crystal size of the CuO nanoparticle was 12.2 nm.

12.4 PREPARATION OF CUO-LOADED CHITOSAN NANOPARTICLES

The modified chitosan was dissolved in acid medium, and CuO nanoparticle was added to the solution and sonicated for proper dispersion of nanoparticle. Then 20 ml of 0.075% TPP solution was added in the solution mixture under vigorous stirring. The existence of cloudy solution is the indicator of the formation of hybrid nanoparticles. The product was centrifuged and freeze-dried. The synthesized hybrid carriers were characterized analytically to ensure the modifications, surface nature, morphology and size.

Samples of modified chitosans and modified chitosan with CuO nanoparticles were characterized by XRD, and their diffraction patterns are presented in Figure 12.6b. The presence of CuO characteristic peaks were observed in all the XRD patterns of modified chitosans with CuO nanoparticle. The CuO peaks are suppressed due to the presence of non-crystalline chitosan and modified chitosan on the CuO nanoparticles.

The chitosan and modified chitosans treated with CuO were characterized by FTIR, as shown in Figure 12.7A and B.

From the figures, the broad peak in the range of 3800 to 2800 cm^{-1} corresponds to –OH group due to the presence of surface adsorbed moisture on the CuO nanoparticles. The characteristic peaks at 1641 cm^{-1} and 513 cm^{-1} are assigned to the stretching vibration of CuO bonds (Dhineshbabu and Rajendran 2016). The presence of peaks at 1641 cm^{-1} and 513 cm^{-1} in the FTIR patterns of all the modified chitosans

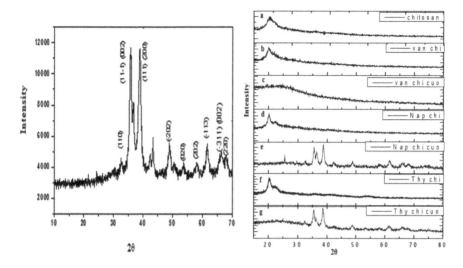

FIGURE 12.6 The XRD patterns of (a) CuO nanoparticle and (b) modified chitosan with CuO nanoparticle.

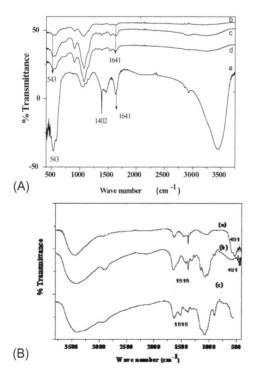

FIGURE 12.7 (A) FTIR spectra of (a) CuO nanoparticle, (b) vanillin-chitosan with CuO, (c) NAA-chitosan with CuO and (d) tyrosine-chitosan with CuO. (B) FTIR spectra of (a) CuO nanoparticle, (b) chitosan-phendione with CuO and (c) chitosan-phenyl benzimidazole with CuO.

with CuO nanoparticles ensures the incorporation of CuO into the modified chitosan nanoparticles.

Figure 12.8a–g shows the SEM images of CuO nanoparticle, chitosan and modified chitosans with CuO nanoparticles. The vanillin chitosan has formed larger nanoparticles (44 nm) when compared with the other modified chitosan and pure chitosan with CuO nanoparticles. The CuO nanoparticles are observed in the size of 15 nm, while the chitosan and modified chitosan with CuO nanoparticles are in the size range of 24 nm to 44 nm, which confirms the biomaterial coating on the

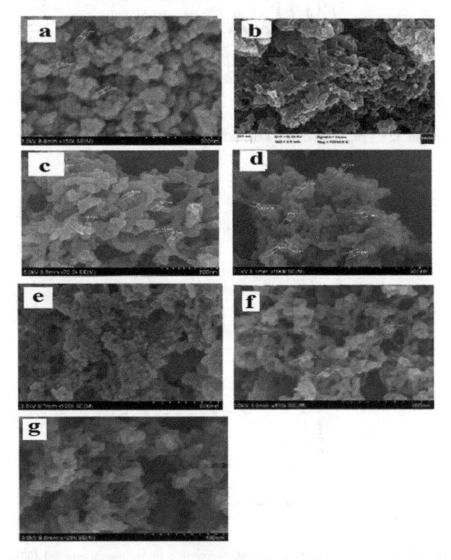

FIGURE 12.8 SEM images of (a) CuO nanoparticle, (b) chitosan-CuO, (c) vanillin-chitosan-CuO, (d) NAA chitosan-CuO, (e) tyrosine-chitosan-CuO, (f) phendione-chitosan-CuO and (g) phbenzim-chitosan-CuO.

CuO metal nanoparticles that in turn increases the size of nanoparticles. The CuO nanoparticle addition helps to lessen the aggregation of nanoparticle formation with modified chitosans.

12.5 BIOCOMPATIBILITY

The cell viability of modified chitosan and the drug-loaded materials were tested using L929 fibroblast cells. The samples were incubated with DMEM (Dulbecco's Modified Eagle Medium) for 24 h and centrifuged, from which the supernatants were subjected to conduct MTT assay for 24 and 72 h in triplicates, which incubated at 37°C under 5% CO_2 using MTT [3-(4, 5-dimethylthiazole-2-yl)-2, 5-diphenyl tetrazolium bromide] reagent. The absorbance was measured at 570 nm using microplate reader (model: Enspire, Perkin Elmer, USA). The percentage of cell viability was calculated using the following equation

$$\text{Cell viability \%} = \frac{\text{Test sample optical density}}{\text{Control sample optical density}} \times 100$$

The chitosan showed high cell viability up to 72 h against L929 fibroblast normal cell lines. The modified chitosans with vanillin, NAA and tyrosine showed biocompatibility until 24 h and then become slightly toxic by 72 h. The phendione-chitosan and phbenzim-chitosan were less biocompatible even by 24 h and were found to be toxic. Table 12.1 presents the cell viability results of chitosan and modified chitosans using L929 mouse fibroblast cell lines for 24 h and 72 h.

12.6 ANTIMICROBIAL ASSAY

Antimicrobial analysis was done by disc diffusion methods (Lumin et al. 2013; Liu et al. 2016). Petri plates (discs) were prepared by pouring 30 ml of nutrient agar and potato dextrose agar for bacteria and fungi cultures, respectively. The inoculums containing the bacteria *Staphylococcus epidermidis* was evenly spread on nutrient agar plates, and the fungi, *Candida albicans*, was evenly spread on potato dextrose agar plates. The plates were allowed to dry for 10 minutes, and the sterile filter papers (6 mm diameter) containing the chitosan compounds to be tested (30 µl) were laid

TABLE 12.1
Biocompatibility Study Results

Samples	% Cell Viability after 24 h	% Cell Viability after 72 h
Chitosan	122	117
Chitosan-vanillin	115	62
Chitosan-NAA	119	49
Chitosan-tyrosine	102	96
Chitosan-phendione	99	32
Chitosan-phbenzim	98	40

down on the surface of the inoculated agar plates. The plates were incubated under the conditions of 24 hours, 37°C for *Staphylococcus epidermidis* and 30°C, 48 hours for *Candida albicans*. Each sample was tested in triplicates. The chloramphenicol and fluconazole were taken as the antibacterial and antifungal standards, respectively. The antimicrobial potential of various modified chitosans were determined by measuring the zone of inhibition (mean diameter) around the disc in millimeters.

The antibacterial and antifungal properties of the following materials, such as pure chitosan and modified chitosans with CuO nanoparticles, were studied using skin infective bacterial (*Staphylococcus epidermidis*) and fungal (*Candida albicans*) strains. The results are as shown in Table 12.2. The activity of pure chitosan is negligible, as it is a proven biocompatible drug carrier material (Khatik et al. 2013). The pure CuO nanoparticles show a zone of inhibition of about 8 and 7 mm respectively towards bacterial and fungal activity, which is the highest inhibition zone among the tested materials. The CuO nanoparticle-added chitosan indicated moderate activity with a zone of inhibition around 5 and 7 mm respectively, with respect to bacterial and fungal activity. Among the modified chitosans, phendione- and phbenzim-substituted

TABLE 12.2
Antimicrobial Properties of Pure and Modified Chitosans With and Without CuO Nanoparticle

Sample (30 µl)	*S. epidermidis* Zone of Inhibition (mm)	*Candida albicans* Zone of Inhibition (mm)
Chitosan	negligible	negligible
CuO	8 ± 0.3	7 ± 0.5
Chitosan with CuO	5 ± 0.3	7 ± 0.5
Vanillin	4 ± 0.3	3 ± 0.2
Vanillin-chitosan	5 ± 0.4	6 ± 0.3
Vanillin-chitosan with CuO	6 ± 0.2	7 ± 0.4
NAA	4 ± 0.2	3 ± 0.2
NAA-chitosan	3 ± 0.7	4 ± 0.7
NAA-chitosan with CuO	7 ± 0.5	5 ± 0.3
Tyrosine	4 ± 0.5	3 ± 0.2
Tyrosine-chitosan	3 ± 0.4	2 ± 0.1
Tyrosine-chitosan with CuO	6 ± 0.5	5 ± 0.4
Phendione	30 ± 0.3	27 ± 0.2
Phendione-chitosan	21 ± 0.4	25 ± 0.3
Phendione-chitosan with CuO	30 ± 0.2	28 ± 0.4
phbenzim	24 ± 0.2	23 ± 0.2
phbenzim-chitosan	18 ± 0.7	22 ± 0.7
phbenzim-chitosan with CuO	26 ± 0.5	24 ± 0.3
Standard (chloramphenicol)	12 ± 0.8	–
Standard (fluconazole)	–	14 ± 0.9

Values were expressed as mean ± SD.

chitosans showed the maximum antibacterial and antifungal activities among all other chitosans, the values ranging from 24 to 30 mm of diameter, as shown in Table 12.2. Next, the vanillin-substituted chitosan was observed a better activity than the NAA and tyrosine-substituted chitosans. Under loading of CuO nanoparticles, the order of inhibition was as follows: phendione-chitosan > phbenzim-chitosan > NAA-chitosan > vanillin-chitosan > tyrosine-chitosan > chitosan. The above results demonstrate that the modified chitosans show higher to moderate antimicrobial properties and are much enhanced with the addition of CuO nanoparticles. Hence, the coordination and the delivery kinetics of CuO nanoparticle-loaded modified chitosan morphology bring out a synergic effect on enhancing the antimicrobial effects. The results were formally compared with the standards of chloramphenicol and fluconazole for antibacterial and antifungal properties, respectively.

12.7 CONCLUSION

The modified chitosans were prepared with the ligands vanillin, NAA, tyrosine, phendione and phbenzim. The modifications were up to 80% in the amino groups of chitosan. The ^{1}H NMR results prove the substitutions on chitosan. The XRD, AFM and SEM characterizations define the materials as more amorphous rather than crystalline. The addition of CuO with the pure and modified chitosans was also amorphous. The CuO-loaded chitosan and modified chitosans show better antimicrobial properties than the unloaded chitosan materials. The CuO nanoparticle-loaded modified chitosans show antimicrobial activities as follows: phendione-CuO > phbenzim-CuO > CuO > NAA-chitosan-CuO > vanillin-chitosan-CuO > tyrosine-chitosan-CuO ≈ chitosan-CuO. The above results clearly demonstrate the improved antimicrobial properties of modified chitosans due to the presence of CuO nanoparticles in comparison to the pure ligands and CuO unloaded modified chitosans. Chitosan can act as a better carrier for ligands of biological importance and can be used with CuO nanoparticles for suitable bio-medical applications.

ACKNOWLEDGMENT

The authors acknowledge the financial support received from DST, India.

REFERENCES

Balan, V., G. Dodi, N. Tudorachi, O. Ponta, V. Simon, M. Butnaru, L. Verestiu. 2015. Doxorubicin-loaded magnetic nanocapsules based on N-palmitoyl chitosan and magnetite: Synthesis and characterization. *Chem. Eng. J.*, 279, 188–197.

Bondarenko, O., K. Juganson, A. Ivask, K. Kasemets, M. Mortimer, A. Kahru. 2013. Toxicity of Ag, CuO and ZnO nanoparticles to selected environmentally relevant test organisms and mammalian cells in vitro: A critical review. *Arch. Toxicol.*, 87(7), 1181–1200.

Cai, J., Q. Danga, C. Liu, T. Wanga, B. Fan, J. Yan, Y. Xua. 2015. Preparation, characterization and antibacterial activity of O-acetyl-chitosan-N-2-hydroxypropyl trimethyl ammonium chloride. *Int. J. Biol. Macromol.*, 80, 8–15.

Dhineshbabu, N.R., V. Rajendran. 2016. Antibacterial activity of hybrid chitosan-cupric oxide nanoparticles on cotton fabric. *IET Nanobiotechnology*, 10(1), 13–19.

Friedman, A.J., J. Phan, D.O. Schairer, J. Champer, M. Qin, A. Pirouz, K. Blecher-Paz, A. Oren, P.T. Liu, R.L. Modlin, J. Kim. 2013. Antimicrobial and anti-inflammatory activity of chitosan–alginate nanoparticles: A targeted therapy for cutaneous pathogens. *J. Invest. Dermatol.*, 133(5), 1231–1239.

Gaspar, V.M., E.C. Costa, J.A. Queiroz, C. Pichon, F. Sousa, I.J. Correia. 2015. Folate-targeted multifunctional amino acid chitosan nanoparticles for improved cancer therapy. *Pharm. Res.*, 32, 562–577.

Gonil, P., W. Sajomsang, U.R. Ruktanonchai, P.N. Ubol, A. Treetong, P. Opanasopit, S. Puttipipatkhachorn. 2014. Synthesis and fluorescence properties of N-substituted 1-cyanobenz[f]isoindole chitosan polymers and nanoparticles for live cell imaging. *Biomacromolecules*, 15(8), 2879–2888.

Holappa, J., T. Nevalainen, J. Savolainen, P. Soininen, M. Elomaa, R. Safin, S. Suvanto, T. Pakkanen, M. Masson, T. Loftsson, T. Järvinen. 2004. Synthesis and characterization of chitosan N-betainates having various degrees of substitution. *Macromolecules*, 37(8), 2784–2789.

Hu, H., C. Tang, C. Yin. 2014. Folate conjugated trimethyl chitosan/graphene oxide nanocomplexes as potential carriers for drug and gene delivery. *Mater. Lett.*, 125, 82–85.

Huang, Y., J. Huang, J. Cai, W. Lin, Q. Linc, F. Wuc, J. Luo. 2015. Carboxymethyl chitosan/clay nanocomposites and their copper complexes: Fabrication and property. *Carbohyd. Polym.*, 134, 390–397.

Khatik, R., R. Mishra, A. Verma, P.D. Vivek Kumar, V. Gupta, S.K. Paliwal, P.R. Mishra, A.K. Dwivedi, A.K. Dwivedi. 2013. Colon-specific delivery of curcumin by exploiting Eudragit-decorated chitosan nanoparticles in vitro and in vivo. *J. Nanopart. Res.*, 15(9), 1893–1908.

Kiparissides, C., O. Kammona. 2012. Nanoscale carriers for targeted delivery of drugs and therapeutic biomolecules. *Can. J. Chem. Eng.*, 91(4), 638–651.

Kowapradit, J., P. Opanasopit, T. Ngawhiranpat, A. Apirakaramwong, T. Rojanarata, U. Ruktanonchai, W. Sajomsang. 2008. Methylated N-(4-N,N-Dimethylaminobenzyl) chitosan, a novel chitosan derivative, enhances paracellular permeability across intestinal epithelial cells (Caco-2). *AAPS PharmSciTech*, 9(4), 1143–1152.

Leceta, I., P. Guerrero, I. Ibarburu, M.T. Dueñas, K. de la Caba. 2013. Characterization and antimicrobial analysis of chitosan-based films. *J. Food. Eng.*, 116(4), 889–899.

Liu, Y., Y. Cai, X. Jiang, J. Wu, X. Le. 2016. Molecular interactions, characterization and antimicrobial activity of curcumin–chitosan blend films. *Food Hydrocolloids*, 52, 564–572.

Lumin, M., S. Iuliana, M. Mihai, D. Valentina, C. Bogdan, B.S. Mihail. 2013. Antifungal vanillin–imino-chitosan biodynameric films. *J. Mater. Chem. B*, 1(27), 3353–3358.

Michael, D.B., A. Merzouki, M. Lavertu, M. Thibault, M. Jean, V. Darras. 2013. Chitosans for delivery of nucleic acids. *Adv. Drug Deliv. Rev.*, 65, 1234–1270.

Muzzarelli, R.A.A., C. Muzzarelli, R. Tarsi, M. Miliani, F. Gabbanelli, M. Cartolari. 2001. Fungistatic activity of modified chitosans against Saprolegnia parasitica. *Biomacromolecules*, 2(1), 165–169.

Ong, S., J. Wu, S.M. Moochhala, M. Tan, J. Lu. 2008. Development of a chitosan-based wound dressing with improved hemostatic and antimicrobial properties. *Biomaterials*, 29(32), 4323–4332.

Pourbeyram, S., S. Mohammadi. 2014. Synthesis and characterization of highly stable and water dispersible hydrogel–copper nanocomposite. *J. Noncryst. Solids*, 402, 58–63.

Rabea, E.I., M.E.I. Badawy, W. Steurbaut, C.V. Stevens. 2009. In vitro assessment of N-(benzyl)chitosan derivatives against some plant pathogenic bacteria and fungi. *Eur. Polym. J.*, 45(1), 237–245.

Rong, L., H. Pei, R. Xuehong, S.D. Worley, T.S. Huang. 2013. Antimicrobial N-halamine modified chitosan films. *Carbohyd. Polym.*, 92(1), 534–539.

Ruparelia, P.J., A.K. Chatterjee, S.P. Duttagupta, S. Mukherji 2013. Strain specificity in antimicrobial activity of silver and copper nanoparticles. *Acta Biomater.*, 3, 707–716.

Saharan, V., G. Sharma, M. Yadav, M.K. Choudhary, S.S. Sharma, A. Pal, R. Raliya, P. Biswas. 2015. Synthesis and in vitro antifungal efficacy of Cu–chitosan nanoparticles against pathogenic fungi of tomato. *Int. J. Biol. Macromol.*, 75, 346–353.

Sajomsang, W., S. Tantayanon, V. Tangpasuthadol, M. Thatte, W.H. Daly. 2008. Synthesis and characterization of N-aryl chitosan derivatives. *Int. J. Biol. Macromol.*, 43(2), 79–87.

Sashiwa, H., Y. Makimura, Y. Shigemasa, R. Roy. 2000. Chemical modification of chitosan: Preparation of chitosan–sialic acid branched polysaccharide hybrids. *Chem. Community*, 11, 909–910.

Shavandi, A., A.A. Bekhit, M.A. Ali, Z. Sun, M. Gould. 2015. Development and characterization of hydroxyapatite/β-TCP/chitosan composites for tissue engineering applications. *Mater. Sci. Eng. C*, 56, 481–493.

Sivaraj, R., K.S.M. Rahman, P. Rajiv, S. Narendhran, R. Venckatesh. 2014. Biosynthesis and characterization of Acalypha indica mediated copper oxide nanoparticles and evaluation of its antimicrobial and anticancer activity. *Spectrochim. Acta A Mol. Biomol. Spectrosc.*, 129, 255–258.

Sriram, K., P. Uma Maheswari, K.M. Meera Sheriffa Begum, G. Arthanareeswaran. 2018a. Functionalized chitosan with super paramagnetic hybrid nanocarrier for targeted drug delivery of curcumin. *Iran. Polym. J.*, 27(7), 469–482.

Sriram, K., P. Uma Maheswari, K.M. Meera Sheriffa Begum, G. Arthanareeswaran, M. Gover Antoniraj, K. Ruckmani. 2018b. Curcumin delivery by vanillin modified chitosan with superparamagnetic hybrid nanoparticles as carrier. *Eur. J. Pharm. Sci.*, 116, 48–60.

Sriram, K., P. Uma Maheswari, A. Ezhilarasu, K.M. Meera Sheriffa Begum, G. Arthanareeswaran. 2017. CuO loaded hydrophobically modified chitosan as hybrid carrier for curcumin delivery and anticancer activity. *Asia Pac. J. Chem. Eng.*, 12(6), 858–871.

Villaa, R., B. Cerroni, L. Viganòa, S. Margheritelli, G. Abolafioa, L. Oddo, B. Paradossi, G.N. Zaffaroni. 2013. Targeted doxorubicin delivery by chitosan-galactosylated modified polymer microbubbles to hepatocarcinoma cells. *Colloids. Surf. B*, 110, 434–442.

Vo, D.T., S. Sabrina, C. Lee. 2017. Silver deposited carboxymethyl chitosan-grafted magnetic nanoparticles as dual action deliverable antimicrobial materials. *Mater Sci Eng C Mater Biol Appl*, 73, 544–551.

Wan Ngah, W.S., K.H. Liang. 1999. Adsorption of gold(III) ions onto chitosan and N-carboxymethyl chitosan: Equilibrium studies. *Ind. Eng. Chem. Res.*, 38(4), 1411–1414.

Wang, Y., Q. Zhang, C. Zhang, P. Li. 2012. Characterisation and cooperative antimicrobial properties of chitosan/nano-ZnO composite nanofibrous membranes. *Food Chem.*, 132(1), 419–427.

Wu, T., S. Zivanovic, F.A. Draughon, W.S. Conway, C.E. Sams. 2005. Physicochemical properties and bioactivity of fungal chitin and chitosan. *J. Agric. Food Chem.*, 53, 3888–3894.

Xu, T., M. Xin, M. Li, H. Huang, S. Zhou. 2010. Synthesis, characteristic and antibacterial activity of N,N,N-trimethyl chitosan and its carboxymethyl derivatives. *Carbohyd. Polym.*, 81(4), 931–936.

Yoksan, R., S. Chirachanchai. 2009. Silver nanoparticles dispersing in chitosan solution: Preparation by γ-ray irradiation and their antimicrobial activities. *Mater. Chem. Phys.*, 115(1), 296–302.

Yurderi, M., A. Bulut, I.E. Ertas, M. Zahmakiran, M. Kaya. 2015. Supported copper–copper oxide nanoparticles as active, stable and low-cost catalyst in the methanolysis of ammonia–borane for chemical hydrogen storage. *Appl. Cat. B Environ.*, 165, 169–175.

Zain, N.M., A.G. Stapley, G. Shama. 2014. Green synthesis of silver and copper nanoparticles using ascorbic acid and chitosan for antimicrobial applications. *Carbohydr. Polym.*, 112, 195–202.

Zhang, F., J. Fei, M. Sun, Q. Ping. 2016. Heparin modification enhances the delivery and tumor targeting of paclitaxel-loaded N-octyl-N-trimethyl chitosan micelles. *Int. J. Pharm.*, 511(1), 390–402.

Zhou, N., X. Zan, Z. Wang, H. Wu, D. Yin, C. Liao, Y. Wan. 2013. Galactosylated chitosan–polycaprolactone nanoparticles for hepatocyte-targeted delivery of curcumin. *Carbohyd. Polym.*, 94(1), 420–429.

13 Fucoxanthin
Biosynthesis, Structure, Extraction, Characteristics, and Its Application

K. Anjana and K. Arunkumar

CONTENTS

- 13.1 Introduction 222
- 13.2 Biological Role of Fucoxanthin 223
- 13.3 Biosynthetic Pathway of Fucoxanthin 223
- 13.4 Structure of Fucoxanthin 224
- 13.5 Spectral Property of Fucoxanthin 226
  - 13.5.1 Absorption Spectrum 226
  - 13.5.2 Emission Spectrum 227
- 13.6 Occurrence and Yield of Fucoxanthin in Algae 228
- 13.7 Solvent Used for Extraction 231
- 13.8 Purification of Fucoxanthin 232
  - 13.8.1 Silica Gel Column Chromatography 232
  - 13.8.2 Thin Layer Chromatography (TLC) 233
  - 13.8.3 High-Performance Liquid Chromatography (HPLC) 234
  - 13.8.4 Reversed-Phase High-Performance Liquid Chromatography (RP-HPLC) 234
- 13.9 Characterization of Fucoxanthin 235
  - 13.9.1 UV-Vis Spectrophotometric Analysis 236
  - 13.9.2 Fourier-Transform Infrared Spectroscopy (FTIR) 236
  - 13.9.3 Fluorescence Spectroscopy 237
  - 13.9.4 Gas Chromatography Mass Spectrometry (GC-MS) Analysis 237
  - 13.9.5 Liquid Chromatography Mass Spectrometry (LC-MS) 237
  - 13.9.6 Nuclear Magnetic Resonance (NMR) 237
  - 13.9.7 High-Performance Liquid Chromatography (HPLC) 238
  - 13.9.8 Reversed-Phase High-Performance Liquid Chromatography (RP-HPLC) 238
- 13.10 Bioactivity of Fucoxanthin 238
- 13.11 Conclusion 241
- References 242

13.1 INTRODUCTION

Algae is grouped into two types based on size: microalgae and macroalgae. Microalgae is the source of omega-3 fatty acid. Omega-3 fatty acids found in fish are actually from microalgae (*Spirulina*, *Chlorella*, and cyanobacteria, etc.) that they feed on, and marine seaweeds are grouped under macroalgae (rich in polysaccharides, pigments, and antioxidants). Macroalgae contribute most of the biomass in intertidal zones (Dawes, 1998). In addition, the production of oxygen by algae (50% of all oxygen production) is another reason for saying "our lives depend on algae" (Chapman, 2013).

Approximately more than 9000 seaweed species have been identified and classified on the basis of their pigments. Vimala and Poonghuzhali (2013) pointed out that those pigments are important both in classification and biodiversity studies. Seaweeds are classified mainly into three groups: brown (*Phaeophyta*), red (*Rhodophyta*), and green (*Chlorophyta*) (Sivagnanam et al., 2015). Recently, much attention has been given to seaweeds as effective biomass resources because of their high carbon dioxide (CO_2) absorption rate relative to that of terrestrial plants (Miyashita et al., 2013). Microalgae are able to transport bicarbonate into cells, which in turn helps them to absorb CO_2. It has been reported that macroalgae can capture CO_2 as high as 90% in open ponds (Sayre, 2010). Seaweed is apparently able to absorb five times more CO_2 (a greenhouse gas contributing to climate change) than land-based plants. As such, seaweed is a good "carbon sink" that can absorb greenhouse gases such as carbon dioxide from the atmosphere. Brown seaweeds, the second most abundant group of marine algae, include approximately 2000 species.

Algae contain a wide variety of pigments that absorb light for photosynthesis. Seven different types of chlorophylls are reported in algae. They are Chl a (present in all groups of algae), Chl b (found in green algae), Chl c (seen in brown algae and diatoms), Chl d (present in red algae and cyanobacterium (Miyashita et al., 1996)), and Chl f (present in cyanobacterium (Chen et al., 2012)). Carotenoid is also a fat-soluble yellow pigment which is present in almost all algal groups. They are found in close association with chlorophyll and protect chlorophyll from photo damage by passing absorbed energy to chlorophyll. Chlorophyll molecules are used in pharmacies as a photosensitizer for cancer therapy (Mishra et al., 2011). Carotenoids are also called accessory pigments. Fucoxanthin is a brown pigment which gives color to brown algae as well as diatoms. Phycobilins are water-soluble pigments which are found in the cytoplasm or in the stroma of the chloroplast. Phycocyanin is responsible for the bluish color of cyanobacteria and phycoerythrin for the red color of red algae. Major pigments in the class Phaeophyceae are β carotene, zeatin, violaxanthin, fucoxanthin, chlorophyll a, and chlorophyll c (Takaichi, 2011). Algal pigments have great commercial value as natural colorants in nutraceutical, cosmetics, and pharmaceutical industry, besides their health benefits (Prasanna et al., 2007).

Fucoxanthin is widely used for several health benefits. Generally, seaweeds are used in salad, jelly, soup, etc. in countries like Indonesia, Malaysia, Singapore, Thailand, Korea, etc. (Dhargalkar and Pereira, 2005). Brown algae are a rich source of iodine and are included as a traditional food in Japan. Brown seaweeds are also used to extract alginic acid, a gel widely used as a stabilizer and emulsifier in ice cream, toothpaste, cosmetics, and thousands of other products. Alginic acid helps to remove radioactive strontium from the body. Insoluble alginate salts are used to waterproof tiles and seal fine paper. Soil erosion can be prevented by spraying sludge of brown alga containing seeds on soil bund (Branch and Branch, 1981; Branch et al., 1994).

Willstatter and Page (1914) first isolated fucoxanthin from the marine brown seaweeds such as *Fucus*, *Dictyota*, and *Laminaria*. The detailed structure and chirality of fucoxanthin was determined by Peng et al. (2011). The Gulf of Mannar region in the southeastern part of India is found to be the richest source of brown algae throughout the season (Sudhakar et al., 2013). The molecular weight of fucoxanthin protein is 24000 g/mol (Anderson and Barrett, 1978), and fucoxanthin is identified as 659.8 and 681.9 corresponding to [M+H]+ and [M+Na]+, respectively (Xia et al., 2013).

Valorization of biomass is a new concept which has been introduced, i.e., extracting or converting more valuable products such as energy, fuels, and other useful materials from wastes (Christopher et al., 2012). Seaweed biomass is valorized for extracting bioactive compounds. Biofuel and bioethanol are extracted from the biomass of microalgae and macroalgae respectively (Murphy et al., 2013). The two most important ways that this algal biomass is converted to valorized product are by biochemical conversion and by thermochemical conversion. Biochemical conversion includes the extraction of hydrocarbons for the production of biodiesels and value-added products by fermentation; ethanol, acetone, and butanol are obtained. Methane and hydrogen are produced by anaerobic digestion. Gasoline aviation fuel is derived by hydroprocessing. Thermochemical liquefaction, pyrolysis, and gasification by thermoconversion are used to produce bio-oil, oil, charcoal, and fuel gas, respectively (Kanes and Forster, 2009).

13.2 BIOLOGICAL ROLE OF FUCOXANTHIN

Fucoxanthin in brown algae helps to capture light for photosynthesis. Mikami and Hosokawa (2013) described fucoxanthin as the main carotenoid produced in brown algae as the component of the light-harvesting complex for photosynthesis and photoprotection (Mikami and Hosokawa, 2013). Martino et al. (2000) isolated fucoxanthin from the algal chloroplasts. The fucoxanthin-containing complexes specifically associated with pigment system (PS)I and PSII reaction centers are designated as light-harvesting complexes containing fucoxanthin (LHCF)-I (Berkaloff et al., 1990) and LHCF-II (Douady et al., 1993), respectively.

13.3 BIOSYNTHETIC PATHWAY OF FUCOXANTHIN

The biosynthetic pathway of fucoxanthin in Phaeophyceae is still unknown. So far, no work has been carried out on the brown algae. Only a hypothetical pathway is predicted by referring the fucoxanthin synthesis in diatoms and land plants (Figure 13.1) (Lohr and Wilhelm, 1999; Takaichi, 2011; Mikami and Hosokawa, 2013; Wang et al., 2014; Kuczynska et al., 2015).

Lohr and Wilhelm (1999) found out that violaxanthin as the major precursor for the synthesis of almost all carotenoids. Fucoxanthin in the diatom *Phaeodactylum tricornutum* was synthesized from violaxanthin through diadinoxanthin. Lichtenthaler (1999) proved that isopentenyl pyrophosphate (IPP) was synthesized

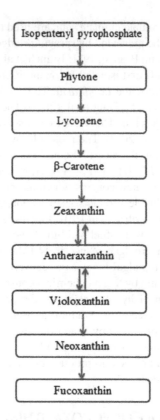

FIGURE 13.1 Predicted biosynthetic pathway of fucoxanthin.

in 1-deoxy-D-xylulose-5-phosphate (DOXP) pathway from pyruvate and glyceraldehyde. Phytoene synthase converts IPP to phytoene, a C40 compound. Oxygenic phototrops require three enzymes for the conversion of phytoene to lycopene. They are phytoene desaturase, δ-carotene desaturase, and cis-carotene isomerase. Bacteria use only phytoene desaturase for the conversion of phytoene to lycopene. Then lycopene is converted to β-carotene by lycopene cyclase (Wang et al., 2014). β-Carotene is hydroxylated by β-carotene hydroxylase to zeaxanthin (Takaichi, 2011).

According to Lohr and Wilhelm (1999), in low light zeaxanthin is converted to violaxanthin through the intermediate antheraxanthin by the enzyme zeaxanthin epoxidase, but in high light the compound violaxanthin is converted back to zeaxanthin by violaxanthin de-epoxidase. Takaichi and Mimuro (1998) explained the above conversion in land plants. The enzyme neoxanthin synthase converts violaxanthin to neoxanthin. Finally, neoxanthin is converted to fucoxanthin. How neoxanthin to converted to fucoxanthin is still unknown (Kuczynska et al., 2015).

13.4 STRUCTURE OF FUCOXANTHIN

Chemical and spectroscopic studies establish its structure as 3′-acetoxy-5,6-epoxy-3,5′-dihydroxy-6′,7′-didehydro-5,6,7,8,5′,6′-hexahydro-β-caroten-8-one

(Bonnett et al., 1969). Fucoxanthin has a unique molecular structure, including an unusual allenic bond (C=C=C) and oxygenic functional group, such as 5,6-monoepoxide, and nine conjugated double bounds (Shang et al., 2011; Zhao et al., 2014a); this allenic bond is mainly present in fucoxanthin, which was not found in other carotenoids in brown seaweeds (Maeda et al., 2006; Sangeetha et al., 2013; Kim et al., 2012a; Mikami and Hosokawa, 2013). See Figure 13.2.

However, the unique structure and its asymmetry, i.e., the structure and its mirror image, are not superimposable, so the fucoxanthin are said to be unstable, which can be decayed easily. It is easily affected by heating, aerial exposure, and illumination (Zhao et al., 2014a; Rajauria et al., 2017). As this compound plays a role of light harvesting and energy transferring, thus it can help marine brown alga survive in shallow coastal waters by offering efficient photosynthesis for acclimatization in their environment, i.e., it is a component of the light-harvesting complex for photosynthesis and photoprotection (Mikami and Hosokawa, 2013). The most common and accepted structure (all-*trans*-fucoxanthin) is shown in Figure 13.3.

Trans and *cis* isomerization is the common feature of carotenoid. This isomerization is due to the presence of conjugated double bonds in their structures. *Tran* isomer is more stable than its *cis* isomer. The factors that contribute to *cis-trans* isomerization are light, thermal energy, chemical reactions, and interaction with biological molecules such as proteins.

All-*trans*-fucoxanthin was the major geometrical isomer (~88%) found in the fresh brown seaweed (*Undaria pinnatifida*), apart from a small amount of 13-*cis* and 13'-*cis* isomers (~9%) (Figure 13.4). Waghmode and Kumbar (2015) and Fung et al. (2013) have determined 13-*cis* and 130-*cis* isomers of fucoxanthin were detected at retention times of 4.3 and 4.5 min, respectively. Zhang et al. (2014) isolated

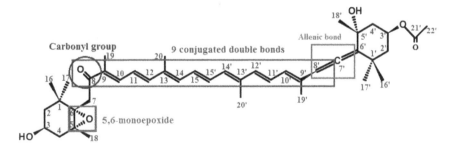

FIGURE 13.2 Detailed structure of fucoxanthin.

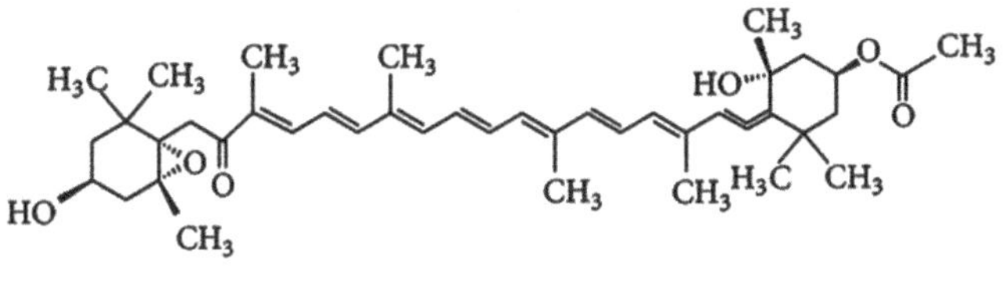

FIGURE 13.3 Chemical structure of fucoxanthin (molecular formula: $C_{42}H_{58}O_6$).

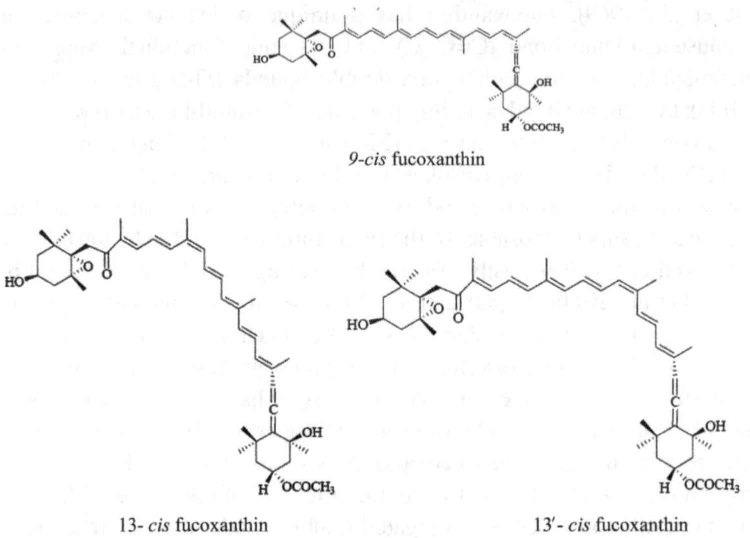

FIGURE 13.4 Different isomers of fucoxanthin.

fucoxanthin, and its stereoisomers 9'cis-Fuco, 13cis-Fuco, and 13'cis-Fuco were extracted from *Laminaria japonica*.

13.5 SPECTRAL PROPERTY OF FUCOXANTHIN

Among the pigments, fucoxanthin is the only naturally occurring stereoisomer in brown algae (Haugan et al., 1992). Fucoxanthin absorbs light significantly in the range of 450 to 540 nm, peaking at around 510–525 nm. Kosumi et al. (2011) reported that two spectroscopic forms of fucoxanthin, blue and red, exist in the polar environment. Fucoxanthin exhibit spectral characteristics attributed to an intramolecular charge transfer (ICT) state that arises in polar solvents due to the presence of the carbonyl group (Figure 13.1).

13.5.1 Absorption Spectrum

Absorption spectroscopy is one of the most widely used techniques employed for determining the concentrations of absorbing species (chromophores) in solutions. It is a nondestructive technique which biologists and biochemists and now systems biologists use to quantify the cellular components and characteristic parameters of functional molecules. This quantification is most relevant in the context of systems biology. For creating a quantitative depiction of a metabolic pathway, a number of parameters and variables are important, and these need to be determined experimentally (Nilapwar et al., 2011). The excitation wavelengths were 477 nm for all-trans-fucoxanthin, and 333 and 477 nm for *cis*-fucoxanthin. All transient absorption spectra of the fucoxanthin isomers in polar solvents show intramolecular charge transfer (ICT) state features, typical for carbonyl carotenoids (Kuznetsova et al., 2017).

The absorption spectrum of fucoxanthin has peaks at 424, 444, and 470 nm in organic solvents (Ke, 2001). When protein binds to fucoxanthin, the absorbance of fucoxanthin is shifted to the red, thus absorbing from 460 nm to 570 nm, a spectral range not used by higher plants and green algae (Veith and Buchel, 2007). Fucoxanthin showed a very good peak in the region between 500–560 nm (Haxo and Blinks, 1946). Sudhakar et al. (2013) confirmed the presence of fucoxanthin at 446 nm in *Sargassum wightii, Sargassum ilicifolium, Sargassum longifolium, Padina sp.*, and *Turbinaria sp*. Some studies confirmed that mostly fucoxanthin is detected at 450 nm (Yan et al., 1999; Mori et al., 2004; Maeda et al., 2006; Kanazawa et al., 2008; Hashimoto et al., 2009; Jaswir et al., 2012; Jaswir et al., 2013; Sivagnanam et al., 2015). Jeffrey et al. (1997) reported that the spectral patterns of fucoxanthin have a minimum absorption at a wavelength of 446 nm and maximum absorption at a wavelength of 468 nm.

According to Papagiannakis et al. (2005), the absorbance of fucoxanthin ranges from 420–470 nm, and they revealed that fucoxanthin was found to be present in *Sargassum binderi* extract, which was detected using UV-vis spectrophotometer at $\lambda max = 420$ nm. Shang et al. (2011) recorded fucoxanthin at 440 nm. Mise et al. (2011) also found that fucoxanthin was detected at 440 nm by comparing with the standard curve obtained with pure fucoxanthin. Ke (2001) found that functionally active fucoxanthin absorbs at 515 and 545 nm, and inactive fucoxanthin absorbs at 428, 448, and 488 nm. But Yip et al. (2014) reported that the absorption peak of standard fucoxanthin was at 423 nm.

Rajauria and Ghannam (2013) gave an excellent idea about the peak of absorption of fucoxanthin by comparing standard and purified fucoxanthin. Both standard and purified fucoxanthin exhibited the same spectroscopic profile with similar λmax (331, 446, and 468 nm). Fucoxanthin absorbs light primarily in the blue green to yellow green part of the visible spectrum, peaking at around 510–525 nm by various estimates and absorbing significantly in the range of 450–540 nm (Narayani et al., 2016). Hugan et al. (1992) reported maximum absorption in two different solvents, i.e., λmax (n-hexane): (428), 446, 475 nm, and λmax (acetone): (420), 444, 467. Wang et al. (2005) reported that DMSO extract showed a strong absorption peak at 452 nm for fucoxanthin. Martino et al. (2000) stated that fucoxanthin has absorption in the wavelength range 500–550 nm; if the peak is not found in this 500–550 nm range, it is suggested that most fucoxanthin is not functionally bound to the proteins, leading to a blue shift in its absorption band towards 460 nm. The absorption bands are located at 448, 476, and 505 nm, as shown by Katoh et al. (1991).

13.5.2 Emission Spectrum

Emission is the ability of a substance to emit light after absorption; emitted light has a longer wavelength than absorbed light. Absorption is the opposite of emission, where energy, light, or radiation is absorbed by the electrons of a particular matter.

Wang et al. (2005), using a fluorescence spectrophotometer (5 nm slit width), observed that fucoxanthin has an excitation peak at 452 nm, and emission spectrum showed peaks at 646 nm. Martino et al. (2000) reported that a fluorescence band at 680 nm indicated fucoxanthin. The emission bands of fucoxanthin were located around 630, 685, and 750 nm (Katoh et al., 1991).

13.6 OCCURRENCE AND YIELD OF FUCOXANTHIN IN ALGAE

Fucoxanthin is present in both macroalgae and microalgae. Peng et al. (2011) reported that fucoxanthin is a carotenoid present in brown algae and diatoms. It is the dominant pigment present in macroalgae, coming under class *Phaeophyceae*, and it also occurs in species coming under the class *Bacillariophyceae* (diatoms). The characteristic color of brown algae is due to the presence of this pigment, and it contributes over 10% of the estimated total production of carotenoid in nature.

Passaquet et al. (1991) reported the ratio of pigments present in some of the brown algae. The relative molar proportions of the four main pigments (Chl a/Chl c/fucoxanthin/violaxanthin) ranged from 100:18:76:6 to 100:30:107:17. They also studied the ratio of pigments in *Fucus serratus, Pelvetia canaliculata, Laminaria sacharina, Dictyota dichotoma,* and *Pylaiella littoralis,* and reported a ratio of 100:18:77:17, 100:6:61:30, 100:30:76:10, 100:30:107:10, and 100:30:85:6, respectively.

Fucoxanthin has been isolated from the following brown algae: *Alaria crassifolia* (Airanthi et al., 2011b), *Analipus japonicus* (Terasaki et al., 2009), *Cystoseira hakodatensis* (Airanthi et al., 2011a), *Eisenia bicyclis* (Kim et al., 2010; Airanthi et al., 2011b), *Fucus serratus* (Strand et al., 1998), *Fucus vesiculosus* (Zaragoza et al., 2008), *Hijikia fusiformis* (Zhang et al., 2014; Yan et al., 1999; Nishino, 1998), *Himanthalia elongate* (Rajauria and Ghannam, 2013; Rajauria et al., 2017), *Himanthalia elongate* (Rajauria and Ghannam, 2013; Rajauria et al., 2017), *Hincksia mitchellae* (Chen et al., 2017), *Ishige okamurae* (Kim et al., 2010; Kang et al., 2014), *Kjellmaniella crassifolia* (Airanthi et al., 2011b), *Laminaria japonica* (Zhang et al., 2014), *Laminaria angustata, Laminaria kjellmaniella, Laminaria longissima* (Kanazawa et al., 2008), *Laminaria ochotensis* (Miyashita et al., 2009), *Laminaria religiosa, Melanosiphon intestinalis* (Terasaki et al., 2009), *Myagropsis myagroides* (Heo et al., 2010), *Padina tetrastromatica* (Sangeetha et al., 2009a,b), *Petalonia binghamiae* (Kang et al., 2012), *Phaeodactylum tricornutum* (Hashimoto et al., 2009), *Sargassum binderi, Sargassum duplicatum* (Noviendri et al., 2011), *Sargassum filipendula* (Zailanie et al., 2015), *Sargassum fusiforme* (Terasaki et al., 2009), *Sargassum heterophyllum* (Afolyan et al., 2008), *Sargassum horneri* (Airanthi et al., 2011a), *Sargassum ilicifolium* (Susanto et al., 2016., Waghmode and Kumbar, 2015), *Sargassum siliquastrum* (Heo and Jeon, 2009), *Sargassum thunbergii* (Terasaki et al., 2009), *Sargassum thunbergii* (Terasaki et al., 2009), *Scytosiphon lomentaria* (Mori et al., 2004), *Undaria pinnatifida* (Kanda et al., 2014; Li et al., 2016; Piovan et al., 2013), and *Chaetoseros sp.* (Iio et al., 2011).

In addition, fucoxanthin has been reported from the following diatoms: *Chaetoceros calcitrans* (Foo et al., 2015), *Chaetoceros gracilis* (Ikeda et al., 2008), *Cladosiphon okamuranus* (Mise et al., 2011), *Cyclotella cryptica* (Westermann and Rhiel, 2005; Guo et al., 2016), *Cyclotella meneghiniana* (Lavanya et al., 2010), *Cylindrotheca closterium* (Rijstenbil, 2003; Vimala and Poonghuzhali, 2013; Erdogan et al., 2016), *Isochrysis aff. Galbana* (Kim et al., 2012a), *Nitzschia closterium* (Strain and Mannin, 1942), *Nitzschia laevis* (Wen and Chen, 2000), *Odontella aurita* (Moreau et al., 2006; Xia et al., 2014), *Phaeodactylum tricornutum* (Nomura et al., 1997; Kim et al., 2012b; Mikami and Hosokawa, 2013; Zhao et al., 2014b), and *Thalassiosira pseudonana* (Ikeda et al., 2013; Nomura et al., 1997).

The amounts of fucoxanthin extracted from various brown seaweeds and diatoms are shown in Table 13.1.

TABLE 13.1
Amount of Fucoxanthin in Different Algae

Species	Yield (mg/g)	Fresh Weight (FW)/Dry Weight (DW)	Reference	Group
Chaetoceros calcitrans	2.02	DW	Foo et al. (2015)	Diatom
Chaetoceros gracilis	2.24	DW	Kim et al. (2012a)	Diatom
Cladosiphon okamuranus	0.27	DW	Mise et al. (2011)	Diatom
Cylindrotheca closterium	5.23	DW	Pasquet et al. (2011)	Diatom
Cystoseira hakodatensis	2.01	DW	Susanto et al. (2016)	Brown algae
Cystoseira hakodatensis	2.4	DW	Terasaki et al. (2009)	Brown algae
Cystoseira indica	3.31	DW	Fariman et al. (2016)	Brown algae
Desmarestia viridi	0.1	DW	Terasaki et al. (2009)	Brown algae
Ecklonia kurome	1.68	DW	Susanto et al. (2016)	Brown algae
Eisenia bicyclis	0.08	FW	Kanazawa et al. (2008)	Brown algae
Eisenia bicyclis	0.26	FW	Kim et al. (2012a)	Brown algae
Eisenia bicyclis	0.42	FW	Shang et al. (2011)	Brown algae
Fucus distichus	0.16	FW	Seely et al. (1972)	Brown algae
Fucus serratus	0.47	DW	Haugan and Liaaen-Jensen (1989)	Brown algae
Fucus serratus	0.74	DW	Haugan et al. (1992)	Brown algae
Hizikia fusiformis	0.02	FW	Kanazawa et al. (2008)	Brown algae
Hydroclathrus clathratus	0.07	DW	Vimala and Poonghuzhali (2013)	Brown algae
Isochrysis aff. Galbana	18.23	DW	Kim et al. (2012a)	Diatom
Isochrysis galbana	1.08	DW	Loredo et al. (2016)	Diatom
Isochrysis galbana	6.04	DW	Kim et al. (2012a)	Diatom
Laminaria japonica	0.03	FW	Xiao et al. (2012)	Brown algae
Laminaria japonica	0.12	FW	Wang et al. (2005)	Brown algae
Laminaria japonica	0.19	FW	Kanazawa et al. (2008)	Brown algae
Laminaria japonica	0.83	FW	Xiao et al. (2012)	Brown algae
Laminaria japonica	1.22	FW	Wang et al. (2005)	Brown algae
Laminaria religiosa	0.24	FW	Mori et al. (2004)	Brown algae
Laminaria saccharina	0.24	FW	Seely et al. (1972)	Brown algae
Leathesia difformis	0.3	DW	Terasaki et al. (2009)	Brown algae
Nitzschia sp.	4.92	DW	Kim et al. (2012a)	Diatom
Nizamuddinia zanardinii	1.65	DW	Fariman et al. (2016)	Brown algae
Odontella aurita	21.7	DW	Xia et al. (2013)	Diatom
Padina australis	0.27	DW	Zailanie and Purnomo (2011)	Brown algae
Padina australis	1.29	DW	Susanto et al. (2016)	Brown algae
Padina gymnospora	0.38	DW	Sudhakar et al. (2013)	Brown algae
Petalonia binghamiae	0.43	DW	Mori et al. (2004)	Brown algae
Petalonia binghamiae	0.58	FW	Mori et al. (2004)	Brown algae
Phaeodactylum tricornutum	1.81	DW	Fuentes et al. (2001)	Diatom

(Continued)

TABLE 13.1 (CONTINUED)
Amount of Fucoxanthin in Different Algae

Species	Yield (mg/g)	Fresh Weight (FW)/Dry Weight (DW)	Reference	Group
Phaeodactylum tricornutum	8.55	DW	Kim et al. (2012a)	Diatom
Phaeodactylum tricornutum	15.33	DW	Kim et al. (2012a)	Diatom
Saccharina japonica	0.48	DW	Sivagnanam et al. (2015)	Brown algae
Saccharina sculpera	4.5	DW	Miyashita et al. (2011)	Brown algae
Sargassum binderi	1	DW	Jaswir et al. (2012)	Brown algae
Sargassum binderi	7.4	FW	Yip et al. (2014)	Brown algae
Sargassum binderi	1.01	DW	Noviendri et al. (2011)	Brown algae
Sargassum binderi	0.73	DW	Jaswir et al. (2012)	Brown algae
Sargassum cinereum	0.16	DW	Zailanie and Purnomo (2011)	Brown algae
Sargassum cinereum	2.02	DW	Narayani et al. (2016)	Brown algae
Sargassum confusum	3.4	DW	Miyashita et al. (2011)	Brown algae
Sargassum crassifolium	1.64	DW	Susanto et al. (2016)	Brown algae
Sargassum cristaefolium	0.14	DW	Kartikaningsih et al. (2017)	Brown algae
Sargassum cristaefolium	0.2	FW	Kartikaningsih et al. (2017)	Brown algae
Sargassum duplicatum	0.73	DW	Jaswir et al. (2013)	Brown algae
Sargassum duplicatum	1.01	DW	Jaswir et al. (2012)	Brown algae
Sargassum duplicatum	0.73	DW	Noviendri et al. (2011)	Brown algae
Sargassum echinocarpum	0.16	DW	Zailanie and Purnomo (2011)	Brown algae
Sargassum filipendula	0.2	DW	Zailanie and Purnomo (2011)	Brown algae
Sargassum filipendula	0.2	FW	Zailanie and Sukoso (2014)	Brown algae
Sargassum fulvellum	0.06	FW	Kanazawa et al. (2008)	Brown algae
Sargassum fusiforme	0.01	DW	Xiao et al. (2012)	Brown algae
Sargassum fusiforme	0.2	DW	Xiao et al. (2012)	Brown algae
Sargassum horneri	0.77	DW	Sivagnanam et al. (2015)	Brown algae
Sargassum horneri	2.12	DW	Susanto et al. (2016)	Brown algae
Sargassum horneri	3.7	DW	Terasaki et al. (2009)	Brown algae
Sargassum ilicifolium	0.01	DW	Vimala and Poonghuzhali (2013)	Brown algae
Sargassum ilicifolium	0.23	DW	Sudhakar et al. (2013)	Brown algae
Sargassum longifolium	0.09	DW	Sudhakar et al. (2013)	Brown algae
Sargassum muricum	0.29	FW	Seely et al. (1972)	Brown algae
Sargassum plagyophyllum	0.71	DW	Jaswir et al. (2012)	Brown algae
Sargassum plagyophyllum	0.79	DW	Jaswir et al. (2013)	Brown algae
Sargassum plagyophyllum	0.79	DW	Jaswir et al. (2012)	Brown algae

(*Continued*)

TABLE 13.1 (CONTINUED)
Amount of Fucoxanthin in Different Algae

Species	Yield (mg/g)	Fresh Weight (FW)/Dry Weight (DW)	Reference	Group
Sargassum polycystum	0.08	DW	Vimala and Poonghuzhali (2013)	Brown algae
Sargassum siliquastrum	1.99	DW	Susanto et al. (2016)	Brown algae
Sargassum thunbergii	5.7	DW	Miyashita et al. (2011)	Brown algae
Sargassum wightii	0.12	DW	Sudhakar et al. (2013)	Brown algae
Scytosiphon lomentaria	0.24	DW	Mori et al. (2004)	Brown algae
Scytosiphon lomentaria	0.5	DW	Terasaki et al. (2009)	Brown algae
Scytosiphon lomentaria	0.56	FW	Mori et al. (2004)	Brown algae
Sphaerotrichia divaricata	0.2	DW	Terasaki et al. (2009)	Brown algae
Turbinaria conoides	0.08	DW	Vimala and Poonghuzhali (2013)	Brown alga
Turbinaria conoides	0.21	DW	Zailanie and Purnomo (2011)	Brown algae
Turbinaria ornata	0.38	DW	Sudhakar et al. (2013)	Brown algae
Turbinaria ornata	1.27	DW	Susanto et al. (2016)	Brown algae
Turbinaria turbinata	0.59	DW	Jaswir et al. (2012)	Brown algae
Turbinaria turbinata	0.71	DW	Jaswir et al. (2012)	Brown algae
Turbinaria turbinata	0.59	DW	Jaswir et al. (2013)	Brown algae
Undaria pinnatifida	0.11	FW	Kanazawa et al. (2008)	Brown algae
Undaria pinnatifida	0.32	DW	Mori et al. (2004)	Brown algae
Undaria pinnatifida	0.39	DW	Kanda et al. (2014)	Brown algae
Undaria pinnatifida	0.73	DW	Xiao et al. (2012)	Brown algae
Undaria pinnatifida	0.87	DW	Kim et al. (2004)	Brown algae
Undaria pinnatifida	1.09	DW	Xiao et al. (2012)	Brown algae
Undaria pinnatifida	2.08	DW	Fung et al. (2013)	Brown algae
Undaria pinnatifida	3.08	DW	Fung et al. (2013)	Brown algae

13.7 SOLVENT USED FOR EXTRACTION

Fucoxanthin is slightly soluble in water and in organic solvents. However, for extraction and isolation, organic solvents are employed. Solvents play an important role in the process of extracting pigments. Mostly polar solvents were used for the extraction of fucoxanthin.

In a few studies, it was found that dimethyl sulfoxide (DMSO) was the best solvent for extraction because it has an ability to penetrate into thallus and rapidly extract much of the fucoxanthin, and it does not require maceration, centrifugation, or filtration (Vimala and Poonghuzhali, 2013). Fucoxanthin was extracted from the intact rhizoid of *Laminaria japonica* Aresch by DMSO, and then recovered from the DMSO extract by partitioning it into ethyl acetate and subsequent evaporation (Wang et al., 2005). Seely et al. (1972) also reported the extraction of the fucoxanthin from the intact thalli of brown algae by dimethyl sulfoxide (DMSO). Wang et al. (2005) also reported that DMSO was a much more effective solvent than acetone for fucoxanthin extraction. DMSO of 6 ml was used to extract fucoxanthin from *Laminaria japonica*; about 88% of fucoxanthin was extracted, but the purity decreased from 0.06 to 0.44. Liquefied dimethyl ether (DME) was said to be the best solvent for extraction in *Undaria pinnatifida* by Kanda et al. (2014). Acetone was used for extraction of fucoxanthin from *Undaria pinnatifida* (Maeda et al., 2006). Sudhakar et al. (2013) found that 90% acetone was a good solvent for extraction of carotenoids, including fucoxanthin in *Sargassum wightii*, *Sargassum ilicifolium*, *Sargassum longifolium*, *Padina gymnospora*, and *Turbinaria ornata*.

According to Mise et al. (2011) and Susanto et al. (2016), absolute methanol is the best solvent for the extraction of fucoxanthin from *Cladosiphon okamuranus* and *Sargassum ilicifolium*. Ethanol was recommended as an efficient solvent for the extraction in *Isochrysis aff. Galbana*, *Laminaria japonica*, and *Sargassum cinereum* (Kim et al., 2012a; Xiao et al., 2012; Narayani et al., 2016). Fucoxanthin extraction was performed using ethanol to a sample ratio of 4:1 mL/g (Lin et al., 2016).

Of the commercial extraction methods adopted for large-scale preparation of fucoxanthin, the best solvent is the mixture of cold acetone:methanol (7:3) (Haugan et al., 1992; Noviendri et al., 2011; Jaswir, et al., 2012). Fucoxanthin was extracted by Sivagnanam et al. (2015) using the mixture of hexane, ethanol, and acetone-mixed methanol (1:1, v/v) at a liquid to solid ratio of 20:1 (v/w) at 25°C and by stirring at 300 rpm using magnetic stirrer for 20 h. Yip et al. (2014) used methanol:chloroform:water (4:2:1, v/v/v) for extracting fucoxanthin from *Sargassum binderi*. Rajauria et al. (2017) and Rajauria and Ghannam (2013) used low-polarity solvents such as n-hexane, diethyl ether, and chloroform (1:1:1) for the extraction of fucoxanthin from *Himanthalia elongata*.

13.8 PURIFICATION OF FUCOXANTHIN

After extraction, purification of fucoxanthin is done in two steps, first partial purification and then complete purification.

13.8.1 SILICA GEL COLUMN CHROMATOGRAPHY

The most common procedure carried out for partial purification is silica gel column chromatography (Sudhakar et al., 2013; Narayani et al., 2016). Xia et al. (2013) extracted crude pigment which was subject to open silica gel column chromatography with n-hexane/acetone (6:4 solution; v/v) ((7:3, v/v) (Kim et al., 2004)) being the eluting solvent system. Noviendri et al. (2011) dissolved condensed crude pigment in benzene, and the benzene-containing residue was loaded to a silica column (Silica 60G, Merck, 0.040–0.063 mm). Elution was initially performed with n-hexane (100%) to remove chlorophyll and carotenoids other than fucoxanthin. Elution was continued with n-hexane:acetone (6:4; v/v) to recover fucoxanthin. Finally, residual fucoxanthin was eluted with acetone solvent. The acetone and hexane:acetone (6:4; v/v) fractions containing fucoxanthin were together evaporated to dryness by a rotary evaporator. The residue from the combined acetone and hexane:acetone (6:4, v/v) evaporation step was redissolved in methanol.

Fucoxanthin

Advantages:

- Useful in purification and isolating of pure materials from mixtures.
- Column chromatography is a preparative technique.
- It is primarily used to separate relatively large samples into pure components.
- Silica gels provide an ideal stationary phase for optimum purification.

Limitations:

- It is time-consuming and tedious, especially for large samples.

13.8.2 Thin Layer Chromatography (TLC)

Another method used for partial purification is TLC (Rajauria et al., 2017). TLC is also used for screening and to assess the purity of fucoxanthin. Dried fucoxanthin pigment was dissolved sufficiently in acetone and spotted onto the bottom of a TLC plate using a micropipette; the process was halted until the plates were dry. Then, the TLC plate was inserted into a glass beaker containing the mobile phase hexane:ethyl acetate (8:2, v/v) (p.a) 10 ml, covered with a petri dish so that the solvent did not evaporate; the process was halted until the solvent reached the upper limit (Kartikaningsih et al., 2017). Thin layer chromatography was used to calculate the retardation factor (Rf) value of the photosynthetic pigments, chlorophyll, carotenoids, and fucoxanthin present in the extracted samples. Silica gel-coated plates (Merck −10×6 cm) and the solvent used for TLC are n-hexane:acetone (7:3 v/v). Initially, the chromatography sheets were pre-saturated with the solvent. 5 µl of the sample was then carefully applied on the plates, and the samples were allowed to dry. The loaded plates were then placed in a pre-saturated tank with caution such that the applied sample does not dip in the solvent system. The set up was left undisturbed and the solvent was allowed to move up until it reached 3/4 of the plate. The plates were then removed from the tank and the color spots were marked immediately (Sudhakar et al., 2013). The Rf values were noted in the TLC plates and calculated by the standard formula given below for fucoxanthin:

[Rf = Distance moved by the pigment/Distance moved by the solvent] (Sudhakar et al., 2013).

Where,

Rf = Retention factor

Advantages:

- Quick, simple, and inexpensive way to analyze small samples.
- TLC is used primarily as an analytical technique.
- It can be used to identify components of a mixture.
- Purity of a sample is checked using TLC.
- TLC is used to find out the best solvents that can be used for column chromatography separation (Maria et al., 2017).

Limitations:

- TLC results are affected by factors like stationary phase, its particle size, type of plate used, and dimension of the development chamber.
- In TLC, path length is limited, and hence separation takes place only up to a certain length.
- Humidity and temperature affect the sample that is loaded on the TLC plate.

13.8.3 HIGH-PERFORMANCE LIQUID CHROMATOGRAPHY (HPLC)

Recently, HPLC analysis was carried out for identifying, quantifying, and purifying the fucoxanthin. The purity of the fucoxanthin was determined by preparative HPLC (ODS double column, 1 mL/min at 450 nm, methanol and acetonitrile were the mobile phase) (Passaquet et al., 1991; Kim et al., 2009; Noviendri et al., 2011; Jaswir et al., 2012; Kim et al., 2012a; Fung et al., 2013; Xia et al., 2013; Yip et al., 2014; Foo et al., 2015; Sivagnanam et al., 2015; Narayani et al., 2016; Erdogan et al., 2016; Susanto et al., 2016). Fung et al. (2013) and Sangeetha et al. (2009a) used absolute methanol for the mobile phase. The mobile phase used was 100% methanol with a flow rate of 1 ml/min, and the sample injection volume was 20 µl.

Advantages:

- Stationary phase is polar and mobile phase non-polar.
- The main purposes for using HPLC are for identifying, quantifying, and purifying the individual components of the mixture.
- HPLC technique is mostly used in the industrial and analytical field and helps in structure elucidation and quantitative determination of impurities and degradation products (Hassan, 2012).
- It requires a lesser sample, and the sample nature is not destroyed.
- HPLC solvents are forced through the column under high pressures of up to 400 atmospheres.
- Less time is required.

Limitations:

- Costly instrument.
- Need skill to run the instrument.
- A large amount of solvent is consumed.

13.8.4 REVERSED-PHASE HIGH-PERFORMANCE LIQUID CHROMATOGRAPHY (RP-HPLC)

Heo and Jeon (2009), Jaswir et al. (2013), and Zhang et al. (2014) used reversed-phase preparative high-performance liquid chromatography (HPLC) methods to quantify and purify fucoxanthin respectively. The extract after partial purification was dissolved in methanol and then combined with acetone and hexane:acetone (6:4; v/v). The concentration of fucoxanthin was then analyzed by HPLC. All-trans-fucoxanthin was further purified by RP-HPLC and filtered with a 0.22-mm membrane filter. The mobile phase used was methanol:acetonitrile (7:3, v/v) at a flow rate of 1.0 mL/min at 28°C (Maeda et al., 2006; Jaswir et al., 2012; Jaswir et al., 2013; Sivagnanam et al., 2015). The detection wavelength was set at 450 nm for detecting fucoxanthin (Mori et al., 2004; Jaswir et al., 2012; Jaswir et al., 2013). All the solvents used were of HPLC grade (Sivagnanam et al., 2015).

Advantages:

- The stationary phase is non-polar and the mobile phase is polar.
- Easier to modify.
- A large number of compounds can be identified.
- RP-HPLC is used for both analytical and preparative applications (Aguilar, 2003).
- Preparative RP-HPLC is also used for the micro-purification of protein fragments for sequencing to large-scale purification of synthetic peptides and recombinant proteins (Aguilar, 2003).
- RP-HPLC has been applied on the nano-, micro-, and analytical scale, and has also been scaled up for preparative purifications to a large industrial scale (Josic and Kovic, 2010).
- RP-HPLC is an indispensable tool in proteomic research (Josic and Kovic, 2010).
- Preparative RP-HPLC is often used for large-scale purification of proteins (Josic and Kovic, 2010).

Limitations:

- To modify polarity, the column must be changed.
- RP-HPLC can cause the irreversible denaturation of protein samples, thereby reducing the potential recovery of material in a biologically active form (Aguilar, 2003).

13.9 CHARACTERIZATION OF FUCOXANTHIN

A characterization study has been mainly carried out for knowing the detailed structure and properties of fucoxanthin. Identification, quantification, and structure elucidation of purified compounds were performed by liquid chromatography with diode array detection electrospray ionization mass spectrometry (LC-DAD-ESI-MS) and nuclear magnetic resonance (NMR) (1H and 13C) by Rajauria et al. (2017) and Yip et al. (2014). The purified fucoxanthin and functional groups and structures were determined using Fourier-transform infrared spectroscopy (FTIR) and proton (^{1}H NMR), respectively.

13.9.1 UV-Vis Spectrophotometric Analysis

UV-vis spectrophotometric analysis is one of the best methods for determination of impurities in organic molecules. Additional peaks can be observed due to impurities in the sample, and it can be compared with that of standard fucoxanthin. By also measuring the absorbance at specific wavelengths, the impurities can be detected. It is also used for the quantitative determination of compounds that absorb UV radiation. UV absorption spectroscopy can characterize those types of compounds which absorb UV radiation. Identification is done by comparing the absorption spectrum with the spectra of known fucoxanthin.

The standard formula used for the estimation of fucoxanthin is shown below:

$$\text{Fucoxanthin (mg g}^{-1}) = A470 - 1.239(A631 + A581 - 0.3 \times A664) - 0.0275 \times A664 / 141$$

Where,
- A = Absorbance at a particular wavelength
- V = Total volume of the pigment extract
- W = Weight of the sample used for extraction (Seely et al., 1972)

Yip et al. (2014) characterized using UV-vis spectroscopy in the wavelength range of 350–750 nm at room temperature and determined the yield of fucoxanthin in the crude extract using a standard fucoxanthin curve (dissolved in methanol:water (1:9, v/v)) with concentrations of 1, 2, 4, 6, 8, and 10 µg/ml, at the wavelength of $\lambda = 420$ nm. Rajauria et al. (2017) and Narayani et al. (2016) characterized purified fucoxanthin using UV-visible spectroscopy. Fucoxanthin content in seaweed samples was expressed as mg/g dry weight of seaweed sample (Jaswir et al., 2012).

13.9.2 Fourier-Transform Infrared Spectroscopy (FTIR)

FTIR spectroscopy helps to identify the presence of specific functional groups that build a compound (Sjahfirdi et al., 2012). One part of extract was mixed with 90 parts of dried potassium bromide (KBr) separately and then compressed to prepare a salt disc of 3 mm in diameter. These discs were subjected to IR-spectrophotometer. The absorption was read between 400 and 4000 cm^{-1} (Narayani et al., 2016). Yip et al. (2014) confirmed the presence of allenic bonds in the fucoxanthin from *Sargassum binderi*, but it showed a very low spectrum. It is believed that the sample contains high amounts of moisture, which interfere with the infrared absorbance.

The FTIR spectra of standard and purified fucoxanthin were analyzed. Mainly, 32 scans were signal-averaged for a single spectrum obtained within the region from 4000 to 500 cm^{-1}. The sample was analyzed as KBr pellet and compared with the fucoxanthin standard (Rajaurai and Ghannam, 2013; Narayani et al., 2016). Yip et al. (2014) compared the spectrum of analytes with fucoxanthin standard. (Rajaurai and Ghannam, 2013). FTIR identification of fucoxanthin crude extract showed the functional groups of OH, CH, C=O, C=C, CH2, C=O, COC, and C=C; trans substituted and isolated fucoxanthin had functional groups of OH, CH, allenic bond, C=O, C=C, CH2, COC, and C=C trans substituted (Zailanie and Purnomo, 2017).

13.9.3 FLUORESCENCE SPECTROSCOPY

A fundamental aspect of fluorescence spectroscopy is the measurement of light absorption. Emission spectra are usually independent of excitation wavelength. Hence it is useful to determine if the emission spectrum remains the same at different excitation wavelengths. Whenever the emission spectrum changes with excitation wavelength, one should suspect an impurity (Lakowicz, 2006; Kartikaningsih et al., 2017).

13.9.4 GAS CHROMATOGRAPHY MASS SPECTROMETRY (GC-MS) ANALYSIS

GC has played a fundamental role in determining how many components a sample has and in what proportion these exist in a mixture. The most used is the mass spectrometric detector (MSD), which allows one to obtain the "fingerprint" of the molecule, i.e., its mass spectrum. Mass spectra provide information on the molecular weight and elemental composition; if a high-resolution mass spectrometer is used, information is provided on functional groups present and in some cases the geometry and spatial isomerism of the molecule (Stashenko and Martinez, 2014).

Narayani et al. (2016) prepared samples for GC-MS with a chloroform solvent mixture. The samples were injected into the gas chromatograph, where high pure helium was used as the carrier gas.

13.9.5 LIQUID CHROMATOGRAPHY MASS SPECTROMETRY (LC-MS)

The LC-MS is an advantageous technique, which first separates fucoxanthin from a mixture, followed by ionization and separation of the ions on the basis of their mass/charge ratio. In a general way, the molecules present on the sample are converted into a gas-phase ionic species by the addition or removal of electrons or protons. Waghmode and Kumbar (2015) and Zailanie and Sukoso (2014) determined pure fucoxanthin by LCMS m/z, having a molecular weight of 659.43 and 658.1 respectively. The optimized parameters such as 120 V fragmentor voltage and 3.5 kV capillary voltages, were selected for liquid chromatography-electrospray ionization mass spectrometry (LC-ESI-MS) analysis of purified compounds (Rajauria et al., 2017). Xia et al. (2013) determined the molecular weight of purified fucoxanthin. Based on the fragment pattern at m/z 659.8 and 681.9 corresponding to [M+H]+ and [M+Na]+, respectively, the molecular weight of fucoxanthin was identified.

13.9.6 NUCLEAR MAGNETIC RESONANCE (NMR)

The NMR analysis is carried out for probing the chemical structure of the molecule. The Proton (^{1}H NMR) and carbon (^{13}C NMR) were performed on purified compound. The spectra were measured at an ambient temperature with 32 K data points and 128–1024 scans. The purified band collected from the preparative TLC plate was dried under nitrogen stream in order to remove traces of TLC-developing solvents.

Xia et al. (2013) subjected purified fucoxanthin to NMR spectroscopy for its structural determination. The complete assignments of the ^{1}H and ^{13}C NMR spectra of fucoxanthin revealed the signals assignable to polyene-containing acetyl, conjugated ketone, olefinic methyl, two quaternary germinal oxygen methyls, two quaternary germinal dimethyls, and allene groups. NMR analysis was conducted to determine the framework of hydrogen molecules of carbon atoms that form the structure of a compound. Functional groups of fucoxanthin in NMR analysis showed the presence of O_2 at 5.6 epoxide groups shown in the C5 and C6 atoms, and the compounds were allene and acetate. ^{13}C-NMR analysis showed the presences of 11 methyl group ($-CH_3$), acetyl group (–COCH3), quaternary C atom (=C=), carbonyl group (–C=O), and hydroxyl group (x-C-OH). And, ^{1}H-NMR showed the presence of 11 methyl singlet, methyl-methyl, and acetyl groups (Zailanie and Purnomo, 2017). Zhang et al. (2014) determined the structure of fucoxanthin, 9′cis-fucoxanthin, 13cis-fucoxanthin, and 13′cis-fucoxanthin by ^{1}H-NMR, ^{13}C-NMR, IR, and UV.

13.9.7 HIGH-PERFORMANCE LIQUID CHROMATOGRAPHY (HPLC)

The fucoxanthin content in the seaweed extracts was determined by HPLC with methanol:acetonitrile (7:3, v/v) as the mobile phase. This extract was dissolved in a mobile phase. Then it was filtered using a 0.22 µm membrane filter. The fucoxanthin detection wavelength was set at 450 nm. A standard curve prepared by using the authentic standard was used for quantification of the fucoxanthin content in the seaweed samples. The fucoxanthin content in the seaweed sample was expressed as milligrams per gram (Maeda et al., 2006; Yip et al., 2014; Sivagnanam et al., 2015). 10 µl pigment solution was injected into the HPLC Shimadzu with the chromolith stationary phase and methanol (p.a) mobile phase, with a flow rate of 1 mL/min at a temperature of 30°C. The next stage was analyzing samples at a wavelength of 430 nm (Kartikaningsih et al., 2017).

13.9.8 REVERSED-PHASE HIGH-PERFORMANCE LIQUID CHROMATOGRAPHY (RP-HPLC)

The fucoxanthin content in the seaweed extracts was determined by reversed-phase HPLC (RP-HPLC), with methanol:acetonitrile (7:3, v/v) as the mobile phase. This extract was dissolved in a mobile phase. Then it was filtered using a 0.22 µm membrane filter. The fucoxanthin detection wavelength was set at 450 nm. HPLC analysis was performed at 30°C using two serially connected reverse-phase (RP) columns with a mixture of methanol and acetonitrile (70:30, v/v) as the mobile phase at a flow rate of 1.0 mL/min. The eluent was monitored at 450 nm with a spectrophotometric detector (Nakazawa et al., 2009). Absolute methanol was used as the mobile for RP-HPLC (Heo et al., 2010; Jaswir et al., 2011; Fariman et al., 2016).

13.10 BIOACTIVITY OF FUCOXANTHIN

Fucoxanthin is used as an excellent feed additive for poultry breeding and in aquaculture. Wang et al. (2005) reported that fucoxanthin increases the yellow color of egg yolk when fed to poultry. Some are most commonly used at the industrial level, such as *Sargassum* spp., *Laminaria* spp., *Ascophyllum* spp., *Fucus* spp., and *Turbinaria* spp. (Woo et al., 2011). Recently, fucoxanthin has been receiving much attention due to its potential benefits to human health (Table 13.2).

TABLE 13.2
Bioactivity of Fucoxanthin

Bioactivity	Reference
Anti-cancer effect	
Lung cancer was treated by arresting cell cycle and induced apoptosis by modulating expression of p53, p21, Fas, PUMA, Bcl-2, and caspase-3/8.	Mei et al. (2017)
Fucoxanthin inhibited cancer by different signaling pathways, including the caspases, Bcl-2 proteins, MAPK, PI3K/Akt, JAK/STAT, AP-1, GADD45, and several other molecules that are involved in cell cycle arrest, apoptosis, antiangiogenesis, or inhibition of metastasis.	Martin (2015)
Fucoxanthin had potential to induce cell apoptosis that occurs in HeLa cells.	Zailanie et al. (2015)
Anticancer effects of fucoxanthin and fucoxanthinol on colorectal cancer cell lines were studied.	Takahashi et al. (2015)
Fucoxanthin induced G1 cell cycle arrest by reducing the expression of pro-apoptotic proteins like cyclin-dependent kinase (CDK) 4, CDK 6, cyclin E, XIAP, Bcl-2, and Bcl-xL.	Rokkaku et al. (2013)
Fucoxanthin inhibited the expression and secretion of MMP-9 and suppressed the expressions of the cell surface glycoprotein CD44 and CXC chemokine receptor-4 (CXCR4).	Chung et al. (2013)
Fucoxanthin inhibited the growth of LNCaP prostate cancer cells in a dose-dependent manner, which in turn effects the induction of GADD45A expression and G1 cell cycle arrest but not apoptosis.	Yoshiko (2012)
Fucoxanthin inhibited the growth of H1299 cells in a dose-dependent manner.	Jaswir et al. (2011)
Stereoisomers of fucoxanthin isolated from *Undaria pinnatifida* on cancer cell lines; these stereoisomers inhibit the growth of HL-60 cells.	Nakazawa et al. (2009)
Intracellular reactive oxygen species (ROS) generated by exposure to UV-B radiation, which was significantly decreased by addition with various concentrations of fucoxanthin.	Heo and Jeon (2009)
Apoptosis induced by fucoxanthin in HL-60 cells was associated with a loss of mitochondrial membrane potential at an early stage, but not with an increase in reactive oxygen species.	Nara et al. (2005)
Anti-pigmentary effect	
Fucoxanthin can be used against sunburn caused by ultraviolet (UV) irradiation. Effect of UV radiation on skin was sunburn and down-regulated filaggrin (Flg). In vitro analysis showed that UV irradiation of human dermal fibroblasts caused production of intracellular reactive oxygen species (ROS) without cellular toxicity. ROS production was suppressed by fucoxanthin.	Matsui et al. (2016)
Anti-pigmentary activity of fucoxanthin on UVB-induced melanogenesis. Fucoxanthin suppressed mRNA expression of cyclooxygenase (COX)-2, endothelin receptor A, p75neurotrophin receptor (NTR), prostaglandin E receptor 1 (EP1), melanocortin 1 receptor (MC1R), and tyrosinase-related protein. Also, oral application of fucoxanthin (10 mg/kg) significantly suppressed expression of COX-2, p75NTR, EP1, and MC1R.	Shimoda et al. (2010)
Anti-hypersensitivity	
Anti-hypersensitivity activity was detected by using Angiotensin I-Converting Enzyme (ACE) inhibitory assay.	Sivagnanam et al. (2015)

(Continued)

TABLE 13.2 (CONTINUED)
Bioactivity of Fucoxanthin

Bioactivity	Reference
Anti-inflammatory	
Fucoxanthin effectively protects against impairments in mice induced by scopolamine by increasing acetylcholinesterase (AChE) activity and by decreasing both choline acetyltransferase activity and brain-derived neurotrophic factor (BDNF) expression.	Lin et al. (2016)
Fucoxanthin significantly inhibited the NO production, which in turn reduced the prostaglandin E2 (PGE2) production.	Heo et al. (2010)
Fucoxanthin inhibited the cytoplasmic degradation of inhibitors of B (IκB)-α, p50, and p65 proteins; as a result, it lowered levels of nuclear factor (NF)-κB transactivation.	Kim et al. (2010)
Antimicrobial activity	
Antimicrobial activity of fucoxanthin against *Listeria monocytogenes* bacteria using TLC bioautography approach. Further, the active compound was purified using preparative TLC. This purified compound showed a strong antimicrobial (inhibition zone: 10.27 mm, 25 μg compound/disc) activities, which were examined by agar disc-diffusion bioassay.	Rajuria et al. (2013)
Anti-obesity effect	
Fucoxanthin-rich diets significantly suppressed body weight and white adipose tissue (WAT) weight gain induced by high-fat diets.	Maeda et al. (2009)
Anti-obesity effects of fucoxanthin were due to inhibition of lipid absorption and metabolism and inhibition of adipocyte differentiation.	Chu and Phang (2016)
Fucoxanthin inhibited accumulation of intercellular lipid during differentiation and regulated adipocytes involved in insulin resistance.	Lelyana (2016)
Fucoxanthin inhibited lipid accumulation during adipocyte differentiation of 3T3-L1 cells.	Maeda et al. (2006)
UCP1 is a mitochondrial protein responsible for thermogenic respiration, i.e., it is used to produce heat to avoid an excess of fat accumulation. Fucoxanthin up-regulates the expression of UCP1 and reduces weight.	Maeda et al. (2005)
Antioxidant activity	
Steaming treatment showed the highest antioxidant activity.	Susanto et al. (2017)
Fucoxanthin showed strong properties of hydroxyl radical scavenging activity.	Li et al. (2016)
Fucoxanthin supplementation in food could reduce oxidative stress.	D'Orazio et al. (2012)
Hydrogen peroxide scavenging (H_2O_2) assay showed the highest antioxidant activity.	Sivagnanam et al. (2015)
Study concluded all three fucoxanthin stereoisomers had stronger scavenging hydroxyl radical activities, but they showed weaker scavenging activities toward DPPH.	Zhang et al. (2014)
Fucoxanthin from *S. ilicifolium* showed the highest DPPH scavenging activity.	Sudhakar et al. (2013)
DPPH radical scavenging activity and CUPRAC assays were partly due to structural variations in the phenolic compounds and the selectivity of the antioxidant reacting to the assays.	Fung et al. (2013)
Fucoxanthin protection against oxidative stress caused by retinol deficiency by suppressing lipid peroxidation and enhancing activities of antioxidant enzymes catalase and glutathione transferase.	Sangeetha et al. (2009b)

(Continued)

TABLE 13.2 (CONTINUED)
Bioactivity of Fucoxanthin

Bioactivity	Reference
Antidiabetic effects	
Antidiabetic effects of fucoxanthin by improving the effect of insulin resistance on the diabetic model of KK-Ay mice. Fucoxanthin shows anti-diabetic effects by changing the adipocyte cell.	Maeda et al. (2015)
Fucoxanthin metabolites accumulate in abdominal adipose tissue at a higher rate than in plasma and other tissues, suggesting that the adipose tissue will be a main target of fucoxanthin metabolites.	Miyashita et al. (2013)
Fucoxanthin improves insulin resistance and decreases blood glucose levels through the regulation of cytokine secretions from white adipose tissue.	Nilapwar (2013)
Anti-wrinkle effect	
Fucoxanthin significantly lessened (1) UVB-induced epidermal abnormality, (2) VEGF (its up-regulation leads to tumor angiogenesis), and (3) MMP-13 (identified member of the matrix metalloproteinase (MMP) family that is expressed in breast carcinomas) expression in the epidermis and thiobarbituric acid reactive substances (TBARS), a byproduct of lipid peroxidation in the skin. These results indicate that topical treatment with fucoxanthin prevents skin photoaging in UV-B-irradiated hairless mice, possibly via antioxidant and antiangiogenic effects.	Urikura et al. (2011)
Vision-protecting effect	
Fucoxanthin inhibits overexpression of vascular endothelial growth factors, resisting senescence, improving phagocytic function, and clearing intracellular reactive oxygen species in retinal pigment epithelium cells. The in vivo experiment also confirmed the superiority of fucoxanthin over lutein in protecting retina against photoinduced damage.	Liu et al., (2016)
Antiangiogenic activity	
Fucoxanthin significantly suppressed human umbilical vein endothelial cells (HUVEC) proliferation and tube formation at more than 10 microM.	Sugawara et al. (2006)
Fucoxanthin suppress the mRNA expression of fibroblast growth factor 2 (FGF-2) and its receptor (FGFR-1), as well as their trans-activation factor, EGR-1. Antiangiogenic effects of fucoxanthin are due to the down-regulation of signal transduction by FGFR-1.	Ganesan et al. (2013)

13.11 CONCLUSION

Fucoxanthin is a valuable natural carotenoid. Presently, much interest has been shown by industry people towards fucoxanthin research as a bioactive compound. Fucoxanthin has a unique molecular structure. It is the one and only compound that contains allenic bond, which makes it highly reactive. Normally, light reflects back when it falls on seawater. Brown algae under the sea has to capture this light before reflecting it back, and this important role is carried out by the pigment present in brown algae, fucoxanthin. It captures light for photosynthesis. Fucoxanthin is present in both brown algae and diatoms. Yield of fucoxanthin varies in different algae. The content of fucoxanthin depends on the quality, quantity, and the methods that are adopted to extract fucoxanthin. Both polar and non-polar solvents are used for extraction. Most of the work is carried out by using polar solvent because it is cheaper and less harmful. Fucoxanthin is extracted from brown algae. Fucoxanthin yield can be increased if large quantities of brown algae are used for extraction. Harvesting algae for a byproduct will be a way of exploiting such a seaweed. So, we can valorize brown algal biomass after the extraction of fucoxanthin. After extracting fucoxanthin, the residue can be used for the extraction of fucoidan and alginate. Fucoxanthin has multiple biological uses such as anti-cancer effects, anti-pigmentary effects, anti-hypersensitivity, anti-inflammatory, antimicrobial activity, antioxidant activity, anti-diabetic effects, anti-wrinkle effects, vision-protecting effects, and antiangiogenic activity. The nutraceutical effects of fucoxanthin are being revealed day by day.

REFERENCES

Afolayan A F, Bolton J J, Lategan C A, Smith P J, Beukes D R, Fucoxanthin, tetraprenylated toluquinone and toluhydroquinone metabolites from *Sargassum heterophyllum* inhibit the in vitro growth of the malaria parasite *Plasmodium falciparum*, *Z. Naturforsch. C* 2008, 63(11–12), 848–852

Aguilar M I, HPLC of peptides and proteins: Methods and protocols, *Methods Mol. Biol.* 2003, 251

Airanthi M K W A, Sasaki N, Iwasaki S, Baba N, Abe M, Hosokawa M, Miyashita K, Effect of brown seaweed lipids on fatty acid composition and lipid hydroperoxide levels of mouse liver, *J. Agric. Food Chem.* 2011a, 59(8), 4156–4163

Airanthi M W A, Hosokawa M, Miyashita K, Comparative antioxidant activity of edible Japanese brown seaweeds, *J. Food Sci.* 2011b, 76(1), C104–C111

Anderson J M, Barrett J, Chlorophyll-protein complexes of brown algae: P700 reaction centre and light-harvesting complexes, *Ciba Found. Symp.* 1978, 61(61), 81–104

Berkaloff C, Caron L, Rousseau B, Subunit organization of PSI particles from brown algae and diatoms ± polypeptide and pigment analysis, *Photosynth. Res.* 1990, 23, 181–193

Bonnett R, Mallams A K, Spark A A, Tee J L, Weedon B C L, Mc Cormick A, Carotenoids and related compounds. Part XX. Structure and reactions of fucoxanthin, *J. Chem. Soc. C* 1969, 0, 429–454

Branch G M, Branch M L, The living shores of Southern Africa. In: *Struik*. Cape Town, Botany Department, University of the Western Cape, 1981

Branch G M, Griffiths C L, Branch M L, Beckley L E, *Two Oceans: A Guide to the Marine Life of Southern Africa*, David Philip, Claremont, Cape Town, 1994.

Chapman R L, Algae: The world's most important "plants"—An introduction, *Mitigation Adapt. Strateg. Glob. Change* 2013, 18(1), 5–12

Chen C R, Lin D M, Chang C M, Chou H N, Wu J J, Supercritical carbon dioxide anti-solvent crystallization of fucoxanthin chromatographically purified from *Hincksia mitchellae* P.C. Silva, *J. of Supercritical Fluids*, 2017, 119, 1–8

Chen M, Li Y, Birch D, Willows R D, A cyanobacterium that contains chlorophyll f—A red-absorbing photopigment, *FEBS Lett.* 2012, 586(19), 3249–3254

Christopher O T, Eduardo P, István T H, Roger A S, Martyn P, Valorization of biomass: Deriving more value from waste, *Science* 2012, 337(6095), 695

Chung T W, Choi H J, Lee J Y, Jeong H S, Kim C H, Joo M, Choi J Y, Han C W, Kim S Y, Choi J S, Ha K T, Marine algal fucoxanthin inhibits the metastatic potential of cancer cells, *Biochem. Biophys. Res. Commun.* 2013, 439(4), 580–585

Dawes C J, *Marine Botany*, 2nd edition, New York, John Wiley and Sons Inc., 1998

Dhargalkar V K, Pereira N, Seaweed: Promising plant of the millennium, *Sci. Cult.* 2005, 71(3–4)

D'Orazio N, Gammone M A, Gemello E, De Girolamo M, Cusenza S, Riccioni G, Marine bioactives: Pharmacological properties and potential applications against inflammatory diseases, *Mar. Drugs* 2012, 10(4), 812–833

Douady D, Rousseau B, Berkaloff C, Isolation and characterization of PS II core complexes from a brown alga, *Laminaria Saccharina*, *FEBS Lett.* 1993 324, 22–26

Erdogan A, Zeliha Demirel Z, Dalay M C, Eroglu A E, Fucoxanthin content of *Cylindrotheca closterium* and its oxidative stress mediated enhancement, *Turk. J. Fish. Aquat. Sci.* 2016, 16, 491–498

Fariman G A, Shastan S J, Zahedi M M, Seasonal variation of total lipid, fatty acids, fucoxanthin content, and antioxidant properties of two tropical brown algae (*Nizamuddinia zanardinii* and *Cystoseira indica*) from Iran, *J. Appl. Phycol.* 2016, 28(2), 1323–1331

Foo S C, Fatimah M Y, Maznah I, Mahiran B, Chan K W, Khong N M H, Yau S K, Production of fucoxanthin-rich fraction (FxRF) from a diatom, *Chaetoceros calcitrans* (Paulsen) Takano 1968, *Algal Res.* 2015, 12, 26–32

Fuentes M M R, Pérez A N, Miras J J R, Guerrero J L G, Biomass nutrients profiles of the microalga Phaeodactylum tricornutum, *J. Food Biochem.* 2001, 25, 57–76

Fung A, Hamid N, Lu J, Fucoxanthin content and antioxidant properties of *Undaria pinnatifida*, *Food Chem.* 2013, 136(2), 1055–1062

Ganesan P, Matsubara K, Sugawara T, Hirata T, Marine algal carotenoids inhibit angiogenesis by down-regulating FGF-2-mediated intracellular signals in vascular endothelial cells, *Mol. Cell. Biochem.* 2013, 380(1–2), 1–9

Guo B, Liu B, Yang B, Sun P, Lu X, Liuand J, Chen F, Screening of diatom strains and characterization of *Cyclotella cryptica* as a potential fucoxanthin producer, *Mar. Drugs* 2016, 14(7), 125

Hashimoto T, Ozaki Y, Taminato M, Das S K, Mizuno M, Yoshimura K, Maoka T, Kanazawa K, The distribution and accumulation of fucoxanthin and its metabolites after oral administration in mice, *Br. J. Nutr.* 2009, 102(2), 242–248

Hassan B A R, HPLC uses and importance in the pharmaceutical analysis and industrial field, *Pharmaceut. Anal. Acta* 2012, 3, e133

Haugan J A, Aakermann T, Jensen S L, Isolation of fucoxanthin and peridinin, *Methods Enzymol.* 1992, 213

Haugan J A, Liaaen-Jensen S, Improved isolation procedure for fucoxanthin, *Phytochemistry* 1989, 28(10), 2797–2798

Haxo F T, Blinks L R, Photosynthetic action spectra of marine algae, *J. Gen. Physiol.* 1946, 33, 389'

Heo S J, Jeon Y J, Protective effect of fucoxanthin isolated from *Sargassum siliquastrum* on UV-B induced cell damage, *J. Photochem. Photobiol. B Biol.* 2009, 95(2), 101–107

Heo S J, Yoon W J, Kim K N, Ahn G N, Kang S M, Kang D H, Affan A, Oh C, Jung W K, Jeon Y J, Evaluation of anti-inflammatory effect of fucoxanthin isolated from brown algae in lipopolysaccharide-stimulated RAW 264.7 macrophages, *Food Chem. Toxicol.* 2010, 48(8–9), 2045–2051

Iio K, Okada Y, Ishikura M, Single and 13-week oral toxicity study of fucoxanthin oil from microalgae in rats, *J. Food Hyg. Soc. Jpn.* 2011, 52, 183–189

Ikeda Y, Komura M, Watanabe M, Minami C, Koike H, Itoh S, Kashino Y, Satoh K, Photosystem I complexes associated with fucoxanthin-chlorophyll binding proteins from a marine centric diatom, *Chaetoceros gracilis*, *BBA Bioenerg.* 2008, 1777(4), 351–361

Ikeda Y, Yamagishi A, Komura M, Suzuki T, Dohmae N, Shibata Y, Itoh S, Koike H, Satoh K, Two types of fucoxanthin-chlorophyll-binding proteins I tightly bound to the photosystem I core complex in marine centric diatoms, *Biochim. Biophys. Acta* 1827, 2013, 529–539

Jaswir I, Noviendri D, Salleh H M, Miyashita Kazuo, Fucoxanthin extractions of brown seaweeds and analysis of their lipid fraction in methanol, *Food Sci. Technol. Res.* 2012, 18(2), 251–257

Jaswir I, Noviendri D, Salleh H M, Taher M, Miyashita K, Isolation of fucoxanthin and fatty acids analysis of *Padina australis* and cytotoxic effect of fucoxanthin on human lung cancer (H1299) cell lines, *Afr. J. Biotechnol.* 2011, 10(81), 18855–18862

Jaswir I, Noviendri D, Salleh H M, Taher M, Miyashita K, Ramli N, Analysis of fucoxanthin content and purification of all-trans-fucoxanthin from *Turbinaria turbinata* and *Sargassum plagyophyllum* by SiO_2 open column chromatography and reversed phase-HPLC, *J. Liq. Chromatogr. Relat. Technol.* 2013, 36, 1340–1354

Jeffrey S W, Mantoura R F C, Wright S W, *Phytoplankton Pigments in Oceanography. Guidelines to Modern Methods*, Paris, Unesco, 1997

Josic D, Kovic S, Reversed-phase high performance liquid chromatography of proteins. *Curr. Protoc. Protein Sci.* 2010; Chapter 8: Unit 8.7

Kanazawa K, Ozaki Y, Hashimoto T, Das S K, Matsushita S, Hirano M, Okada T, Komoto A, Mori N, Nakatsuka M, Commercial-scale preparation of bio functional fucoxanthin from waste parts of brown sea algae *Laminalia japonica*, *Food Sci. Technol. Res.* 2008, 14(6), 573–582

Kanda H, Kamo Y, Machmudah S, Wahyudiono E Y, Goto M, Extraction of fucoxanthin from raw macroalgae excluding drying and cell wall disruption by liquefied dimethyl ether, *Mar. Drugs* 2014, 12(5), 2383–2396

Kanes S, Forster D, *The Choice of Next-Generation Biofuels (Algae Excerpt); Equity Research Industry Report*, New York, Scotia Capital, 2009

Kang M C, Lee S H, Lee W W, Kang N, Kim E A, Kim S Y, Lee D H, Kim D, Jeon Y, Protective effect of fucoxanthin isolated from *Ishige okamurae* against high-glucose induced oxidative stress in human umbilical vein endothelial cells and zebrafish model, *J. Funct. Foods* 2014, 11, 304–312

Kang S I, Shin H S, Kim H M, Yoon S A, Kang S W, Kim J H, Ko H C, Kim S J, *Petalonia binghamiae* extract and its constituent fucoxanthin ameliorate high-fat diet-induced obesity by activating AMP-activated protein kinase, *J. Agric. Food Chem.* 2012, 60(13), 3389–3395

Kartikaningsih H, Deviana Mufti E, Eko Nurhanief A, Fucoxanthin from brown seaweed *Sargassum cristaefolium* tea in acid ph., 2017, *Therapeutic 7th International Conference on Global Resource Conservation AIP Conf. Proc. 1844*, American Institute of Physics

Katoh T, Nagashima U, Mimuro M, Fluorescence properties of the allenic carotenoid fucoxanthin: Implication for energy transfer in photosynthetic pigment systems, *Photosynth. Res.* 1991, 27(3), 221–226

Ke B, *Photosynthesis: Photobiochemistry and Photobiophysics*, Kluver Academic Publisher, Vol. 10: Chapter 13, 239, 2001

Kim K N, Heo S J, Yoon W J, Kang M, Ahn G, Hoo Y T, Jeon Y J, Fucoxanthin inhibits the inflammatory response by suppressing the activation of NF-κB and MAPKs in lipopolysaccharide-induced RAW 264.7 macrophages., *Eur. J. Pharmacol.* 2010, 649(1–3), 369–375

Kim S J, Kim H J, Moon J S, Kim J M, Kang S G, and Jung S T, Characteristic and extraction of fucoxanthin pigment in *Undaria pinnatifida*, *Journal of Korean Society for Food Science and Nutrition*, 2004, 33, 847–851

Kim S M, Kang S W, Kwon O N, Chung D, Pan C H, Fucoxanthin as a major carotenoid in *Isochrysisaff. galbana*: Characterization of extraction for commercial application, *J. Korea Soc. Appl. Biol. Chem.* 2012a, 55(4), 477–483

Kim S M, Jung Y J, Kwon O N, Cha K H, Um B H, Chung D, Pan C H, A potential commercial source of fucoxanthin extracted from the microalga *Phaeodactylum tricornutum*, *Appl. Biochem. Biotechnol.* 2012b, 166(7), 1843–1855

Kosumi D, Kusumoto T, Fujii R, Sugisaki M, Iinuma Y, Oka N, Takaesu Y, Taira T, Iha M, Frank H A, Hashimoto H, Ultrafast excited state dynamics of fucoxanthin: Excitation energy dependent intramolecular charge transfer dynamics, *Phy. Chem. Chem. Phys.* 2011, 13(22), 10762–10770

Kuczynska P, Rzeminska M J, Strzalka K, Photosynthetic pigments in diatoms, *Mar. Drugs* 2015, 13(9), 5847–5881

Kuznetsova V, Chábera P, Litvín R, Polívka T, Fuciman M, Effect of isomerization on excited-state dynamics of carotenoid fucoxanthin, *J. Phys. Chem. B* 2017, 121(17), 4438–4447

Lakowicz J R, 2006, *Principles of Fluorescence Spectroscopy*, Springer, XXVI, 954

Lavanya R, Maheshwari S U, Harish G, Raj J B, Kamali S and Hemamalani D, Investigation of in vitro anti-inflammatory anti-platelet and anti-arthritic activities in the leaves of *Anisomeles malabarica* Linn, Research *Journal of Pharmaceutical, Biological and Chemical Sciences* 2010 1(4), 745–752

Lee C S, Chong M F, Robinson J, Binner E, Preliminary study on extraction of bio-flocculants from okra and Chinese yam. *ASEAN Journal of Chemical Engineering* 2015, 15(1), 41–51

Lelyana R, Role of marine-natural ingredient fucoxanthin on body's immune response of obesity, *J. Nano. Med. Nano. Technol.* 2016, 7, 5

Li Y, Liu Y, Wang P Y, Yu Y, Zeng Y, Li L, Wang L, The bioactivity of fucoxanthin from *Undaria pinnatifida* invitro, *Am. J. Biochem. Biotechnol.* 2016, 12(2), 139–148

Lichtenthaler H K, The 1-deoxy-D-xylulose-5-phosphate pathway of isoprenoid biosynthesis in plants, *Annu. Rev. Plant Physiol. Plant Mol. Biol.* 1999, 50, 47–65

Lin J, Huang L, Yu J, Xiang S, Wang J, Zhang J, Yan X, Cui W, He S, Wang Q, Fucoxanthin, a marine carotenoid, reverses scopolamine-induced cognitive impairments in mice and inhibits acetylcholin esterase in vitro, *Mar. Drugs* 2016, 14(4), 67

Liu Y, Liu Meng, Zhang Xichun, Chen Qingchou, Chen Haixiu, Sun Lechang, Liu Guangming, Protective effect of fucoxanthin isolated from *laminaria japonica* against visible light-induced retinal damage both in vitro and in vivo, *J. Agric. Food Chem.* 2016, 64(2), 416–424

Lohr M, Wilhelm C, Algae displaying the diadinoxanthin cycle also possess the violaxanthin cycle, *Proc. Natl. Acad. Sci. U.S.A.* 1999, 96(15), 8784–8789

Loredo A G, Benavides J, Palomares R M, Growth kinetics and fucoxanthin production of Phaeodactylum tricornutum and Isochrysis galbana cultures at different light and agitation conditions, *Journal of Applied Phycology* 2016, 28(2), 849

Maeda H, Hosokawa M, Sashima T, Funayama K, Miyashita K, Fucoxanthin from edible seaweed, *Undaria pinnatifida*, shows antiobesity effect through UCP1 expression in white adipose tissues. *Biochem. Biophys. Res. Commun.* 2005, 332(2), 392–397

Maeda H, Hosokawa M, Sashima T, Murakami-Funayama K, Miyashita K, Anti-obesity and anti-diabetic effects of fucoxanthin on diet-induced obesity conditions in a murine model, *Mol. Med. Rep.* 2009, 2(6), 897–902

Maeda H, Hosokawa M, Sashima T, Takahashi N, Kawada T, Miyashita K, Fucoxanthin and its metabolite, fucoxanthinol, suppress adipocyte differentiation in 3T3-L1 cells, *Int. J. Mol. Med.* 2006, 18(1), 147–152

Maeda H, Kanno S, Kodate M, Hosokawa M, Miyashita K, Fucoxanthinol, metabolite of fucoxanthin, improves obesity-induced inflammation in adipocyte cells, *Mar. Drugs* 2015, 13(8), 4799–4813

Maria F C, Abesamis M E C P, Acosta F M, Agustin M C G, Aquitania M J H, Bagsican. Formal report prepared by Marilu Jane H Bagsican Group 1 2E Medical Technology, 2017, Organic Chemistry Laboratory

Martin L J, Fucoxanthin and its metabolite fucoxanthinol in cancer prevention and treatment, *Mar. Drugs* 2015, 13(8), 4784–4798

Martino A D, Douay D, Szely M Q, Rousseau B, Âpineau F C, Apt K, Caron L, The light-harvesting antenna of brown algae: Highly homologous proteins encoded by a multi-gene family, *Eur. J. Biochem.* 2000, 267, 5540–5549

Mei C H, Zhou S C, Zhu L, Ming J X, Zeng F D, Xu R, Antitumor effects of *laminaria* extract fucoxanthin on lung cancer, *Mar. Drugs* 2017, 15(2), 39

Mikami M, Hosokawa K, Biosynthetic pathway and health benefits of fucoxanthin, an algae-specific xanthophyll in brown seaweeds, *Int. J. Mol. Sci.* 2013, 14(7), 13763–13781

Mise T, Ueda M, Yasumoto T, Production of fucoxanthin-rich powder from *Cladosiphon okamuranus*, *Adv. J. Food Sci. Technol.* 2011, 3(1), 73–76

Mishra V K, Bachet R K, Husen A, Medicinal uses of chlorophyll: A critical overview. In: *Chlorophyll: Structure, Function and Medicinal Uses*, Hua Le and Elisa eSalcedo, ed., Hauppauge, Nova Science Publishers, Inc., 177–196, 2011

Miyashita H, Ikemoto H, Kurano N, Adachi K, Chihara M, Miyachi S, Chlorophyll d as a major pigment, *Nature* 1996, 383(6599), 402

Miyashita K, Mikami N, Hosokawa M, Chemical and nutritional characteristics of brown seaweed lipids: A review, *J. Funct. Food* 2013, 5, 1507–1517

Miyashita K, Nishikawa N, Beppu F, Tsukui T, Abe M, Hosokawa M, The allenic carotenoid fucoxanthin, a novel marine nutraceutical from brown seaweed, *J. Sci. Food Agric.* 2011, 91, 1166–1174

Miyashita Y, Saiki A, Endo K, Ban N, Yamaguchi T, Kawana H, Nagayama D, Ohira M, Oyama T, Shirai K, Effects of olmesartan, an angiotensin II receptor blocker, and amlodipine, *J Atheroscler Thromb.* 2009,16(6), 912

Moreau D, Tomasoni C, Jacquot C, Kaas R, Le Guedes R, Cadoret J P, Muller-Feuga A, Kontiza I, Vagias C, Roussis V, Roussakis C, Cultivated microalgae and the carotenoid fucoxanthin from Odontella aurita as potent anti-proliferative agents in bronchopulmonary and epithelial cell lines, *Environ. Toxicol. Pharmcol.* 2006, 22(1), 97–103

Mori K, Ooi T, Hiraoka M, Oka N, Hamada H, Tamura M, Kusumi T, Fucoxanthin and its metabolites in edible brown algae cultivated in deep seawater, *Mar. Drugs* 2004, 2, 63–72

Murphy F, Devlin G, Deverell R, McDonnell K, Biofuel production in Ireland—An approach to 2020 targets with a focus on algal biomass, *Energies* 2013, 6(12), 6391–6412

Nakazawa Y, Sashima T, Hosokawa M, Miyashita K, Comparative evaluation of growth inhibitory effect of stereoisomers of fucoxanthin in human cancer cell lines, *J. Funct. Foods* 2009, 1, 88–97

Nara E K, Terasaki M, Nagao A, Characterization of apoptosis induced by fucoxanthin in human promyelocytic leukemia cells., *Biosci. Biotechnol. Biochem.* 2005, 69(1), 224–227

Narayani S S, Saravanan S, Bharathiaraja S, Mahendran S, Extraction, partially purification and study on antioxidant property of fucoxanthin from *Sargassum cinereum* J. Agardh, *J. Chem. Pharm. Res.* 2016, 8(3), 610–616

Nilapwar S M, Nardelli M, Westerhoff H V, Verma M, Absorption spectroscopy, *Methods Enzymol.* 2011, 500, 59–75

Nishino H, Cancer prevention by carotenoids, *Mutat. Res.* 1998, 402(1–2), 159–163

Nomura T, Kikuchi M, Kubodera A, Kawakami Y, Proton-donative antioxidant activity of fucoxanthin with 1,1-diphenyl-2-picrylhydrazyl (DPPH), *Biochem. Mol. Biol. Int.* 1997, 42(2), 361–370

Noviendri D, Jaswir I, Salleh H M, Taher M, Fucoxanthin extraction and fatty acid analysis of *Sargassum binderi* and *S. duplicatum*, *J. Med. Plants Res.* 2011, 5(11), 2405–2412

Papagiannakis E, van Stokkum H M, Fey H, Buchel C, van Grondelle R, Spectroscopic characterization of the excitation energy transfer in the fucoxanthin-chlorophyll protein of diatoms, *Photosynth. Res.* 2005, 86(1), 241–250

Pasquet V, Cherouvrier J R, Farhat F, Thiery V, Piot J M, Berard JB, Kaas R, Serive B, Patrice T, Cadoret J P, Study on the microalgal pigments extraction process: performance of microwave assisted extraction, *Process Biochem* 2011, 46(1), 59–67

Passaquet C, Thomas J C, Caron L, Hauswirth N, Puel F, Berkuloff C, Light-harvesting complexes of brown algae. Biochemical characterization and immunological relationships, *FEBS Lett.* 1991, 280(1), 21–26

Peng J, Yuan J P, Wu C F, Wang J H, Fucoxanthin, a marine carotenoid present in brown seaweeds and diatoms: Metabolism and bioactivities relevant to human health, *Mar. Drugs* 2011, 9(10), 1806–1828

Piovan A, Seraglia R, Bresin B, Caniato R, Filippini R, Fucoxanthin from Undaria pinnatifida: photostability and coextractive effects, *Molecules* 2013 May 29, 18(6), 6298–6310

Prasanna R, Sood A, Suresh A, Nayak S, Kaushik S, Potentials and applications of algal pigments in biology and industry, *Acta Bot. Hung.* 2007, 49(1–2)

Rajauria G, Ghannam N A, Isolation and partial characterization of bioactive fucoxanthin from *Himanthalia elongata* brown seaweed: A TLC-based approach, *Int. J. Anal. Chem.* 2013, Article ID 802573, 6 pages

Rajauria G, Foley B, Ghannam N A, Characterization of dietary fucoxanthin from *Himanthalia elongate* brown seaweed, *Food Res. Int.* 2017, 99, 995–1001

Rijstenbil J W, Effects of UVB radiation and salt stress on growth, pigments and antioxidative defence of the marine diatom *Cylindrotheca closterium*, *Mar. Ecol. Prog. Ser.* 2003, 254, 37–48

Rokkaku T, Kimura R, Ishikawa C, Yasumoto T, Senba M, Kanaya F, Mori N, Anticancer effects of marine carotenoids, fucoxanthin and its deacetylated product, fucoxanthinol, on osteosarcoma., *Int. J. Oncol.* 2013, 43(4), 1176–1186

Sangeetha R K, Bhaskar N, Baskaran V, Comparative effects of β-carotene and fucoxanthin on retinol deficiency induced oxidative stress in rats, *Mol. Cell. Biochem.* 2009a, 331(1–2), 59–67

Sangeetha R K, Bhaskar N, Divakar S, Baskaran V, Bioavailability and metabolism of fucoxanthin in rats: structural characterization of metabolites by LC-MS (APCI), *Mol. Cell. Biochem*, 2009b, 333, 299–310

Sangeetha R K, Hosokawa M, Miyashita K, Fucoxanthin: A marine carotenoid exerting anticancer effects by affecting multiple mechanisms, *Mar. Drugs* 2013, 11(12), 5130–5147

Sayre R, Microalgae: The potential for carbon capture, *BioScience* 2010, 60(9), 722–727

Seely G R, Duncan M J, Vidaver W E, Preparative and analytical extraction of pigments from brown algae with dimethyl sulfoxide, *Mar. Biol.* 1972, 12, 184

Shang Y F, Kim S M, Lee W J, Um B H, Pressurized liquid method for fucoxanthin extraction from *Eisenia bicyclis* (Kjellman) Setchell, *J. Biosci. Bioeng.* 2011, 111(2), 237–241

Shimoda H, Tanakaa J, Shanaand S-J, Maoka T, Anti-pigmentary activity of fucoxanthin and its influence on skin mRNA expression of melanogenic molecules, *J. Pharm. Pharmacol.* 2010, 62(9), 1137–1145

Sivagnanam S P, Yin S, Choi Y H, Park Y B, Woo H C, Chun B S, Biological properties of fucoxanthin in oil recovered from two brown seaweeds using supercritical CO_2 extraction, *Mar. Drugs* 2015, 13(6), 3422–3442

Sjahfirdi L, Mayangsari and Nasikin M, Protein identification using Fourier Transform Infrared (FTIR), *IJRRAS* 2012, 3, 10

Stashenko E, Martinez J R, Gas chromatography-mass spectrometry, *Intact* 2014, Chapter 1

Strain H H, Manning W M, Chlorofucine, *(Chlorophyll R), A Green Pigment of Diatoms and Brown Algae*, Carnegie Institution of Washington, Division of Plant Biology, Stanford University, California, 1942

Strand A, Herstad O, Liaaen-Jensen S, Fucoxanthin metabolites in egg yolks of laying hens, *Comp. Biochem. Phys. A Mol. Integr. Physiol.* 1998, 119(4), 963–974

Sudhakar M P, Ananthalakshmi J S, Nair Beena B, Extraction, purification and study on antioxidant properties of fucoxanthin from brown seaweeds, *J. Chem. Pharm. Res.* 2013, 5(7), 169–175

Sugawara T, Matsubara K, Akagi R, Mori M, Hirata T, Antiangiogenic activity of brown algae fucoxanthin and its deacetylated product, fucoxanthinol, *J. Agric. Food Chem.* 2006, 54(26), 9805–9810

Susanto E, Fahmi A S, Agustini T W, Rosyadi S, Wardani A D, Effects of different heat processing on fucoxanthin, antioxidant activity and colour of Indonesian brown seaweeds, *IOP Conf. Ser. Earth Environ. Sci.* 2017, 55, 012063

Susanto E, Fahmia A S, Abeb M, Hosokawa M, Miyashita K, Lipids, fatty acids, and fucoxanthin content from temperate and tropical brown seaweeds, *Aquat. Procedia* 2016, 7, 66–75

Takahashi K, Hosokawa M, Kasajima H, Hatanaka K, Kudo K, Shimoyama N, Miyashita K, Anticancer effects of fucoxanthin and fucoxanthinol on colorectal cancer cell lines and colorectal cancer tissues, *Oncol. Lett.* 2015, 10(3), 1463–1467

Takaichi S, Carotenoids in algae: Distributions, biosyntheses and functions, *Mar. Drugs* 2011, 9(6), 1101–1118

Takaichi S, Mirauro M, Distribution and geometric isomerism of neoxanthin in oxygenic phototrophs: 9′-cis, a sole molecular form, *Plant Cell Physiol.* 1998, 39(9), 968–977

Terasaki M, Hirose A, Narayan B, Baba Y, Kawagoe C, Yasui H, Saga N, Hosokawa M, Miyashita K, Evaluation of recoverable functional lipid components of several brown seaweeds of Japan with special reference to fucoxanthin and fucosterol contents, *J. Phycol.* 2009, 45(4), 974–980

Urikura S T, Hirata T, Hirata T, Protective effect of fucoxanthin against UVB-induced skin photoaging in hairless mice, *Biosci. Biotechnol. Biochem.* 2011, 75(4), 757–760

Veith T, Buchel C, The monomeric photosystem I-complex of the diatom *Phaeodactylum tricornutum* binds specific fucoxanthin chlorophyll proteins (FCPs) as light-harvesting complexes, *Biochemical et Biophysical Acta* 2007, 1767, 1428–1435

Vimala T, Poonghuzhali T V, Estimation of pigments from seaweeds by using acetone and DMSO, *International Journal of Science and Research (IJSR)*, ISSN (Online) Index Copernicus Value 2013, 6(14) | Impact Factor (2014): 5.611, 2319–7064

Waghmode A V, Kumbar R R, Phytochemical screening and isolation of fucoxanthin content of *Sargassum ilicifolium*, *Int. J. Pure App. Biosci.* 2015, 3(6), 218–222

Wang C, Kim J-H, Kim S-W, Synthetic biology and metabolic engineering for marine carotenoids: New opportunities and future prospects, *Mar. Drugs* 2014, 12(9), 4810–4832

Wang L, Zeng Y, Liu Y, Hu X, Li S, Wang Y, Li L, Lei Z, Zhang Z, Fucoxanthin induces growth arrest and apoptosis in human bladder cancer 2014 T24 cells by up-regulation of p21 and down-regulation of mortalin, *Acta Biochimica et Biophysica Sinica* 2014, 6, 877–884

Wang W J, Wang G C, Zhang M, Tseng C K, Isolation of fucoxanthin from the rhizoid of *Laminaria japonica* Aresch, *J. Integr. Plant Biol.* 2005, 47(8), 1009–1015

Wen Z Y, Chen F, Production potential of eicosapentaenoic acid by the diatom, *Biotechnol. Lett.* 2000, 22(9), 727–733

Westermann M, Rhiel E, Localisation of fucoxanthin chlorophyll a/c-binding polypeptides of the centric diatom *Cyclotella cryptica* by immuno-electron microscopy, *Protoplasma* 2005, 225(3–4), 217–223

Willstatter R, Page H J, Chlorophyll. XXIV. The pigments of the brown algae, *Justus Liebigs Ann. Chem.* 1914, 404(3), 237–271

Woo H C, Chun B S, Jasso R, Mussatto S I, Pastrana L, Aguilar C N, Teixeira J A, Microwave-assisted extraction of sulfated polysaccharides (fucoidan) from brown seaweed, *Carbohydr. Polym.* 2011, 86(3), 1137–1144

Xia S, Gao B, Li A, Xiong J, Ao Z, Zhang C, Preliminary characterization, antioxidant properties and production of chrysolaminarin from marine diatom *Odontella aurita*, *Mar. Drugs* 2014, 12(9), 4883–4897

Xia S, Wang K, Wan L, Li A, Hu Q, Zhang C, Production, characterization, and antioxidant activity of fucoxanthin from the marine diatom *Odontella aurita*, *Mar. Drugs* 2013, 11(7), 2667–2681

Xiao X, Si X, Yuan Z, Xu X, Li G, Isolation of fucoxanthin from edible brown algae by microwave-assisted extraction coupled with high-speed countercurrent chromatography, *J. Sep. Sci.* 2012, 35(17), 2313–2317

Yan X, Chuda Y, Suzuki M, Nagata T, Fucoxanthin as the major antioxidant in *Hijikia fusiformis*, a common edible seaweed, *Biosci. Biotechnol. Bioeng.* 1999, 63(3), 605–607

Yip W H, Joe L S, Mustapha W A W, Maskat M Y, Said M, Characterisation and stability of pigments extracted from *Sargassum binderi* obtained from Semporna, Sabah, *Sains Malays.* 2014, 43(9), 1345–1354

Zailanie K, Kartikaningsih H, Kalsum U, Effect of *Sargassum filipendula* fucoxanthin against HeLa cell and lymphocyte proliferation, *J. Life Sci. Biomed.* 2015, 5(2): 53–59

Zailanie K, Purnomo H, Study of fucoxanthin content and identification from three types of brown seaweeds (Sargassum cinereum, Sargassum echinocarpum and Sargassum filipendula) from Padike, Talongo, Sumenep, Madura. *Berkala Penelitian Hayati Edisi Khusus*, 2011, 7(A), 143–147

Zailanie K, Purnomo H, Identification of fucoxanthin from brown algae (*Sargassum filipendula*) from Padike village, Talango district, Sumenep regency, Madura islands, using nuclear magnetic resonance (NMR), *Int. Food Res. J.* 2017, 24(1), 372–378

Zailanie K, Sukoso, Study on of fucoxanthin content and its identification in brown algae from Padike Vilage Talango District, Madura Islands, *J. Life Sci. Biomed.* 2014, 4(1), 01–03

Zaragoza M C, Lopez D, Sáiz M P, Poquet M, Perez J, Puig-Parellada P, Marmol F, Simonetti P, Gardana C, Lerat Y, Burtin P, Inisan C, Rousseau I, Besnard M, Mitjavila M T, Toxicity and antioxidant activity in vitro and in vivo of two *Fucus vesiculosus* extracts, *J. Agric. Food Chem.* 2008, 56(17), 7773–7780

Zhang Y, Fang H, Xie Q, Sun S, Liu R, Hong Z, Yi R, Wu H, Comparative evaluation of the radical-scavenging activities of fucoxanthin and its stereoisomers, *Molecules* 2014, 19(2), 2100–2113

Zhao D, Kim S M, Pan C H, Chung D, Effects of heating, aerial exposure and illumination on stability of fucoxanthin in canola oil, *Food Chem.* 2014a, 145, 505–513

Zhao P, Zanga Z, Xie X, Huanga A, Wanga G, The influence of different flocculants on the physiological activity and fucoxanthin production of *Phaeodactylum tricornutum*, *Proc. Biochem.* 2014b, 49(4), 681–687

Index

A

Absorption spectrum, 226–227
Acid-base catalysis, 142
Adsorption-coupled flocculation, 170–172
AD, *see* Anaerobic digestion
Advanced oxidation processes (AOPs), 22
AF, *see* Anaerobic filters
Algae, 8–9, 228–231
Algal biomass
   biofuel production
      biochemical conversion (BCC), 132–134
      chemical conversion, 130–131
      thermochemical conversion, 131–132
   biomass feedstock, 124–125
      various feedstocks, 124
   biorefinery, 134–135
   cultivation and nutrients of
      closed-loop system, 128–130
      open pond, 127–128
   macro-algae, advantage of, 125–127
Anaerobic digestion (AD), 133–134
Anaerobic filters (AF), 30
Anaerobic sequential batch reactor (ASBR), 39
   continuous study in, 46–48
   F/M ratio in, 50
   gas production, 50
   mixed liquor volatile suspended solid concentration, 49
   optimization of process variables in, 44–46
   optimization studies in, 39–41
   sludge volume index (SVI), 49
   for textile dyeing effluent treatment, 41–44
   volatile fatty acid, 49–50
AOPs, *see* Advanced oxidation processes
ASBR, *see* Anaerobic sequential batch reactor
Autocatalysis, 141
Azo dye
   biological degradation
      advancements, 31
      bioaugmentation of microbes, 25–27
      by microbes, 25
      nanotechnology in biological remediation, 27–28
      by plants, 24
      reactors for, 28–30
   degradation of
      advanced oxidation processes (AOPs), 22
      biosorption, 22
      enzymatic degradation, 22–23
      enzymatic methods, 23–24
      ozonation, 22
      physico-chemical degradation, 22
   environmental concerns, 19–21
   industrial application of, 18–19

B

Bacteria, 8, 183–184
BCC, *see* Biochemical conversion
Bioaugmentation of microbes, 25–27
Biocatalysis, 141–142; *see also* Enzymatic catalysis
Biocatalytic reactions
   diffusional limitations
      hydrogels, 145
      non-porous supports, 146–147
      sensitive matrices, 145–146
   factors influencing biocatalytic action
      diffusion, 143–145
      enzyme concentration, 143
      substrate concentration, 142–143
      surface area, 143
   types of
      acid-base catalysis, 142
      autocatalysis, 141
      electrocatalysis, 141
      enzymatic catalysis (biocatalysis), 141–142
      heterogeneous catalysis, 140–141
      homogeneous catalysis, 140
      nanocatalysis, 141
      photocatalysis, 141
Biochemical conversion (BCC), 132–134
   anaerobic digestion (AD), 133–134
   fermentation, 134
Biodiesel synthesis, 56
Biofuel production
   advancements, 100–101
   applicability, 100
   benefits, 99–100
   biochemical conversion (BCC), 132–134
   biorefineries in India, 90–91
   biorefinery
      approach, 92–93
      phases, 93–94
      prevailing technologies for, 94
   chemical conversion, 130–131
   conceptualization, 93

intensified biorefinery processes
  modifying the equipment, 98–99
  with novel synthetic routes, 98
  sophisticated and prominent technology, 95–98
 thermochemical conversion, 131–132
Biological azo dye degradation
 advancements, 31
 bioaugmentation of microbes, 25–27
 by microbes, 25
 nanotechnology in biological remediation, 27–28
 by plants, 24
 reactors for, 28–30
Biomass feedstock, 124–125
Biopolymer synthesis, 55
Biorefinery, 134–135
 approach, 92–93
 phases, 93–94
 prevailing technologies for, 94
 processes
  modifying the equipment, 98–99
  with novel synthetic routes, 98
  sophisticated and prominent technology, 95–98
Biosorption, 22
 algae, 8–9
 bacteria, 8
 biosorbents, 10–11
  analytical techniques, 11–12
  conventional methodologies
   coagulation and flocculation, 4
   electrochemical techniques, 5–6
   ion exchange, 4–5
   membrane filtration, 5
   precipitation process, 5
  factors affecting
   characteristics, 10
   concentration of biomass, 10
   initial metal ion concentration, 10
   pH, 9
   temperature, 9
 fungi, 8
 heavy metals, 2–3
 isotherms
  Freundlich adsorption isotherm, 13
  Freundlich isotherm, 13
  Langmuir isotherm, 12–13
 kinetics study, 13–14
 mechanism, 7–8
 process, 6–7
 water pollutants, 2

C

Carbon nanotube-based adsorption, 157
Chemical conversion, 130–131

Chitosans
 antimicrobial assay, 215–217
 biocompatibility, 215
 copper and copper oxide nanoparticles, 204
  antimicrobial studies, 205–206
  hybrid functionalized drug carrier, 206
 CuO-loaded chitosan nanoparticles, 212–215
 hydrophobically-modified chitosans
  CuO nanoparticles, 210–212
  FM analysis, 207
  FTIR analysis, 207
  ^{1}H NMR analysis, 207, 208
  SEM analysis, 209–210
 nanocarrier for drug delivery, 204
 surface functionalization of, 204
Coagulation and flocculation, 4
Conventional biosorption methodologies
 coagulation and flocculation, 4
 electrochemical techniques, 5–6
 ion exchange, 4–5
 membrane filtration, 5
 precipitation process, 5
CuO-loaded chitosan nanoparticles, 212–215
CuO nanoparticles, 210–212

D

Drug delivery, 204
Dry torrefaction, 109

E

Electrocatalysis, 141
Electrochemical techniques, 5–6
Emission spectrum, 227
Enzymatic azo dye degradation, 22–23
Enzymatic catalysis, 141–142

F

Fluorescence spectroscopy, 237
FM analysis, 207
Fourier-transform infrared spectroscopy (FTIR), 207, 236
Freundlich adsorption isotherm, 13
Freundlich isotherm, 13
FTIR, *see* Fourier-transform infrared spectroscopy
Fucoxanthin
 absorption spectrum, 226–227
 in algae, 228–231
 bioactivity of, 238–241
 biological role of, 223
 biosynthetic pathway of, 223–224
 characterization of
  fluorescence spectroscopy, 237

Index

Fourier-transform infrared spectroscopy (FTIR), 236
gas chromatography mass spectrometry (GC-MS) analysis, 237
high-performance liquid chromatography (HPLC), 238
liquid chromatography mass spectrometry (LC-MS), 237
nuclear magnetic resonance (NMR), 237–238
reversed-phase high-performance liquid chromatography (RP-HPLC), 238
UV-vis spectrophotometric analysis, 236
emission spectrum, 227
occurrence and yield of, 228–231
purification of
  high-performance liquid chromatography (HPLC), 234
  reversed-phase high-performance liquid chromatography (RP-HPLC), 234–235
  silica gel column chromatography, 232–233
  thin layer chromatography (TLC), 233–234
solvent used for extraction, 231–232
structure of, 224–226
Fungi, 8

G

Gas chromatography mass spectrometry (GC-MS) analysis, 237
Gasification, 108
GC-MS, *see* Gas chromatography mass spectrometry analysis
Genetic engineering, 27
The Godavari Sugar Mills Ltd., 91
Groundnut shell powder, 39

H

Heavy metals, 2–3
Heterogeneous catalysis, 140–141
High-performance liquid chromatography (HPLC), 234, 238
^{1}H NMR analysis, 207, 208
Homogeneous catalysis, 140
HPLC, *see* High-performance liquid chromatography
HRT, *see* Hydraulic retention time
HTC, *see* Hydrothermal carbonization
Hydrogels, 145
Hydraulic retention time (HRT), 185–186
Hydrothermal carbonization (HTC)
conversion of biomass, 106–108
  dry torrefaction, 109
  gasification, 108
  hydrothermal carbonization, 109–110
  pyrolysis, 108–109
decomposition reactions and mechanisms, 111
hydrochar applications
  activated carbon adsorbent, 114
  additional applications, 114
  carbon sequestration, 114
  renewable energy resource, 113–114
  soil amendment, 113
influence of reaction parameters
  operating pressure, 112
  pH, 112
  reaction temperature, 112
  reaction time, 112
  solid load, 113
of rice husk, 114–116
  thermogravimetric analysis, 116–117
two-step kinetic model for, 118

I

India
  biorefineries in, 90–91
  energy crisis in, 181–182
  energy options in, 182
Ion exchange, 4–5
Isotherms
  Freundlich adsorption isotherm, 13
  Freundlich isotherm, 13
  Langmuir isotherm, 12–13

L

Langmuir isotherm, 12–13
Liquid chromatography mass spectrometry, 237

M

Material properties, MFC
  anode, 186–188
  cathode, 188–190
  chambers, 190–191
  separator, 190
Membrane filtration, 5
MFC, *see* Microbial fuel cells
Microbes, azo dye biological degradation
  bioaugmentation, 25–27
  microbial dye decoloration, 25
  microbial dye degradation, 25
Microbial fuel cells (MFC), 30
  bacteria, extra cellular electron transfer by, 183–184
  factors affecting
    material properties, 186–191
    operating factors, 185–186

large-scale architecture
separate electrode modules, 193–194
stacked MFC, 193
tubular MFCs, 192
power harvesting in, 194
real-time testing, 195–197

N

Nanocatalysis, 141
Nanomembranes, 157–158
Nanosorbents, 156–157
Nanotechnology
in biological remediation, 27–28
wastewater treatment
carbon nanotube-based adsorption, heavy metals, 157
heavy metal removal, 156–157
nanofiltration, 157–158
photocatalytic oxidation, 158
water disinfection, 158
Nuclear magnetic resonance, 237–238

O

Operating factors, MFC
electrochemically active biofilm formation, 186
hydraulic retention time (HRT), 185–186
organic loading rate, 185
pH, 185
temperature, 186
Operating pressure, 112
Organic flocculation
composite preparation, 169–170
natural sources
biopolymers, 167–169
bio-sludge, 165
microbes, 166–167
plants, 167
technological advances
adsorption-coupled flocculation, 170–172
ultrasonic-assisted flocculation, 172
UV-coupled flocculation, 172–173
Ozonation, 22

P

pH, 9, 112
Photocatalysis, 141
Physico-chemical azo dye degradation, 22
Power production, *see* Microbial fuel cells
Pyrolysis, 108–109

R

Reaction temperature, 112
Reaction time, 112

Reactors, azo dye biological degradation
anaerobic filters (AF), 30
continuously stirred anaerobic digester, 29
expanded bed reactors, 30
fluidized, 30
microbial fuel cell (MFC), 30
up-flow anaerobic sludge blanket reactor, 29–30
Renewable energy resource, 113–114
Reversed-phase high-performance liquid chromatography, 234–235, 238

S

SBR, *see* Sequencing batch reactor
SEM analysis, 209–210
Sequencing batch reactor (SBR)
anaerobic sequential batch reactor (ASBR), 39
continuous study in, 46–48
F/M ratio in, 50
gas production, 50
mixed liquor volatile suspended solid concentration, 49
optimization of process variables in, 44–46
optimization studies in, 39–41
sludge volume index (SVI), 49
for textile dyeing effluent treatment, 41–44
volatile fatty acid, 49–50
groundnut shell powder, 39
screening of sorbent, 41
sorbent and support media, 38–39
Silica gel column chromatography, 232–233
Sludge volume index (SVI), 49
Soil amendment, 113
Solid load, 113
SVI, *see* Sludge volume index

T

Temperature, 9
Textile dyeing effluent treatment, 41–44
Thermochemical conversion, 131–132
gasification, 132
liquefaction, 132
pyrolysis, 131–132
torrefaction, 132
Thin layer chromatography, 233–234

U

Ultrasonic-assisted flocculation, 172
UV-coupled flocculation, 172–173
UV-vis spectrophotometric analysis, 236

Index

W

Waste sea shells
  beneficial uses of
    adsorbent, 56
    astaxanthin extraction, 55
    biodiesel synthesis, 56
    biopolymer synthesis, 55
    constructions, 55–56
    wastewater treatment, 56
  biodiesel production, 71–81
  catalyst preparation, 65–71
  characterization of, 56–64
  environmental impacts of, 54–55
  modification technique, 65–71

Wastewater
  characteristics of, 153
  conventional treatment
    adsorption, 155
    coagulation and flocculation, 153
    electro-chemical methods, 154
    ion exchange, 154
    membrane separation, 154–155
    nanotechnology, 155–156
    precipitation, 153–154
    pros and cons of, 155
  organic flocculation, 163–164
  treatment
    carbon nanotube-based adsorption, 157
    heavy metal removal, 156–157
    nanofiltration, 157–158
    photocatalytic oxidation, 158
    water disinfection, 158

Water pollutants, 2

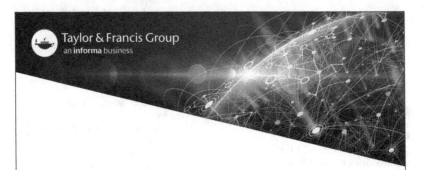